WITHDRAWN

**PLUMBER'S
LICENSING EXAM**

PLUMBER'S LICENSING EXAM

Second Edition

Copyright © 2014 LearningExpress, LLC.

All rights reserved under International and Pan-American Copyright Conventions.
Published in the United States by LearningExpress, LLC, New York.

Printed in the United States of America

9 8 7 6 5 4 3 2 1

Second Edition

ISBN 978-1-57685-999-5

Regarding the Information in This Book
We attempt to verify the information presented in our books prior to publication. It is always a good idea, however, to double check such important information as the test format, application and testing procedures, and deadlines, as such information may change from time to time.

For more information on LearningExpress, other LearningExpress products, or bulk sales,
please write to us at:
    80 Broad Street
    4th Floor
    New York, NY 10004

Or visit us at:
    www.learningexpressllc.com

CONTENTS

ABOUT THE AUTHOR		xiii
HOW TO USE THIS BOOK		1
PREPARING FOR THE EXAM		3
PRETEST		7
CHAPTER 1	Different Plumbing Codes and Abbreviations	25
	Chapter Summary	25
	Learning Objectives	26
	Key Terms	26
	General Regions Served by the IPC and the UPC	26
	Differences between the IPC and the UPC	27
	Drainage Vents	27
	Dishwashers	28
	Water Heater	29
	Water Distribution	29
	Drainage Systems	29
	Vents	29
	Abbreviations	29
	Trade-Related Associations and Governing Bodies	29
	Materials and Procedure Abbreviations	30
	Chapter Review	31

CONTENTS

CHAPTER 2	General Regulations	33
	Chapter Summary	33
	Learning Objectives	34
	Key Terms	34
	General Principles of Plumbing Regulations	34
	Connections to Building Sanitary Drainage Systems and Water Supply Systems	34
	Materials Not Allowed in a Sewer System	35
	Vermin Protection	35
	Protecting the Plumbing System	36
	Protecting the Building from the Plumbing System	37
	Underground Piping Practices	38
	Bathroom Design Regulations	40
	Testing New Plumbing Systems	40
	Drainage System Testing	40
	Supply System Testing	41
	Chapter Review	42
CHAPTER 3	Math for Plumbers	45
	Chapter Summary	45
	Learning Objectives	46
	Key Terms	46
	Basic Facts	46
	Tape Measure Basics	46
	Fitting Allowance Basics	48
	Basic Calculations	50
	Simple Geometry	50
	Chapter Review	53
CHAPTER 4	Fixtures	59
	Chapter Summary	59
	Learning Objectives	60
	Key Terms	60
	Fixture Material	60
	Fixture Requirements in Various Buildings	61
	Fixture Locations	69
	Bathroom Design Relative to Fixtures	70
	Approved Fixture Connections	71
	Regulations for Specific Fixtures	72
	Automatic Clothes Washers	72
	Bathtubs	73

CONTENTS

	Bidets	73
	Dishwashers	73
	Drinking Fountains	73
	Floor Drains	73
	Food Waste Grinders	74
	Garbage Can Washers	74
	Laundry Trays	74
	Lavatories	74
	Showers	74
	Sinks	75
	Urinals	75
	Water Closets	76
	Whirlpool Baths	76
	Specialized Healthcare Fixtures	76
	Specialty Fixtures	77
	Emergency Eye Wash and Shower	77
	Chapter Review	78
CHAPTER 5	Faucets and Fixture Fittings	81
	Chapter Summary	81
	Learning Objectives	82
	Key Terms	82
	Water Consumption	83
	Design	83
	Temperature Control	84
	Chapter Review	86
CHAPTER 6	Water Heaters	89
	Chapter Summary	89
	Learning Objectives	90
	Key Terms	90
	Location	91
	Accessibility	91
	Room as a Plenum	91
	Garages	92
	Temperature Control	92
	Type of Control	92
	Control Range	92
	Installation Requirements	93
	Piping Requirements	93
	Drain Pans	94

CONTENTS

	Safety	94
	Antisiphon Strategies	95
	Shutdown	96
	Relief Valves	96
	Relief Valve Outlet	97
	Chapter Review	98
CHAPTER 7	Gas Piping Installations	101
	Chapter Summary	101
	Learning Objectives	102
	General Rules	102
	Materials	102
	Pressure	103
	Sizing	103
	Chapter Review	103
CHAPTER 8	Water Supply and Distribution	107
	Chapter Summary	107
	Learning Objectives	108
	Key Terms	108
	General Rules	108
	Quantity	109
	Water Service	109
	Sizing	111
	Installation	114
	Distribution System	114
	Sizing	114
	Pressure	114
	Materials	116
	Water Hammer Controls	117
	Installation Criteria	117
	Hot Water	117
	Backflow Prevention	118
	Expansion Accommodation	118
	Chapter Review	118
CHAPTER 9	Sanitary Drainage	123
	Chapter Summary	123
	Learning Objectives	124
	Key Terms	124
	General Requirements	124
	Allowable Sewage	125

	Piping Materials	125
	Aboveground Piping	125
	Belowground Piping	125
	Piping Joints	129
	Connections between Similar Fittings and Pipe	129
	Connections between Dissimilar Fittings and Pipe	129
	Use of Fittings	130
	Cleanouts	131
	Manholes	131
	Sizing	131
	Sumps and Ejectors	135
	Healthcare Drainage Piping	136
	Backwater Valves	136
	Chapter Review	137
CHAPTER 10	Indirect and Special Wastes	141
	Chapter Summary	141
	Learning Objectives	142
	Key Term	142
	Objective of Indirect Waste	142
	Specific Waste (Special Wastes)	145
	Specific Fixtures	145
	Sinks	145
	Dishwashing Machines	146
	Swimming Pools	146
	Chapter Review	147
CHAPTER 11	Vents	151
	Chapter Summary	151
	Learning Objectives	152
	Key Term	152
	Objective of Vents	153
	Venting Theory	153
	Chemical Waste System Vents	154
	Misuse of Vents	154
	Materials	154
	Installation	154
	Vent Terminations	154
	Vent Connections	155
	Sizing	155
	Vent Patterns	157
	Combination Drain and Vent System	158

CONTENTS

	Multistory Buildings	160
	Air Admittance Valves	161
	Chapter Review	162
CHAPTER 12	Traps, Interceptors, and Separators	167
	Chapter Summary	167
	Learning Objectives	168
	Key Term	168
	Trap Requirements	168
	Trap Configurations	169
	Interceptors and Separators	172
	Design	172
	Sizing	173
	Other Interceptor Requirements	174
	Materials	174
	Chapter Review	174
CHAPTER 13	Storm Drainage	177
	Chapter Summary	177
	Learning Objectives	178
	Key Terms	178
	General Requirements	178
	Roof Drains	180
	Subsoil	181
	Building Subdrains	182
	Chapter Review	182
CHAPTER 14	Special Piping and Storage Systems	185
	Chapter Summary	185
	Learning Objectives	186
	Key Terms	186
	Nonflammable Medical Gasses	186
	Medical Oxygen	188
	Nonmedical Oxygen	190
	Vacuum System Piping	190
	Medical Air Compressors	190
	Chapter Review	191
CHAPTER 15	Gray-Water Recycling	193
	Chapter Summary	193
	Learning Objectives	194
	Key Term	194
	What Is Gray-Water Recycling?	194
	Connections	195

CONTENTS

	Water Closets	196
	Storage	196
	Disinfection	196
	Dye	196
	Makeup Water	197
	Landscape Irrigation	197
	Chapter Review	198
CHAPTER 16	Vacuum Drainage Systems	201
	Chapter Summary	201
	Learning Objectives	202
	Key Term	202
	What Is a Vacuum Drainage System?	202
	Chapter Review	205
POSTTEST 1		207
POSTTEST 2		223
ANSWERS AND EXPLANATIONS		239
APPENDIX A: CORRELATION GUIDE		285
APPENDIX B: PROJECT MANAGEMENT		289
APPENDIX C: PLUMBING ELEMENTS AND FACILITIES		291
APPENDIX D: STATE LICENSING RESOURCE LIST		317
GLOSSARY		323
ADDITIONAL ONLINE PRACTICE		329

ABOUT THE AUTHOR

Jack Burgess began his career as a plumbing apprentice in southern New Jersey, where he worked for a small jobber and then for a large construction company, building housing projects. After his apprenticeship ended, he moved to upstate New York and worked at a manufacturing plant, doing extensive electrical, heating, plumbing, and industrial machine maintenance.

He started his first plumbing business in 1980 and three years later, merged it with the service departments of two local fuel oil companies to create a new, larger plumbing, heating, installation, and service company. He served as the president and general manager for this company for eight years, directing a field crew of eight technicians. Following the sale of this business, Mr. Burgess became a sales engineer with the new company. His position eventually evolved into a branch manager position. He later accepted an offer to be an assistant manager at a plumbing and HVAC wholesaler that consistently placed in the top 100 in size in the country.

In 2004, he became a faculty member at the State University of New York at Delhi. Mr. Burgess, specifically, teaches the IPC Plumbing codes course, in addition to teaching other courses for the Plumbing and Heating curriculum.

Additional contributions made by:

Dan Andrews has been in the plumbing and heating trade since 1970. From 1991 to the present, he has been on the faculty at SUNY Delhi College of Technology, teaching in the Plumbing, Heating, and Pipe Fitting Program.

Dean Bliven is a licensed master plumber and lead service technician, working for a large plumbing contractor based in the Albany, New York, area. His areas of expertise are in residential and commercial plumbing, heating, air-conditioning, and indoor air quality controls.

Rob Kehn is a freelance writer from Troy, New York. A former ad executive and project manager for a national consulting firm, Rob has written on a number of topics including Language Arts, Veterinary Science, Paralegalism, Cosmetology, Early Childhood Education, Trucking, Medical Assisting, and Fire Science.

PLUMBER'S LICENSING EXAM

HOW TO USE THIS BOOK

This reference guide combines the theory and direction of traditional textbooks with the educational and testing material of conventional workbooks. The result is a unique, unmatched reference that offers you the guidance you need to familiarize yourself with the type of content on the plumber's licensing exam, as well as the test material you can use to experience an exam's approach firsthand.

Because standards and test questions can differ from state to state and from municipality to municipality, this book takes an unparalleled, two-pronged approach: to guide *and* to test. This approach centers on two areas: 1) general information that applies to plumbing knowledge throughout the country, and 2) how to locate and reference this information in the UPC and IPC 2006 codebooks. To ensure that you gain real-world testing experience—wherever you live—questions mimic those found on actual exams. Questions may test specific information not covered in a chapter, but the underlying principles and approach have already been made clear.

Before reviewing the topics surrounding the Plumber's Journeyman and Master Licensing Exam, test your existing skills by taking the pretest at the beginning of the book. You will encounter 150 questions that may be similar to the types you will find on a state or local licensing exam. When you are finished, check your answers on page 241 carefully to assess your results. Each answer in the Answers and Explanations section provides a cross reference to the plumbing code addressed by the question. Use the correlation guide in Appendix A to connect these codes to specific chapters in this book as well as in the IPC or UPC codebooks. Your pretest outcome will help you determine how much preparation you need and in which areas you need the most careful review and practice.

HOW TO USE THIS BOOK

Next, begin your review of the chapters. The content of this book is set up to align with the plumbing codebook, and each chapter contains a review section so that you can further test your knowledge at your own pace.

The following special feature boxes, present in each chapter, will provide further clarification of key content areas:

Hints and Tips
Bolstered by the experience of plumbers well versed in the trade, this feature helps new professionals navigate plumbing challenges. It also points out how standards can differ by state and/or locale.

JOB CONNECTION
Through the experiences of actual plumbers on the job, these real-world scenarios illustrate the importance of having the right professional knowledge and experience.

DID YOU KNOW?
Helpful facts round out the book's guidance information, providing material that is both useful and interesting.

ALERT!
These boxes highlight, in clear detail, the potential pitfalls facing all plumbers.

Once you have explored each chapter and feel you have increased your knowledge in those areas indicated by the pretest, go ahead and take one of the posttests found at the end of the book. Three posttests have been provided to ensure you are comfortable with the types of question you may encounter on your licensing exam, and to ensure you have as much practice with your codebook as possible prior to taking your exam.

PREPARING FOR THE EXAM

You've been working in the plumbing industry for a while now, and you think you're ready to take "the test"—the plumber's licensing exam. Have you completed all of your apprenticeship courses? Did you attend the classes? Take notes? Do the homework? Has your job exposed you to various plumbing installations? Okay, then—maybe you are ready.

Even if you feel you're *truly* ready for the plumber's exam, you still *don't* know it all. That's right. You still need answers. And that's where your plumbing codebook comes in, and where this review book will help you get the answers you need.

When you begin using this *Preparation Guide for the Plumber's Licensing Exam*, you should also have your copy of the local plumbing code with you. As you study the chapters of this book, you should review the corresponding section of your plumbing code. The chapters of this book are roughly in the same order as they are found in the International Plumbing Code (IPC) and the Uniform Plumbing Code (UPC). The specific code sections referenced in this book will point you to the relevant sections of your codebook.

Not all of the information in the UPC and the IPC plumbing codes is covered in this book. *Plumber's Licensing Exam* is intended to highlight the portions of the code that more than likely will be on your test, and to clarify those sections where many students find confusion. Have both books open at the same time and make comparisons. Every chapter has a review section at the end that will help to increase your understanding of the chapter as well as give you practice taking the plumber's licensing exam. Some of the questions in this book have not been discussed within the chapter. This was done purposely to make you look up information in your codebook and to increase your familiarity with your local code.

PREPARING FOR THE EXAM

Starting with the Right Approach

Instead of trying to learn every aspect of every conceivable plumbing situation, a near-impossible task, use the resources available to you—resources like the plumbing code used in your area. As you work, refer to the codebook whenever you need it. You know which areas of plumbing are difficult for you. You may struggle with a type of plumbing to which you haven't been exposed. Maybe you run into the same problem repeatedly and try to find a way around it because you don't know the solution. When you're in these situations, open your codebook. Each time you do, you not only gain insight into the solutions you need, you learn more about plumbing code and the way the codebook presents information.

The codebook can be an invaluable resource during the plumber's licensing exam, as many states allow for an open-book examination depending on local rules and regulations. Nevertheless, there is no replacement for exam study and preparation. Don't expect to walk into the test with your codebook, open it for the second time, and pass the test. You still need to know where to go to find the right answers. So, open the codebook often beforehand and learn where to find answers. The more prepared you are for the test, the more you'll relax, and the more clearly you'll think during the exam.

This guide, along with your local plumbing code, will go a long way toward preparing you for the plumber's licensing exam. If you don't pass the test the first time, don't worry. Consider it a dress rehearsal, and get ready for your next performance. You'll already know what to expect. But don't rush to the first available test. Give yourself time to review slowly and learn. Familiarize yourself with all the chapters in your codebook and learn where to find the charts and diagrams. Again, you don't have to know all the answers. You just have to know where to find them.

Preparing for the Exam Administratively

The purpose of the plumber's licensing exam is to ensure that qualified professionals do plumbing work in your community and protect your health and the health of all citizens. The goal is *not* to deny you a career. The club of licensed plumbers welcomes all qualified members, so accept the challenge and prepare.

Taking Initial Steps

The challenge of exam preparation begins long before the day of the test. Depending on where you live, the plumber's licensing exam may be administered by the state, county, city, or town. When you first decide to take the exam, contact the appropriate agency. Find out which forms you must submit and when they are due. Provide all needed information, and allow time for processing.

Also, be sure to learn what fees are involved. There may be a fee to apply for the test, a fee to take the test, a fee to apply for a license, and a fee to obtain the license. Be sure to file all forms and pay all fees on time. Write things down to avoid surprises. Don't learn that you filed your application three days late and must wait six months for the next test. To stay on top of scheduling or fee changes, stay in contact with your licensing agency. Adhering to all agency requirements will help ensure your success. Also, find out what materials you can and cannot bring into the test.

Meeting Exam and Licensing Requirements

To take the exam, you'll have to prove that you've met certain experience and education requirements. Find out what proof your administering agency requires to accept your application. To prove that you have the proper experience, you may have to provide pay stubs or a notarized statement from the master plumber who employed you. To prove you have the right

PREPARING FOR THE EXAM

education, an official transcript from your trade school will usually suffice.

Be sure to allow your employer time to provide any needed documents. Files must be pulled, letters written, and papers mailed—all employers have busy workday schedules. Planning, therefore, is key to meeting application requirements. Request employer information at least eight weeks in advance.

Demonstrating Proper Study Techniques

When you set out to study for the plumber's licensing exam, make sure you're referencing the correct and complete code. Two major plumbing codes form the basis for most plumbing codes adopted in the United States. First is the Uniform Plumbing Code (UPC), created by the International Association of Plumbing and Mechanical Officials (IAPMO). This code is popular in western parts of the United States. The second code, the International Plumbing Code (IPC) created by the International Code Council (ICC), is used widely across the eastern and southern United States.

If you have any question about which plumbing code is used in your area or you're unsure how to obtain it, ask your local administering agency. Often, states, counties, or cities will add items that make the code unique to that locality. While administering agencies can make plumbing codes more stringent, they cannot relax them.

> **Hints and Tips**
> The IAPMO's *Uniform Plumbing Code Illustrated Training Manual* is a good study resource when your state or local agency is using the UPC.
>
> The *International Plumbing Code: Code & Commentary* is a helpful resource for locations using the IPC.

Conducting a Thorough Code Review

As you train in the industry and familiarize with the proper techniques of competent plumbing professionals, you'll come to recognize much of the basic plumbing code. When it comes time to take the licensing test, though, details are important. And knowledge of details comes from a thorough code review.

Move thoroughly and systematically through your review. Understand each section's concepts—and the intentions of the specific code—before moving to the next section. To maximize your review time, always take your codebook and this guidebook with you. When you take your study resources with you, you can read and ask your coworkers questions. Many times, experienced plumbers will be able to clarify difficult concepts for you. Your coworkers, like this guide, are excellent study resources.

You are also a valuable study resource, so be sure to take extensive review notes. As you read, write down important facts, as well as any questions you might have. Your notes will serve two purposes. First, they'll serve as a personalized study guide. Second, they'll help reinforce information in your mind. According to one learning theory, the more of your senses that you involve in learning, the more learning that you'll do. So read it, write it, say it, and hear it—as much as you can.

Tackling Questions Strategically

Most of your plumber's licensing exam—and in some instances all of it—will consist of multiple-choice questions. Many administering agencies are now giving computerized tests, but don't let that scare you. Such tests are designed to be fair, and you need minimal computer skills. In addition, you can make each question easier by eliminating obviously wrong answers and then using the resource you brought with you—your codebook— to determine the correct answer. When you're really stumped on a question, move to the next one. Don't let one tough question stop you. If you have time once you've answered all the other questions, go back and try the tough ones again. When you've prepared well, you'll have plenty of time to complete the exam and return to any sticky questions.

Setting Yourself Up for Success

A good night's sleep for two or three nights before the test will help you feel relaxed and mentally ready. Do not try to cram for the test the night before. Staying up late, staring at code, will only serve to increase your anxiety. Your exam results will be better when you are calm and well rested.

Bringing the Process Full Circle

Once you've passed the plumbing exam and obtained your license, there's generally one more requirement: You'll have to show your administering agency you have contractor's liability insurance. If you can't prove your ability to protect the properties you work in from damages, many agencies will cancel your license. And you'll have worked far too hard to let that happen.

In a sense, professional working relationships are insurance for a successful plumbing career. Savvy master plumbers forge good relationships with the people who inspect their jobs: their code enforcement officers. Friendly and open relationships with these professionals are real assets in the plumbing world. Once inspectors learn your work will be thorough and accurate, your inspections will become smoother and more productive—even enjoyable. So once you're licensed and working on the job, keep your inspectors' trust by always doing work that deserves admiration, not scrutiny.

Good luck!

PRETEST

Before reviewing the plumbing topics included in *Plumber's Licensing Exam*, test your existing skills by taking the pretest that follows. You will encounter 150 questions that may be similar to the types you will find on a state or local licensing exam.

Your licensing exam will be based on the plumbing code used in your area, and you should have your codebook with you when you take this pretest. Answer every question; however, if you are not sure of an answer, put a question mark by the question number to note that you are making a guess. On the official exam, an unanswered question is counted as incorrect, so making a good guess is an important skill to practice. When you are finished, check your answers on page 241 carefully to assess your results. Your pretest score will help you determine how much preparation you need and in which areas you need the most careful review and practice.

PRETEST

Note: Answers may vary based on whether the IPC or UPC is applicable.

1. What is an access panel?
   a. a cleanout plug
   b. an alternate door to mechanical room
   c. a fastened cover with pipes or valves behind it
   d. a small covered tank

2. What is a building drain?
   a. any part of the piping system except a riser, main, or stack
   b. a vent connecting one or more individual vents
   c. the lowest portion of a drainage system extending within the building to 30 in. outside the building
   d. a drain that carries only sewage

3. Which letter in the following illustration shows the depth of the water seal?

   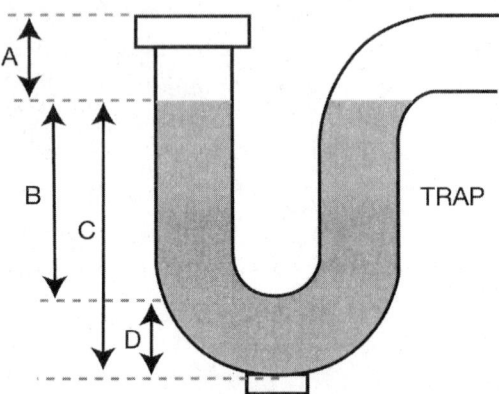

   a. A
   b. B
   c. C
   d. D

4. What is a conductor?
   a. a vertical pipe carrying rainwater down the outside of a building
   b. a horizontal pipe connecting to a soil stack
   c. a pipe carrying rainwater inside a building
   d. the pipe outside a building connecting the building drain to the sewer main

5. Define *code official*.
   a. code enforcement officer
   b. building inspector
   c. plumbing inspector
   d. all of the above

6. The definition of a fixture branch is
   a. a drain serving two or more fixtures that discharges to another drain or to a stack.
   b. the drain from the trap of a fixture to a junction with any other drainpipe.
   c. a branch that is terminated with a cap, plug, or other closing device.
   d. the length of a pipeline measured along the centerline of the pipe and fittings.

7. The term *toxic* means
   a. not hazardous to human health.
   b. bedpan washer and sterilizer drain.
   c. hazardous to human health.
   d. a fixture drain for a laboratory.

8. The effective opening is
   a. the outside diameter of the smallest orifice leading to a fixture.
   b. the insider diameter of the smallest orifice leading to a fixture.
   c. the smallest opening required to supply water to a fixture.
   d. an opening where a valve is located.

PRETEST

9. A backflow preventer is a device
   a. with a compression washer.
   b. that prevents backflow.
   c. that promotes backflow.
   d. that ensures proper flushing of a fixture.

10. Any pressure less than that exerted by the atmosphere is a(n)
   a. head pressure.
   b. unacceptable pressure.
   c. vacuum.
   d. vapor.

11. A tank or pit that receives sewage or liquid waste, located below the normal grade of the gravity system, and that must be emptied by mechanical means (i.e., a pump) is a(n)
   a. septic.
   b. sump.
   c. trap.
   d. interceptor.

12. A fitting or device that provides a liquid seal to prevent sewer gasses from entering the building is a
   a. union.
   b. trap.
   c. backflow preventer.
   d. pump.

13. Water with a temperature higher than 110°F is called
   a. potable.
   b. warm.
   c. hot.
   d. disinfected.

14. A pressure-actuated valve held closed by a spring or other means and designed to relieve pressure automatically at the pressure at which such valve is set is a(n)
   a. emergency valve.
   b. check valve.
   c. T-P valve.
   d. relief valve.

15. Water that is safe for drinking, cooking, and personal use as determined by the board of health is
   a. potable.
   b. purified.
   c. clean.
   d. healthy.

16. The edge of the receptacle from which water overflows is the
   a. threshold.
   b. overflow.
   c. flood-level rim.
   d. support rim.

17. That part of the lowest piping of a drainage system that receives the discharge from soil, waste, and other drainage pipes inside, and that extends 30 in. in developed length of pipe beyond the exterior walls of the building and conveys the drainage to the building sewer is the
   a. building drain.
   b. house drain.
   c. building sewer extension.
   d. main drainpipe.

PRETEST

18. A removable plate, usually secured by bolts or screws, to permit access to a pipe or pipe fitting (e.g., valve) for the purposes of inspection, repair, or cleaning is a(n)
   a. cleanout.
   b. access cover.
   c. saddle.
   d. partition.

19. What is the highest temperature water allowed to be discharged into a building drain?
   a. 120°F
   b. 125°F
   c. 140°F
   d. 180°F

20. How many drinking fountains are required in a factory?
   a. 6 per 200 workers
   b. 1 per 400 workers
   c. 3 per 100 workers
   d. 2 per 150 workers

21. This figure shows an approved horizontal dry vent.

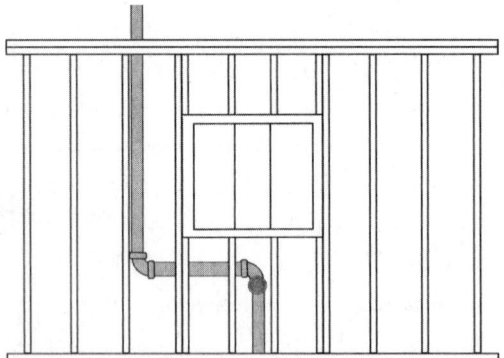

   a. true
   b. false

22. Which of the following are allowed positions for using a drainage tee to carry wastewater?
   a.
   b.
   c.
   d.

23. What is the maximum distance allowed between manholes on a sanitary sewer?
   a. 100 ft.
   b. 200 ft.
   c. 300 ft.
   d. 400 ft.

24. Slip joints are only allowed where they are easily accessible.
   a. true
   b. false

25. The smallest size allowed for a building sewer is 4 in.
   a. true
   b. false

26. A shower has a DFU of 2.
    a. true
    b. false

27. If a roof has a use besides weather protection, such as a roof garden or patio, how high above the roof surface must all plumbing vents end?
    a. 1 ft.
    b. 5 ft.
    c. 7 ft.
    d. 10 ft.

28. A tee-wye is used in a venting-only application. Which of the following are acceptable placement positions?
    a.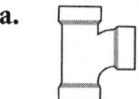
    b.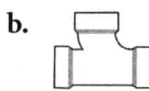
    c.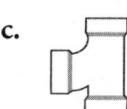
    d.

29. S-traps are allowed only in residential plumbing.
    a. true
    b. false

30. Valves can be concealed in a wall as long as a drawing showing the location is submitted to the building owner.
    a. true
    b. false

31. A fixture can be vented with an air admittance valve as long as there is one vent in the system that goes through the roof.
    a. true
    b. false

32. A building drain is the main vertical stack that extends from the lowest point in the drainage system to the end of the vent above the roof.
    a. true
    b. false

33. The building sewer is the pipe that extends from 30 in. outside the building to the municipal sewer main or to the private sewer system.
    a. true
    b. false

34. Contaminated water is less toxic than polluted water.
    a. true
    b. false

35. A conductor is a pipe carrying rainwater down the *outside* of a building.
    a. true
    b. false

PRETEST

36. A cistern prevents backflow in a water supply system.
 a. true
 b. false

37. Describe the air gap on a faucet.
 a. the device on the end of the spout that prevents splashing
 b. the vertical distance between the end of the spout and the drain
 c. the part that shuts off the spout and sends water to the rinser
 d. the vertical distance between the end of the spout and the flood-level rim of the fixture

38. Define *hot water*.
 a. water warmer than 100°F
 b. water at 120°F or warmer
 c. water warmer than 85°F
 d. water at 110°F or warmer

39. Define *indirect waste pipe*.
 a. a pipe receiving waste through an air gap
 b. waste piping not directly connected with the drain system, discharging through an air gap into a trap or fixture
 c. the outlet pipe from a grease receptor
 d. a pipe that is under a backflow condition

40. A stack vent is the extension of a soil stack.
 a. true
 b. false

41. A plumbing appurtenance uses additional water to operate.
 a. true
 b. false

42. Polluted water is not hazardous to the public health.
 a. true
 b. false

43. A vacuum is any pressure less than normal atmospheric pressure.
 a. true
 b. false

44. A vertical pipe is any pipe angled 45 degrees or more above the horizontal.
 a. true
 b. false

45. What happens inside a water heater that makes a relief valve necessary?

_____
_____
_____
_____

46. Define *yoke vent*.

_____
_____
_____
_____

PRETEST

47. When would a reduced pressure principle backflow preventer most likely be used?

_____

_____

_____

_____

48. Mechanically formed tees (T-drill) must be brazed.
   a. true
   b. false

49. PVC pipe is commonly used for hot water.
   a. true
   b. false

50. The minimum diameter of water service pipe must be 1 in.
   a. true
   b. false

51. A water storage tank is 38 in. high and 23 in. in diameter. How many gallons of water will it hold when it is $\frac{3}{4}$ full?
   a. 38.33 gal.
   b. 51.23 gal.
   c. 59.40 gal.
   d. 44.23 gal.

52. A 12 ft. high soil stack made of 4-in. PVC is full of water due to a blockage. How much water will flow out when the cleanout is opened at the base of the stack?
   a. 10.20 gal.
   b. 13.44 gal.
   c. 7.83 gal.
   d. 3.61 gal.

53. Water flow in a stack in North America is
   a. clockwise.
   b. counterclockwise.
   c. down the center.
   d. straight down the pipe wall.

54. How many DFUs can a 2-in. horizontal fixture branch carry?
   a. 4
   b. 6
   c. 5
   d. 7

55. How many DFUs can a 2-in. stack of three branch intervals or less carry?
   a. 6
   b. 10
   c. 8
   d. 12

56. A _____ horizontal branch interval can carry 20 DFUs.
   a. 2-in.
   b. 3-in.
   c. 4-in.
   d. 6-in.

57. What size must this building drain be if the slope is $\frac{1}{4}$-in. per foot?

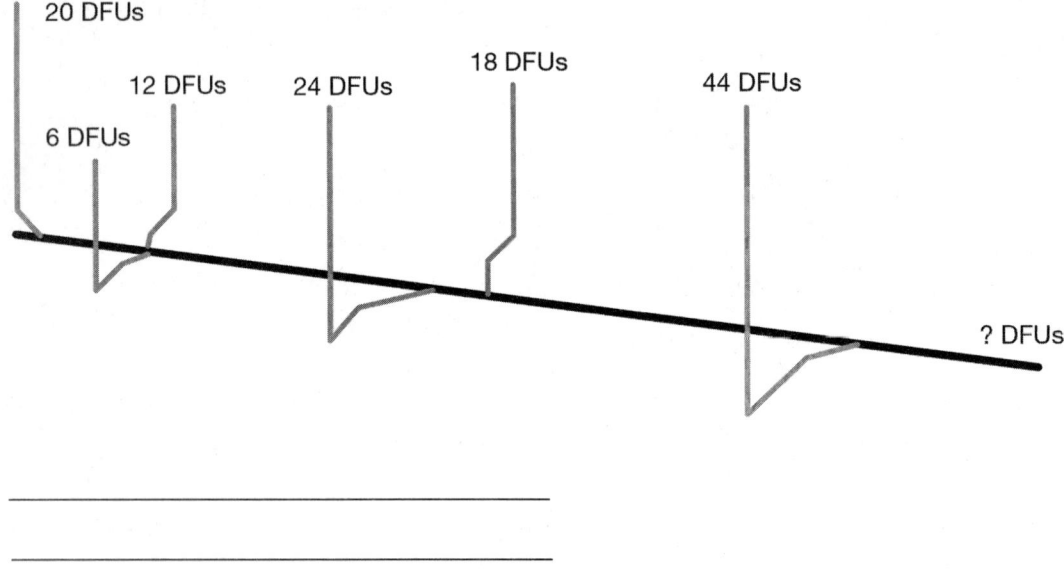

_____

_____

_____

_____

58. An automatic clothes washer requires a _____ drain.
a. $1\frac{1}{2}$-in.
b. 2-in.
c. 3-in.
d. 4-in.

59. A shower has a DFU of
a. 1.
b. 2.
c. 3.
d. 4.

60. A bathroom group has a DFU of
a. 3.
b. 4.
c. 5.
d. 6.

61. Bituminous fiber pipe is permissible for use
a. as a building drain.
b. as a building sewer.
c. as a vent stack.
d. never.

62. Asbestos cement pipe is permissible for use
a. as a water main.
b. as a vent.
c. as aboveground drainage piping.
d. never.

63. A cleanout must be located within _____ of the junction between the building drain and the building sewer.
a. 10 ft.
b. 3 ft.
c. 5 ft.
d. 8 ft.

64. An automatic clothes washer must drain through
 a. a 1½-in. trap.
 b. a tailpiece.
 c. a laundry tray.
 d. an air break.

65. A vent terminal cannot be located within _____ horizontally of any door, window that can be opened, or air intake in a building.
 a. 5 ft.
 b. 20 ft.
 c. 8 ft.
 d. 10 ft.

66. A crown vent is allowed
 a. when it is used for a 1 DFU fixture.
 b. when cast iron is the drain material.
 c. on the top floor.
 d. never.

67. An air admittance valve must be installed a minimum of _____ above the horizontal drain being vented.
 a. 1 ft.
 b. 4 in.
 c. 6 in.
 d. 36 in.

68. The vertical distance from a fixture outlet to the weir of the trap cannot be greater than
 a. 1 ft.
 b. 24 in.
 c. 18 in.
 d. 6 in.

69. A leader conducts storm water
 a. on the outside of a building.
 b. to the sanitary sewer.
 c. on the inside of a building.
 d. underground to the storm sewer.

70. Slip-joint connections
 a. are prohibited.
 b. have unrestricted uses.
 c. can be used at a trap outlet.
 d. must use plastic washers.

71. A showerhead can flow a maximum of _____ GPM.
 a. 1.6
 b. 2.2
 c. 2.0
 d. 2.5

72. A water closet with a flushometer valve requires the water supply pipe to be a minimum size of
 a. 1 in.
 b. $\frac{3}{4}$ in.
 c. $\frac{1}{2}$ in.
 d. $\frac{3}{8}$ in.

73. A grid strainer can have slots that are no wider than
 a. $\frac{1}{4}$ in.
 b. $\frac{3}{4}$ in.
 c. $\frac{1}{2}$ in.
 d. $\frac{1}{8}$ in.

PRETEST

74. Piping, other than cast iron and steel, that passes through drilled or notched holes in framing members must be protected by shield plates if it is within _____ of the nearest edge of the framing member.
   a. $\frac{1}{2}$ in.
   b. $1\frac{1}{2}$ in.
   c. 1 in.
   d. 2 in.

75. The maximum spacing for supports on $\frac{3}{4}$-in. copper pipe is
   a. 6 ft.
   b. 10 ft.
   c. 3 ft.
   d. 4 ft.

76. A compartment for a water closet must be a minimum of _____ wide.
   a. 30 in.
   b. 28 in.
   c. 36 in.
   d. 24 in.

77. The minimum size water supply pipe to a wall hydrant is
   a. $\frac{3}{4}$ in.
   b. $\frac{1}{2}$ in.
   c. 1 in.
   d. $\frac{3}{8}$ in.

78. One psig of pressure will create _____ of head.
   a. 1.6 ft.
   b. 12 ft.
   c. 2.3 ft.
   d. 2.5 ft.

79. A little-used floor drain must have
   a. a diameter of 2 in.
   b. a double trap.
   c. a trap primer.
   d. a backwater valve.

80. When sizing and designing a roof drain system, any wall that diverts water onto the roof must be sized as part of the roof, but only _____ of its area is included in the calculation.
   a. 25%
   b. 10%
   c. 50%
   d. 75%

81. One gallon per flush is the flushing rate for a
   a. bedpan washer.
   b. water closet.
   c. urinal.
   d. bidet.

82. A valve is required on
   a. the cold water inlet of a water heater.
   b. the discharge pipe of a relief valve.
   c. the hot water outlet of a water heater.
   d. none of the above.

83. Every building that is equipped with plumbing fixtures on a single lot must have an individual connection to the public sewer.
   a. true
   b. false

84. CPVC is approved for aboveground drainage and vent piping.
   a. true
   b. false

85. Drainage from a kitchen sink is considered gray water.
  a. true
  b. false

86. Gray water needs to be disinfected when it is used for
  a. flushing fixtures.
  b. disposal directly into the sanitary sewer.
  c. underground irrigation.
  d. none of the above.

87. A bell trap is approved for floor drains.
  a. true
  b. false

88. The minimum slope allowed for a 6-in. horizontal drainpipe is
  a. $\frac{1}{8}$-in per ft.
  b. $\frac{1}{4}$-in. per ft.
  c. $\frac{3}{8}$-inch per ft.
  d. $\frac{1}{2}$-in. per ft.

89. Solvent-welded PVC joints are permitted in
  a. vent piping.
  b. aboveground piping.
  c. underground drainage piping.
  d. all of the above.

90. A double tee-wye can be used to drain back-to-back water closets if
  a. 4-in. pipe is used.
  b. the horizontal developed length from the water closet to the connection with the double tee-wye is 18 in. or greater.
  c. they have a common vent.
  d. the water closets are a rear flush design that is mounted on closet carriers.

91. In drainage piping, the position for a sanitary tee can be
  a. horizontal to vertical.
  b. vertical to horizontal.
  c. horizontal to horizontal.
  d. all of the above.

92. A 2-inch short sweep is permitted for a horizontal-to-horizontal change of direction in a drainage pipe
  a. when the 2-in. drain serves a single fixture.
  b. always.
  c. never.
  d. when there are less than 10 DFUs flowing through it.

93. A residential automatic clothes washer has a DFU value of
  a. 1.
  b. 2.
  c. 3.
  d. 4.

94. The discharge pipe for a temperature and pressure relief valve must terminate
  a. within 6 in. of the floor.
  b. into a floor drain through an air gap.
  c. within 12 in. of the floor.
  d. outdoors.

95. A food preparation sink must drain
  a. through an air gap.
  b. through an air break.
  c. directly to the sanitary drainage system.
  d. through a 2-in. drain.

PRETEST

96. The primary purpose of a vent is to
   a. discourage corrosion in the drainage piping.
   b. relieve pressure differentials in the public sewer.
   c. maintain the trap seals.
   d. carry sewer gas out of the drainage system.

97. Any horizontal dry vent must
   a. terminate within 10 ft. of the fixture served.
   b. not be installed because it is not permitted.
   c. rise vertically 6 in. above the flood-level rim of the highest trap or trapped fixture being vented.
   d. have a diameter one size larger than that which the venting fixture units demand for the fixture served by the vent.

98. Vitrified clay tile drainage piping
   a. is prohibited.
   b. must be joined with an elastomeric seal.
   c. must be joined with cement and tar.
   d. can only be used for subsoil drainage.

99. A full-opening valve is required in a drainage system
   a. that serves a grease interceptor.
   b. after the last fixture draining into a gray-water recycling system.
   c. never.
   d. when three or more commercial automatic clothes washers are served.

100. A backflow preventer on a water supply pipe to a boiler
   a. must have a vent between the check valves.
   b. is necessary only when chemicals have been added to the heating system.
   c. must have the same effective opening as the supply pipe.
   d. cannot be mounted vertically.

101. At sea level, how much pressure is created by 100 ft. of head?
   a. 43.3 psig
   b. 433 psig
   c. 14.14 psig
   d. 23 psig

102. How many gallons of water can a 4-in. diameter pipe that is 100 ft. long contain?
   a. 1,256.00
   b. 31.40
   c. 12.56
   d. 65.25

103. The minimum slope permitted for a $1\frac{1}{2}$-in. drain is
   a. $\frac{1}{8}$-in. per ft.
   b. $\frac{1}{16}$-in. per ft.
   c. $\frac{1}{4}$-in. per ft.
   d. $\frac{1}{2}$-in. per ft.

104. In a public bathroom, urinals must be spaced _____ inches apart.
   a. 24
   b. 18
   c. 30
   d. 36

PRETEST

105. There must be _____ inches of clearance in front of a water closet.
 a. 16
 b. 21
 c. 18
 d. 30

106. _____ inches of clearance are required between the centerline of a water closet and any wall that may be on either side of the water closet.
 a. 12
 b. 18
 c. 30
 d. 21

107. An overflow built into a lavatory
 a. is optional.
 b. is mandatory.
 c. must have the same cross-sectional area as a $1\frac{1}{4}$-in. drain.
 d. must be made of vitreous china.

108. No plumbing system can be installed in
 a. an elevator shaft.
 b. a tunnel.
 c. an unheated area.
 d. an area prone to annual flooding.

109. A building trap is required
 a. when the Authority Having Jurisdiction has determined its necessity.
 b. never.
 c. on all commercial buildings.
 d. when cast iron is used for the building drain.

110. A roof or floor truss can be cut
 a. never.
 b. if the notch is less than one-third the depth of the support member.
 c. with written approval from a registered design professional.
 d. when it interferes with a roof penetration for a vent.

111. Copper tubing with a $1\frac{1}{2}$-in. diameter that is horizontally installed must have hangers installed every
 a. 6 ft.
 b. 10 ft.
 c. 8 ft.
 d. 3 ft.

112. Each drain outlet of a laundry tray must have a minimum diameter of
 a. $1\frac{1}{4}$ in.
 b. $1\frac{1}{2}$ in.
 c. 2 in.
 d. whatever the plumber decides.

113. The minimum width of a shower floor is
 a. 30 in.
 b. 28 in.
 c. 36 in.
 d. 24 in.

114. Water closets with flushometer tanks
 a. must flush no more than 1.6 GPF.
 b. are required in public bathrooms.
 c. are prohibited.
 d. can flush up to 2.5 GPF.

PRETEST

115. The maximum outlet water temperature to a potable hot water distribution system is
   a. 120°F.
   b. 140°F.
   c. 110°F.
   d. 160°F.

116. The temperature setting of a temperature and pressure relief valve installed in a water heater can be no more than
   a. 190°F.
   b. 212°F.
   c. 140°F.
   d. 210°F.

117. A _____-inch diameter water supply pipe is required for a water closet with a flush valve instead of a tank.
   a. $\frac{1}{2}$
   b. $\frac{3}{4}$
   c. 1
   d. $\frac{3}{8}$

118. A water hammer arrestor is required
   a. where quick-closing valves are used.
   b. on every distribution main.
   c. never.
   d. at the furthest fixture from the main valve.

119. The spout of a faucet must be a minimum of _____ above the flood-level rim of the fixture.
   a. 1 in.
   b. 2 in.
   c. 4 in.
   d. no requirement

120. A barometric loop is
   a. a term referring to atmospheric pressure.
   b. part of a sump.
   c. a vacuum breaker.
   d. a vent stack.

121. How many water services does a hospital require?
   a. 1
   b. 2
   c. 3
   d. 0

122. Manholes are required as cleanouts when a building sewer is _____ inches diameter or larger.
   a. 6
   b. 8
   c. 10
   d. 12

123. A drinking fountain has a DFU of
   a. 10.
   b. 1.
   c. 2.
   d. 0.5.

124. A 2-in. horizontal fixture branch has a DFU capacity of
   a. 6.
   b. 5.
   c. 10.
   d. 4.

125. Wastewater can be no hotter than
   a. 120°F.
   b. 140°F.
   c. 130°F.
   d. no limit.

126. In a combination drain and vent system, a 3-in. floor drain could be installed _____ feet away from a vent.
   a. 6
   b. 8
   c. 10
   d. unlimited

127. The constant used to find the length of a 45° offset is
   a. 3.14.
   b. 0.707.
   c. 1.414.
   d. 231.

128. If a combination drain and vent is used for a kitchen sink, what size must the vertical portion be?
   a. $1\frac{1}{4}$ in.
   b. $1\frac{1}{2}$ in.
   c. 2 in.
   d. 3 in.

129. In a jurisdiction that uses the International Plumbing Code, how high above the drain served must an air admittance valve be located?
   a. 4 in.
   b. 12 in.
   c. 4 ft.
   d. 10 ft.

130. How many vents through the roof are needed in a building where air admittance valves are permitted (IPC only)?
   a. 1
   b. 2
   c. 3
   d. 0

131. A grease interceptor requires
   a. a flow control device on the inlet side of the interceptor.
   b. a flow control device on the outlet side of the interceptor.
   c. no flow control device.
   d. an inlet and outlet flow control device.

132. A secondary roof drain must
   a. drain directly to the sanitary sewer.
   b. be one-half the capacity of the primary roof drain.
   c. drain through conductors to the subsoil drainage system.
   d. drain where the discharge will be noticed.

133. An approved material for building storm sewer pipe is
   a. CPVC.
   b. malleable iron.
   c. acrylonitrile butadiene styrene.
   d. ductile iron.

134. A storm sewer can
   a. be laid side by side with a sanitary sewer.
   b. receive the discharge of a sump pump.
   c. drain directly to a waterway.
   d. all of the above.

PRETEST

135. What is the minimum number of roof drains required for a roof that is 10,000 square feet or less?
 a. 1
 b. 2
 c. 3
 d. 4

136. The minimum pipe diameter for a subsoil drain is
 a. 2 in.
 b. 3 in.
 c. 4 in.
 d. 6 in.

137. What fitting is required on the discharge pipe of a sump pump?
 a. check valve
 b. globe valve
 c. backflow preventer
 d. male adapter

138. The material used for medical vacuum piping can be
 a. ABS.
 b. galvanized steel.
 c. PVC.
 d. CPVC.

139. Recycled gray water that is used for flushing fixtures must be
 a. dyed blue or green.
 b. supplied through piping that is $1\frac{1}{2}$ times the standard pipe diameter.
 c. dyed red or orange.
 d. drained from lavatories only.

140. The operating vacuum in a vacuum drainage system is
 a. 19 in. WC.
 b. 16 to 20 in. WC.
 c. 23 in. mercury column.
 d. 16 to 20 in. mercury column.

141. Where a storm drain connects to a combination sewer, the storm drain must be trapped.
 a. true
 b. false

142. A grease interceptor with a flow-through rating of 10 GPM must have a grease retention capacity of 10 lbs.
 a. true
 b. false

143. The trap weir of a urinal can be an unlimited distance from a vent.
 a. true
 b. false

144. A bathtub can be vented using a combination drain and vent system.
 a. true
 b. false

145. A vent stack only carries gray-water waste.
 a. true
 b. false

146. A high heel tee installed as the first fitting at the outlet of a water closet is an acceptable way to connect a vent.
    a. true
    b. false

147. In 6-in. drainage pipes, the minimum cleanout size is 4 in.
    a. true
    b. false

148. The minimum slope allowed in a 4-in. drain is $\frac{1}{8}$-in. per foot.
    a. true
    b. false

149. The maximum horizontal hanger spacing for 10-ft. lengths of cast-iron pipe is 10 ft.
    a. true
    b. false

150. Polluted water is not considered a hazard to public health.
    a. true
    b. false

Readiness

This test has given you an idea of the kinds of questions you will be asked on your plumber's licensing test. Remember, most testing authorities allow you to bring a copy of the plumbing code that has been adopted in your area, and sometimes you will be allowed to bring additional materials as well. Be sure to find out well in advance of the test date which materials are allowed and order them right away because you *will* want to take them with you to the test. Become familiar with the book(s), especially with the table of contents and the index and/or glossary at the end of the book(s). It will be much faster for you to use the book(s) to answer the questions when you are confident that you can find the information you need.

If you found that you were pretty much clueless on this test, don't worry about it. You are not the first, and it's OK to admit to yourself that you are not ready. If your test date is within two months, maybe you would be better off waiting for the next test six months later in the year. Get studying. You cannot walk into this test cold with nothing but a few years of field experience. Use your codebook and study guides like this one to help give meaning to some of the more difficult topics in the code. Study a little every couple of days. If you keep your nose in these books for an hour or so, four or five nights a week, you will be in good shape for exam day.

CHAPTER 1: DIFFERENT PLUMBING CODES AND ABBREVIATIONS

CHAPTER SUMMARY

It is important at the outset to note some of the differences between the International Plumbing Code and the Uniform Plumbing Code. The principles of hydraulics and physics work the same everywhere on Earth, but the opinions and experience of the folks who write these codes vary greatly. A single code written by the same body will differ from one edition to the next as changes in administrative practices, updates of equipment, and the results of recent studies shed new light on the operation of plumbing systems. Following any valid plumbing code will result in a well-designed plumbing system. However, it is important that you have a copy of your local code with you and follow it so that you work within the boundaries of your jurisdiction.

DIFFERENT PLUMBING CODES AND ABBREVIATIONS

Learning Objectives

- Recognize code differences
- Understand the reasons for inconsistencies
- Become aware of your local code

Key Terms

air admittance valve
backflow preventer
hoar frost
$97\frac{1}{2}$% design temperature
stack vent
vent stack
weir

General Regions Served by the IPC and the UPC

Many states have adopted the International Plumbing Code (IPC) created by the International Code Council, which is headquartered in Washington, D.C. The IPC is used primarily throughout the eastern, midwestern, and southern United States.

The International Association of Plumbing and Mechanical Officials (IAPMO) is the creator of the Uniform Plumbing Code. Home base for IAPMO is in Southern California. The UPC has been adopted in many western, northwestern, and northern midwestern states.

The adoption of different codes in different geographical regions is simply a result of the growth of the country from east to west, as well as the changing influences over time in the differing regions. Cities grew large must faster in the east, so the necessity of codes was evident more quickly in the east. Western cities grew much later and the technology of plumbing was already evolving. Cities and states in the west started new codes with fresh research. The eastern cities and states continued to build on their established code that was already being used successfully. In spirit, the codes achieve the same end of creating plumbing systems that protect the health of the population. The differences are in the details.

Areas of some states have adopted mixtures of the two codes and others have their own unique code, usually based on either the IPC or the UPC. Code adoption is not necessarily consistent within a state. Many individual municipalities and counties have their own version of a code. That is why it is important to check with your local government office and obtain a copy of the code on which you will be tested. There is no substitute for getting the correct information direct from the source.

JOB CONNECTION

After completing the renovation of several apartment buildings in a medium-sized city in Ohio, Ace Plumbing Company will be ready to move on to the townhouse construction job—won through the efforts of the owner, Carlos Mendez, and the company estimator, Sophia. The project happens to be just across the state line in Indiana. Sophia had the additional challenge of applying the UPC code, which is used in Indiana, rather than the IPC code, with which she is more accustomed.

DIFFERENT PLUMBING CODES AND ABBREVIATIONS

> **ALERT**
>
> Even if you believe you have followed the code on a particular feature of an installation, the "Jurisdiction Having Authority" makes the final judgment about whether the installation meets code. Many sections of a code are subject to interpretation. The Jurisdiction Having Authority, which may be a building inspector, a plumbing inspector, a code enforcement officer, or another inspector with a different title, makes the final decision.

Differences between the IPC and the UPC

In addition to geographic variations between the IPC and UPC, there are differences that have been formed by hard research and also differences that are based on the experience and opinions of those charged with writing a code. This section provides a broad outline of some of the most important areas of difference.

Drainage Vents

There is one dramatic difference between the codes concerning venting. The UPC requires all vents in a system to terminate through the roof one way or another. Vents may pass through and end above the roof individually, or they can be connected to a **stack vent** or **vent stack** that goes up through the roof. The IPC allows the use of **air admittance valves** (Figure 1.1) throughout a system as long as at least one vent goes up through the roof. An air admittance valve is an air inlet valve that is installed where a vent pipe normally would be installed. They are commonly located under cabinets, as in the case of vanities and kitchen cabinets, or they can be installed within a wall behind a grille to allow the intake of air and access for servicing. Air admittance valves allow the plumber to easily vent a fixture without running a pipe all the way through framing members to penetrate the roof. This results in significant savings in labor and material, especially when plumbing is added to an existing home where all the surfaces are finished. (Refer to IPC 917.3 and 917.7; UPC Chapter 9 Introduction)

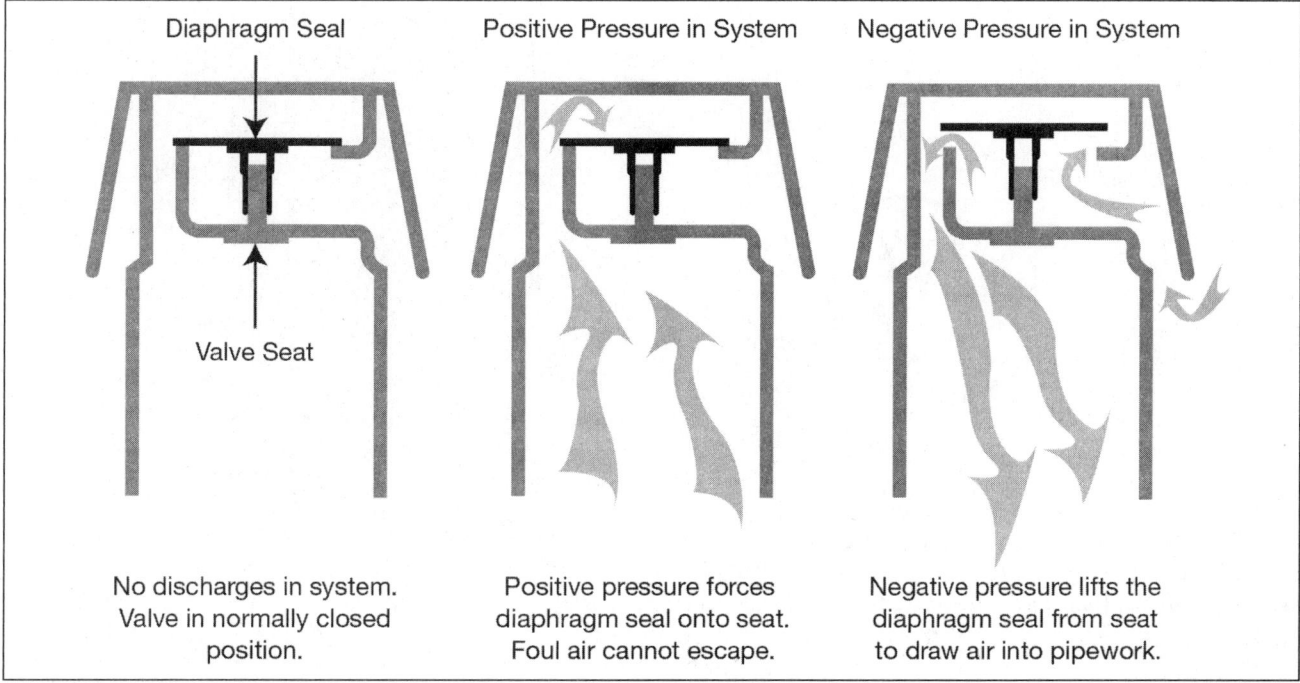

Figure 1.1 Air Admittance Valve

Air admittance valves are different from mechanical vents, which are commonly referred to as cheater vents. Air admittance valves are closed by gravity pulling a disc closed. Mechanical vents are closed by a lightweight spring. Gravity is far more dependable than a spring. Therefore, mechanical vents are not allowed under any plumbing code.

Dishwashers

Under the UPC, a dishwasher drain hose is required to connect to an air gap device above the flood level rim of the associated fixture before it continues to the drainage piping on the inlet side of the trap (Figure 1.2). This is to prevent the backflow of sewage into a dishwasher in the event of a blockage in the drainage piping. The IPC recognizes that currently manufactured dishwashers are fitted with an integral **backflow preventer** in the water supply line and no further air gap is required (Figure 1.3). (Refer to IPC 802.1.6; UPC 807.4)

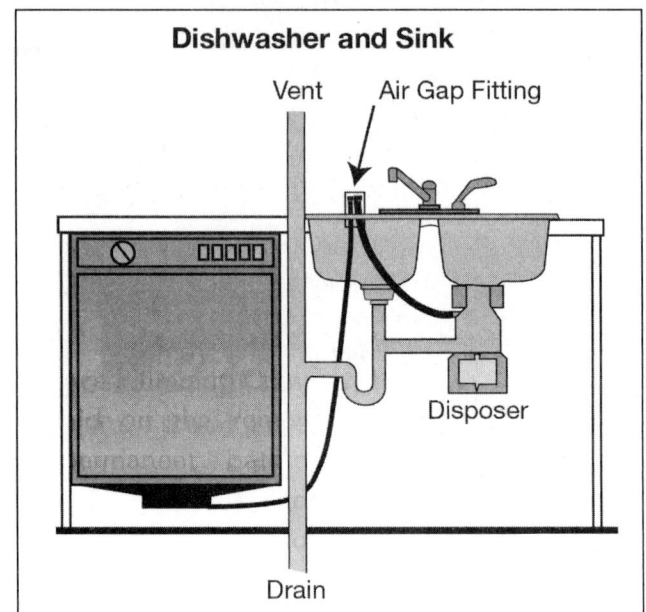

Figure 1.2 Air Gap for a Dishwasher Drain

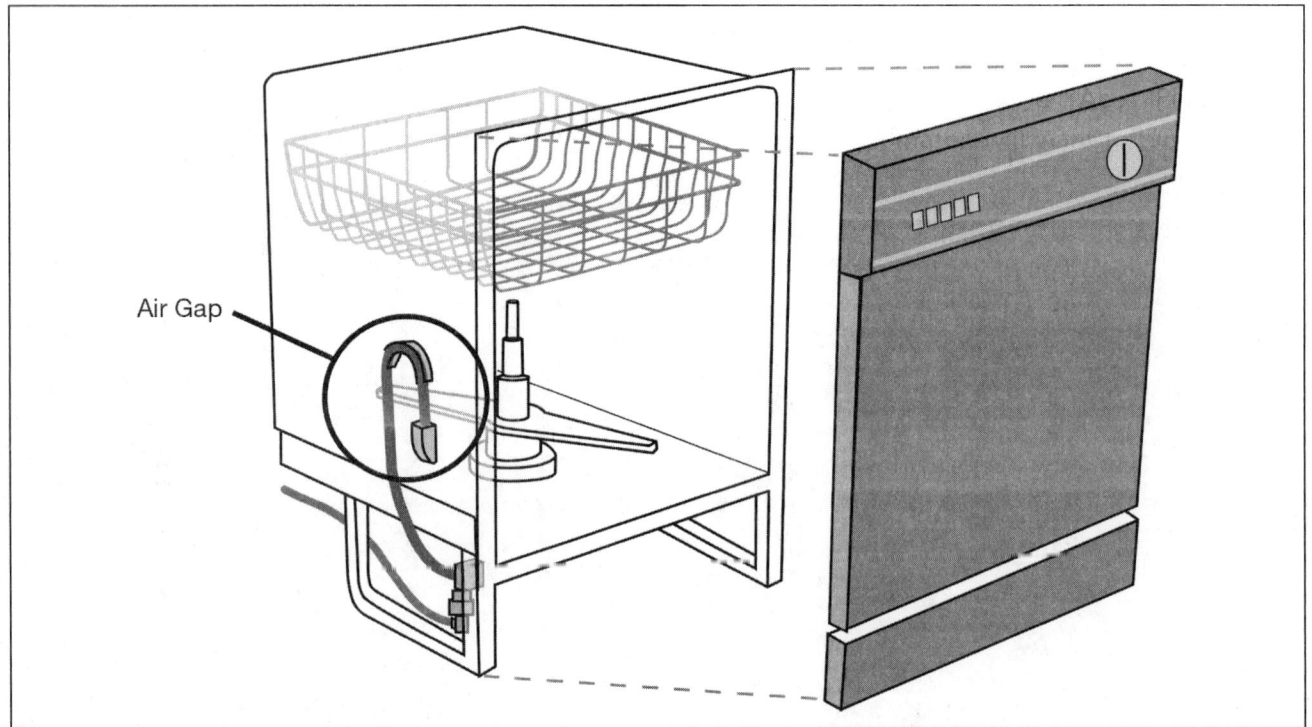

Figure 1.3 Air Gap for the Dishwasher Supply

DIFFERENT PLUMBING CODES AND ABBREVIATIONS

Water Heater

The UPC goes into great detail concerning the proper venting of water heaters. If the UPC is the code used in your area, make sure to study Chapter 5 in your codebook. The IPC doesn't cover the venting of water heaters. The IPC leaves that to another ICC code. (Refer to UPC Chapter 5)

Water Distribution

Both codes are equally conscientious in regard to water distribution systems. The primary goal is to ensure the safety of the public. The prevention of cross-connections and backflow are covered carefully in both codes. The secondary concern of both the UPC and the IPC is the adequate supply of water to all of the fixtures in a building. (Refer to IPC Chapter 6; UPC Chapter 6)

Drainage Systems

The requirements for drainage systems in both the IPC and the UPC generally are similar, but there are a few differences. Both codes allow a kitchen sink and a shower to have a $1\frac{1}{2}$-inch trap. However, the IPC allows the drain for a kitchen sink to be $1\frac{1}{2}$ inches, while the UPC requires a 2-inch drain for a kitchen sink. The IPC allows a $1\frac{1}{2}$ inch drain for a shower, but the UPC requires a 2-inch drain.

The IPC calls for an individual vent for all trap arms, except when venting bathroom groups. The UPC states maximum distances that a fixture may be from a vent according to the size of the drain, regardless of the type of fixture with which it is associated. The distances allowed for a horizontal drain from a trap **weir** to a vent pipe are different in each code. These will be explained later in Chapter 11.

Vents

The UPC has no provision for the increase of the diameter of a vent termination through the roof when it is located in a cold climate. In the IPC, if a vent through the roof is in an area where the $97\frac{1}{2}$% **design temperature** is 0°F or below, the vent size must be increased to 3 inches. An increaser must be installed at least 12 inches down inside the building and then the 3-inch pipe must extend out of the roof by whatever distance is stated in your local code. The purpose of this increase is to prevent the vent pipe from being plugged up with ice at the top by a buildup of **hoar frost**.

DID YOU KNOW?

As a general practice, plumbers use a 2-inch drain for showers unless there is a construction issue that doesn't allow for the depth of a 2-inch trap. Then, a plumber may decide to use the $1\frac{1}{2}$-inch trap.

Abbreviations

This section will acquaint you with some of the acronyms and abbreviations you will frequently encounter.

Trade-Related Associations and Governing Bodies

AGA—American Gas Association
ANSI—American National Standards Institute
ASME—American Society of Mechanical Engineers
ASPE—American Society of Plumbing Engineers
ASSE—American Society of Sanitary Engineering

DIFFERENT PLUMBING CODES AND ABBREVIATIONS

ASTM—American Society for Testing and Materials
AWWA—American Water Works Association
BOCA—Building Officials Conference of America
IAPMO—International Association of Plumbing and Mechanical Officials
ICC—International Code Council
IFC—International Fire Code
IFGC—International Fuel Gas Code
IPC—International Plumbing Code
IRC—International Residential Code
NFPA—National Fire Protection Association
NSF—National Sanitation Foundation
UL—Underwriters Laboratories, Inc.
UPC—Uniform Plumbing Code

Hints and Tips

It is important that you keep your plumbing codebook handy and refer to it often while studying for your test. You will be asked to look up certain information from the codebook during the test that would be very difficult to memorize. Part of the goal of a plumbing licensing test is to determine if you can effectively use your codebook as a reference.

Materials and Procedure Abbreviations

ABS—acrylonitrile butadiene styrene. A plastic drainage material.
B & S—bell and spigot. A pipe design for connecting one piece of pipe or fitting to the next (Figure 1.4).

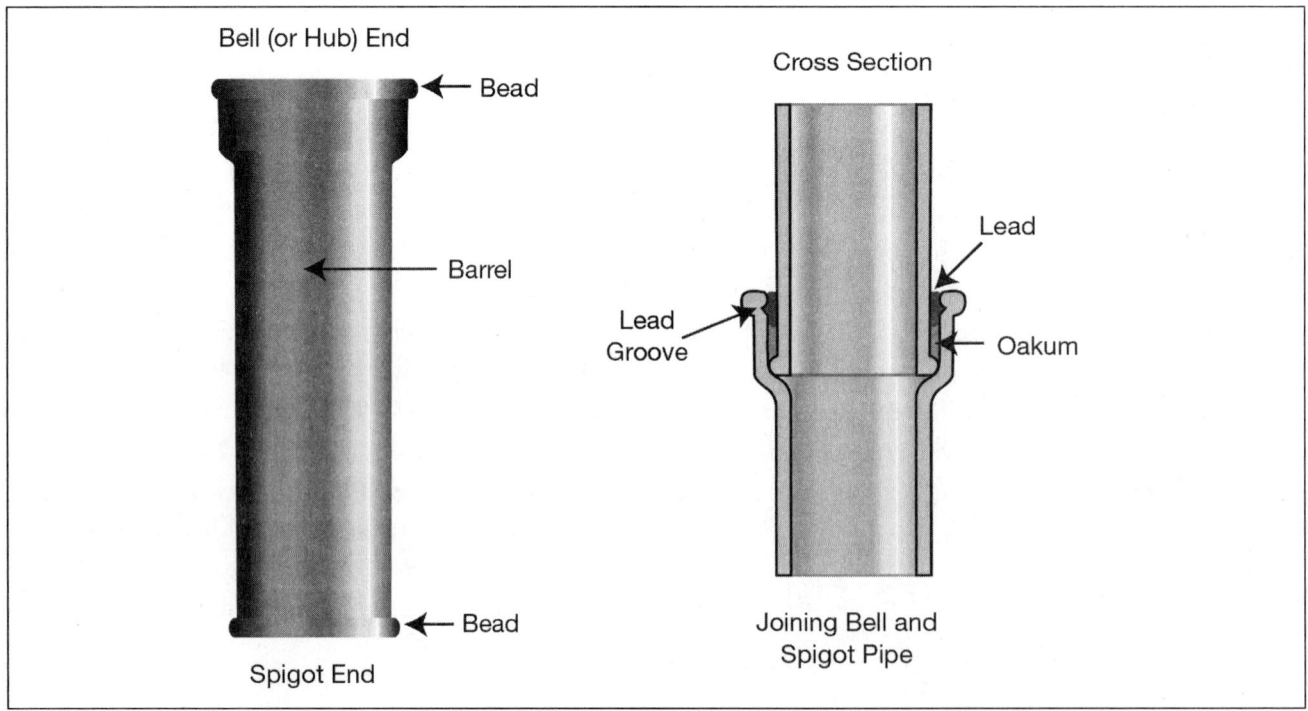

Figure 1.4 Bell and Spigot Connection

DIFFERENT PLUMBING CODES AND ABBREVIATIONS

BTU—British thermal unit. The amount of heat needed to raise one pound of water one degree Fahrenheit.

CPVC—chlorinated polyvinyl chloride. A type of plastic water distribution pipe (Figure 1.5).

DFU—drainage fixture unit. A unit of measure used to determine the drainage load a certain fixture will place on the drainage system.

DWV—drainage, waste, and vent. This refers to a weight and wall thickness of copper piping that is used for drainage, waste, and vent piping. It is not appropriate for pressure applications.

FPS—feet per second. FPS is used commonly in plumbing to describe the speed of water traveling in a pipe.

GPM—gallons per minute. A reference to the flow rate of a pipe or fixture fitting.

NPT—National Pipe Thread Tapered. Standard pipe threads.

PPM—parts per million. A measurement of waterborne materials.

VTR—vent through roof.

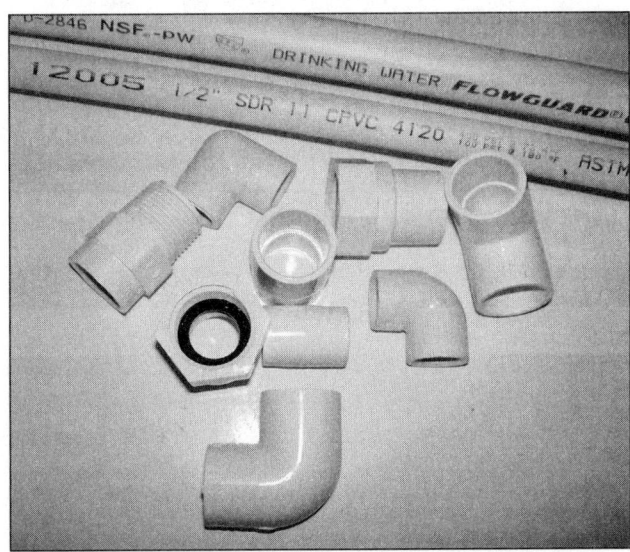

Figure 1.5 CPVC Water Distribution Piping

Chapter Review

Note: Answers may vary based on whether the IPC or UPC is applicable.

1. In which code(s) are air admittance valves allowed?
   a. both the IPC and the UPC
   b. IPC
   c. UPC
   d. neither the IPC or the UPC

2. What is the difference between the UPC and the IPC concerning the regulations for shower drains?

   _____
   _____
   _____
   _____

3. In which code(s) are mechanical vents allowed?
   a. both the IPC and the UPC
   b. IPC
   c. UPC
   d. neither the IPC or the UPC

4. Hoar frost can close a plumbing vent when
   a. the outdoor winter $97\frac{1}{2}$% design temperature is below 32°F.
   b. the outdoor winter $97\frac{1}{2}$% design temperature is below 0°F and the vent is smaller than 3-inch diameter.
   c. the outdoor winter $97\frac{1}{2}$% design temperature is below 0°F and the vent is smaller than 2-inch diameter.
   d. the outdoor winter $97\frac{1}{2}$% design temperature is below 0°F and the vent is 3-inch diameter or larger.

DIFFERENT PLUMBING CODES AND ABBREVIATIONS

5. DWV copper tubing can be used for
   a. distribution of potable cold water at pressures under 100 psi.
   b. irrigation piping.
   c. building drains.
   d. all aboveground plumbing applications.

6. The abbreviation that designates the type of threads that are cut in pipes, which form a tapered thread used commonly in plumbing is
   a. NPT.
   b. PSI.
   c. CPT.
   d. TPT.

7. In a jurisdiction using the UPC in its entirety, a kitchen drain must
   a. have an S-trap.
   b. be made of PVC.
   c. be 2 in. in diameter.
   d. have a dishwasher connection in the tailpiece.

8. A DFU is used
   a. to pressure test a drain and vent system.
   b. to determine drainage pipe size.
   c. to cap a pipe that went to a dead-end fixture.
   d. only by plumbing engineers.

9. In the _____, a deck-mounted air gap is required for dishwasher drains.
   a. UPC
   b. commercial kitchen
   c. IPC
   d. event a food waste disposer is installed

10. In both the UPC and the IPC, a vent stack carries
    a. air and sewer gas only.
    b. air and wastewater only.
    c. sewer gas and wastewater only.
    d. water vapor and gray water only.

11. The smallest diameter fixture branch allowed for a lavatory in the UPC and the IPC is
    a. $1\frac{1}{4}$ in.
    b. $1\frac{1}{2}$ in.
    c. 1 in.
    d. 2 in.

12. CPVC is
    a. a material used for drainage pipes.
    b. a material used for supply pipes.
    c. a third-party testing agency.
    d. a type of solvent cement.

13. UL is
    a. a material used for underground sewer pipe.
    b. a third-party testing agency.
    c. an appendix in the UPC code.
    d. an agency of the federal government that regulates plumbing.

14. In which state are both the UPC and the IPC accepted as a part of the state building code.
    a. Kansas
    b. Pennsylvania
    c. Utah
    d. none of the above

15. The UPC has been adopted primarily in the
    a. northeastern states.
    b. western states.
    c. mid-Atlantic states.
    d. eastern states.

CHAPTER 2: GENERAL REGULATIONS

CHAPTER SUMMARY

This chapter covers all those rules about how to install piping without discussing pipe sizes, capacities, or any of the hydraulics involved with designing and installing a plumbing system. We must protect the building, the plumbing system, and the people who will use it. In this chapter, we will consider some of the practical rules for actually getting the piping into the ground and the building.

The plumbing system is only a portion of a complete building; so, when plumbers are working, they must look beyond "the pipes" to the other aspects of a building's construction. The integrity of the building must be maintained and the safety and comfort of the occupants must be kept in mind.

GENERAL REGULATIONS

Learning Objectives

- Connect to municipal systems
- Identify materials not allowed in a sewer system
- Provide vermin protection
- Protect a plumbing system
- Protect a building from the plumbing system
- State underground piping practices
- Explain bathroom design regulations
- Test a new drainage system
- Test a new supply system

Key Terms

adhesion
ambient
effluent
gray water
green
head
ID
potable
psi
psig
sanitary drainage system
WC

General Principles of Plumbing Regulations

Connections to Building Sanitary Drainage Systems and Water Supply Systems

This portion of the regulations simply states that all plumbing fixture drains must be connected to the building **sanitary drainage system**, and all fixtures that need a water supply must be connected to the water supply system of the building. This seems obvious, but occasionally you will find situations where this has not been done. Many older homes may have the kitchen sink and/or the **gray-water** fixtures connected to a separate drain from the sanitary sewer. This drain may lead to a manufactured or homemade seepage pit located on the property. It may even simply lead out to a drainage ditch (Figure 2.1). (Refer to IPC 301.3, 301.4; UPC 305.1)

JOB CONNECTION

Ace Plumbing Company is bidding on the contract for the new addition to the Springfield Elementary School. There are many design features that Sophia, the Ace estimator, must consider. The existing school was built in 1952. Trenching will be necessary next to the old school's foundation. What must be done to protect the old foundation as well as the new building drain? Is the earth in the area of the new addition stable enough for bedding a pipe in a trench? Sophia knows from the plans of the old school that lots of fill was necessary because the school was built on the old Bigler swamp. Can pipe connections be made to the old school's building drain, or will a new connection to the sewer main be needed? Do we have the necessary equipment for testing the new system?

The motivation to make these drains separate was to reduce the quantity of water flowing into an inadequate septic system. When automatic clothes washers became popular after 1947, many homeowners thought the excess water would reduce the effectiveness of the bacteria in their septic tank.

GENERAL REGULATIONS

Figure 2.1 Seepage Pit

The purpose of the requirement to connect all water-using fixtures to the water distribution system of a building is to prevent the flushing of fixtures with some outside and uncontrolled water source, like using a bucket to flush a toilet rather than connecting the building water supply to the toilet tank.

In locations where water usage must be carefully restricted due to limited supply or when a building owner chooses to build **green**, gray water is collected and stored, and then reused for the flushing of water closets and urinals, which do not require **potable** water to operate properly.

Materials Not Allowed in a Sewer System

You can probably think of a number of materials that should not be put into a sewer system. Think of what could happen if a certain material or liquid were introduced into a sewer system. That way, you don't have to memorize a list of disallowed materials. Think of just one item; got it? Now, think of what might happen to the people working on the plumbing system (like you!) if they were unaware of or unprepared for that harmful material. Consider the ability of the sewage disposal system to handle this material; could the system render the material harmless to the environment? Think about the pipes, the building, and the main sewer system. Could they safely carry this material without a blockage occurring or the piping being aggressively corroded? What about the final destination of the material, be it a river, ocean, or leach field? Would plants, animals, and the air be able to tolerate the introduction of this material? (Refer to IPC 302; UPC 303)

A sewage disposal system is designed to receive organic waste and turn it into harmless **effluent** that is ready to be put back into our environment—you know, the stuff that is normally flushed down the toilet and washed down the kitchen sink. The system is also designed to handle the extra water, soap, and detergents that go down the same building drain. If the material is highly acidic, caustic, radioactive, flammable, or poisonous, it does not belong in a sanitary sewer. In the case of industrial waste, many varieties of exotic and harmful substances are routinely disposed of. Such waste must be treated in a manner that renders it harmless before it can be put into a sewer system. If it cannot be made harmless, it cannot be put down a drain; it's as simple as that. In some cases, this can be as minor as cooling the water to 140°F (60°C), which is the highest temperature of wastewater allowed in a building drainage system.

Vermin Protection

The most notable vermin we as plumbers must deal with is the rat. Rats can carry a wide array of diseases and they effectively transmit them to other animal populations, including humans, through their own body fluids, dead carcasses, and the fleas they carry. The most famous disease to which they have subjected the world is bubonic plague, or the Black Death. We have all learned about the plague's devastation of Europe during the Middle Ages. (Refer to IPC 304; UPC 313.12)

GENERAL REGULATIONS

There have been other plague epidemics throughout history. Plague still roams the earth in the fleas on the backs of the rats we hate to see, but cannot eliminate.

That is why we must do our part to control rats. The IPC and the UPC are almost identical on the steps required for making a plumbing system rodent-proof. The smaller dimension of any opening of drain grates can be no larger than $\frac{1}{2}$ inch. For example, a floor drain can have openings that extend 4 inches or more in one direction, but the opposite dimension, which creates a rectangle, can be no larger than $\frac{1}{2}$ inch. All pipes that go through floors, ceilings, or walls must have the space around them covered with a metal plate or collar. These prevent rats from getting out of the sewer system if they are already in it, and prevent them from getting into the sewer system if they are inside a building.

> **DID YOU KNOW?**
>
> The first documented plague epidemic was long before the Middle Ages. From about 500 A.D. to 700 A.D., an epidemic killed about a quarter of the world's population from Ethiopia, up through the Middle East, and into Europe.

Protecting the Plumbing System

We want to protect our good work from manmade and natural forces. Wherever metallic pipes pass through concrete, or through a soil that is corrosive, the pipes must have a protective sheath, or covering, that is at least 0.025-inch thick. That's about as thick as this line:

_____

The covering also has to be flexible enough to allow for expansion and contraction to prevent scraping of the pipe. (Refer to IPC 305, 306; UPC 313.0–313.11, 314.0–315.4)

When a pipe passes under or through a wall, it must be protected from damage that could occur if the wall shifts and the weight of the building bears down on the pipe, or creates some other force to be transferred to the pipe due to movement of the building. This is accomplished by running the pipe through a supporting sleeve or by building an arch over the pipe.

Piping must be protected against damage from expansion and contraction due to temperature changes. This can occur with changes in the **ambient** temperature, or when the temperature of the liquid flowing through the pipe changes. If a straight section of pipe is rigidly supported on both ends, great stress and probable damage would occur when its temperature changed, generating expansion or contraction. Either the design of the piping or the piping supports must accommodate this natural movement. Expansion joints or loops in the piping are used to allow such movement. Imagine a worm trying to sink back into its hole (like a pipe that suddenly has chilled liquid flowing through it) but a robin is firmly holding on to it. The worm has to let go of the earth or rip apart; either way the worm dies. The point is, something's got to give.

Freezing is a common nemesis of water-conducting pipes. Those who live in Miami and San Diego may not have to be too concerned about this, but most of the rest of us do. If you live in a colder climate, you have seen the results of the irresistible force of ice expanding inside a pipe. Broken pipes and building damage can be quite expensive. Any pipe intended to conduct a liquid that will freeze under a normal weather condition must be protected. This can be done with freeze-proof hydrants, pipe burial at least 6 inches below the local frost line in the case of supply piping, and insulation with a heat source. Sewer pipes do not have to meet the "frost line" rule because they are not normally filled with water and they only flow when water runs down a drain. Generally, the minimum depth required is 6 inches, but this can vary by region and by a specific situation. You must check the local code.

GENERAL REGULATIONS

During new construction or remodeling, pipes are installed that must pass through the framing members of the building in order to conceal them. Unless those pipes are made of cast iron or galvanized steel, they must be protected from the possibility of damage when finish nails or screws are used to fasten the wall and ceiling panels to the framing members. This is done with what are commonly called nail plates (Figure 2.2). They are fastened to the frame at any point where the surface of a pipe passing through a framing member is closer than $1\frac{1}{2}$ inches to the face of the frame. A little-known and less followed rule concerning nail plates is that they must extend 2 inches beyond the edge of the framing member in the direction where panels are to be applied to protect against the likelihood of a fastener missing the stud.

Figure 2.2 Protective Shield Plate or Nail Plate

Protecting the Building from the Plumbing System

Probably the most common crime plumbers commit is overzealous drilling, cutting, and notching of the support members of a building. If you adhere to the regulations of the plumbing code, this will not happen. To begin with, the code says in a general way that you must leave all parts of a building in a safe structural state when your work is done. Some general rules include not notching a bearing stud any more than 25% and a nonbearing stud any more than 40%. You should avoid notching joists whenever possible. No notching is allowed in the middle third of a joist and very limited notching should occur in other areas. You must refer to your local building code for specifics. The size and location of drilled holes are also limited by the building code. Never cut any part of a truss. A truss is designed to bear specific loads as a complete, intact structure. A truss cannot bear its designed load if any part of it is cut or drilled. (Refer to IPC 307; UPC 313.7, 313.10.3, 313.11)

> **Hints and Tips**
>
> Have you ever heard the old myth that hot water freezes faster than cold water? Obviously, hot water will not freeze faster than cold water if they are both subjected to freezing temperatures at the same time. Say you put a coffee can full of hot water and one full of cold water in your freezer. The hot water must lose all that extra heat and cool down before it will begin to freeze. The can of cold water will have a head start on freezing. However, water that *has been heated* and cooled to the same temperature as cold water *will* freeze faster than the cold water. The reason for this is that microscopic air bubbles are driven out of water when it is heated. The little bubbles act like insulation (minimal as it may be) for the cold water. This often occurs in plumbing systems when piping that has had no water drawn from it for a long period of time, such as overnight, is subjected to frigid temperatures and freezes.

> **ALERT**
>
> Cutting and notching of the structural members of a building beyond acceptable limits not only renders the structure unsafe, but also becomes a very expensive repair for your company. Notching and drilling regulations must be strictly adhered to.

Underground Piping Practices

The UPC makes an important reference to trenching near buildings at or below the grade of the footing. If a trench is parallel to the footing, any settling of the building could damage the pipe, but more importantly the integrity of the structure could be seriously compromised. (Refer to IPC 306; UPC 315)

Installing underground pipe by jacking or tunneling methods is allowed. Generally, this is done by contractors who specialize in this unique piping practice. These methods are efficient when piping must be installed under a paved area, roadway, or another improved surface that would be expensive to repair or inconvenient for interruption of service. If true tunneling is the method used, the tunnel must be reinforced to prevent future cave-in of the tunnel on the pipe. Jacking and tunneling pipe-laying methods should be limited to areas where it is clearly the best route to go. Trenching and backfilling is the most controllable method of burying pipe and it is arguably the most reliable (Figure 2.3).

Figure 2.3 Large-Scale Trenching Project

Pipe in a trench must be continuously supported on unexcavated earth whenever possible. If you are installing a type of pipe with hub and spigot joints, holes must be dug where each hub sits to allow the body of the pipe to rest continuously on the bottom of the trench, rather than the pipe being suspended by the hubs. Now, we know that it is a rare backhoe or excavator operator who can dig a trench at the exact depth and exact pitch needed to meet the "unexcavated earth" requirement. So, what to do?

GENERAL REGULATIONS

Wherever the earth is dug deeper than necessary, sand or fine gravel must be placed in the trench and compacted at every 6 inches of depth. The UPC states that any material that can pass through a $\frac{3}{16}$-inch screen is appropriate for back fill. All rocks, frozen soil, concrete, and other large, heavy fill items must be kept away from the pipe. The single point pressure that those kinds of items can place on a pipe may push the pipe out of alignment or break the pipe. After 12 inches of fine fill has been backfilled and compacted every 6 inches, large items may be placed in the trench. It's better if they can be kept out of the trench altogether, but that isn't always possible.

> **Hints and Tips**
> When laying a pipe in a trench, the top half of the pipe must be left visible for the inspector to examine before completing the backfilling operation. If it is covered, the inspector may compel you to uncover it for inspection.

Occasionally, and in some regions more than others, an excavated trench will have a bottom that is soft and cannot dependably support the pipe. This must be remedied by overexcavating the trench to two times the inside diameter, or **ID**, of the pipe and placing appropriate fill, such as gravel, crushed stone, or, in extreme cases, concrete, to provide a solid base to support the pipe.

After a pipe has been laid and backfilled around the bottom and sides, the top half must be left exposed until after the appropriate authority has inspected and approved the installation.

Keep in mind that you must protect this pipe that you are installing underground, anticipating that it could stay in service for fifty years, eighty years, or perhaps even longer (Figure 2.4).

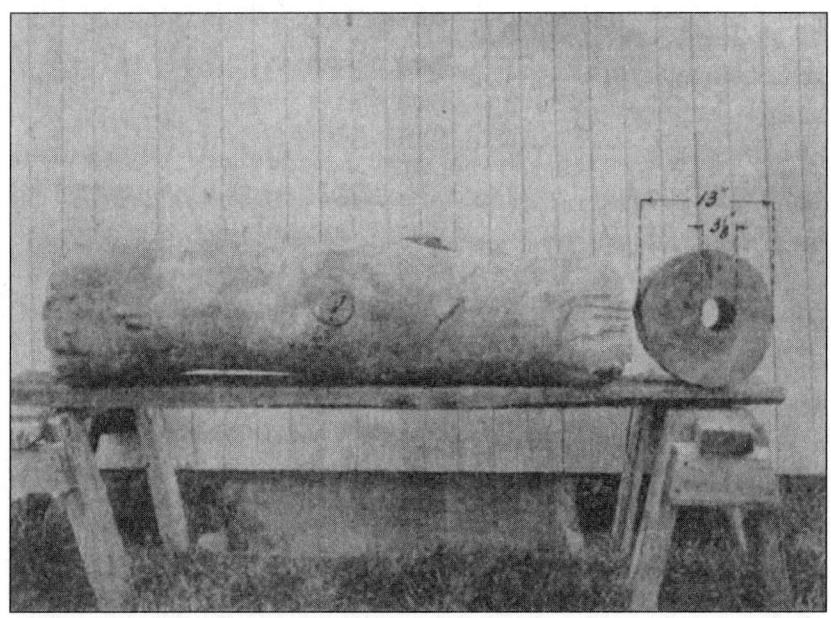

Figure 2.4 Centuries-Old Water Main *(Photo courtesy of the American Society of Civil Engineers [ASCE])*

GENERAL REGULATIONS

Bathroom Design Regulations

The IPC has regulations governing the design of bathrooms. The UPC leaves these design considerations to the local building code. (Refer to IPC 310.1–311.1)

Interference with the normal operation of any doors or windows by any piping or fixtures is prohibited. So, if your cellar door opens halfway and then runs into an overhead soil pipe, the pipe must be rerouted. If the bathroom door swings open and then clunks into the toilet bowl, the water closet must be moved or the door must be moved.

In public restrooms, every toilet must have its own compartment with a door for privacy. All urinals must have partitions that extend a minimum of 18 inches from the wall or 6 inches from the front rim of the urinal, whichever is furthest out from the wall. The walls around a urinal must be easily washable. The partition must start no more than 12 inches above the floor and must go no higher than 60 inches. The exception to these privacy provisions is in childcare facilities where children are likely to need assistance. There are size provisions for these compartments, which we will look at in Chapter 5.

Section 311 requires that toilet facilities are available to construction workers who are in an area where plumbing is not yet finished or will not be included in the project. Refer to your copy of the plumbing code to find the number of toilets required based on the number of workers at a site. Portable toilets are allowed, and are far more prevalent on jobsites than permanent facilities. This rule gives workers the dignity of using a proper toilet rather than having to leave the site to find a bathroom or search for a bush to hide behind.

DID YOU KNOW?

Wooden water mains buried in Philadelphia before the Industrial Revolution were replaced in the 1970s, and that was only because they were too small to handle the amount of water needed. Those original pipes had been carefully supported, bedded, and backfilled!

Testing New Plumbing Systems

Gauges used for testing must be in a range that is reasonable for the pressure being measured. This is an important point. Specifically, you should never use a gauge that has a range of more than twice the test pressure. For pressure tests up to 10 **psi**, the gauge increments should be of a pound. For measurements between 10 and 100 psi, the increments should be $\frac{1}{10}$ psi. Pressure tests at higher ranges should have increments that are 2% of the test pressure, and no gauge used should have an upper limit that is greater than twice the test pressure. (Refer to IPC 312, 107.1–107.5; UPC 319, 103.5.3–103.5.3.3)

Drainage System Testing

An important aspect of the IPC 107 section and the UPC 103.5 section is that the obligation to test plumbing systems falls entirely on the permit owner. This includes having all the equipment and labor necessary on site to perform the tests. The inspector simply observes and comments. The inspector's job is to inspect only; he or she plays no role in performing the tests. The plumbing codes put the responsibility on the permit holder because that is where the legal recourse lies should there be any problems arising on the job. The permit holder, if it is not the plumber, will undoubtedly transfer the responsibility of testing to the plumber through the owner's contract with the

GENERAL REGULATIONS

plumber. On very small jobs, such as a water heater replacement, the plumber may very well be the permit holder. On larger jobs where construction is involved, the building owner will probably be the permit holder. (Refer to IPC 312.1–312.9.2; UPC 319.1 and 319.4)

If you are testing a drainage system, the whole system must be filled with water so that it runs out of the roof vent. This puts the entire system under a pressure of at least 10 feet of **head**, except for the top 10 feet, which is the code requirement. In the case of a multistory building where filling the system would create unusual and excessive pressure in the lower parts of the system, testing is permitted to be done in sections. The tests must be maintained for a minimum of 15 minutes to satisfy the requirements of inspection.

The pressure created by 10 feet of head is 4.33 **psig**. An alternative to testing the drainage system with water is an air pressure test. Air is pumped into the system, after all the openings are plugged, up to 5 psig. Again, this pressure must be maintained for 15 minutes. It is far more difficult to find leaks in a system tested with air pressure, but the risks of freezing and water damage from leaks are eliminated.

If there is any doubt that a system is watertight, the inspector may require a smoke test (Figure 2.5). A smoke test is done by filling a drainage system with smoke and air, then capping all openings and creating a pressure of only 1-inch **WC**. Keeping the pressure this low is critical because too much pressure will cause the smoke to blow through the trap seals and fill the rooms with smoke, making the test useless for finding leaks. The procedure would have to be repeated after the smoke is totally cleared, resulting in a great waste of time and money.

Supply System Testing

Once again, the complete responsibility of testing lies with the permit holder. When testing a water supply system with water, potable water must be used. It may be convenient to pump water from a bucket, ditch, or pond for testing before a water main is connected, but this would contaminate the new piping system you just installed, which would then require sanitizing and testing for bacteria. Only use water from a potable supply. (Refer to IPC 312.1, 312.1.1, and 312.5; UPC 319.2–319.4)

The minimum pressure for testing water supply systems is 50 psi. If the operating pressure of the system will be greater than 50 psi, the test must be at least at the working pressure of the system.

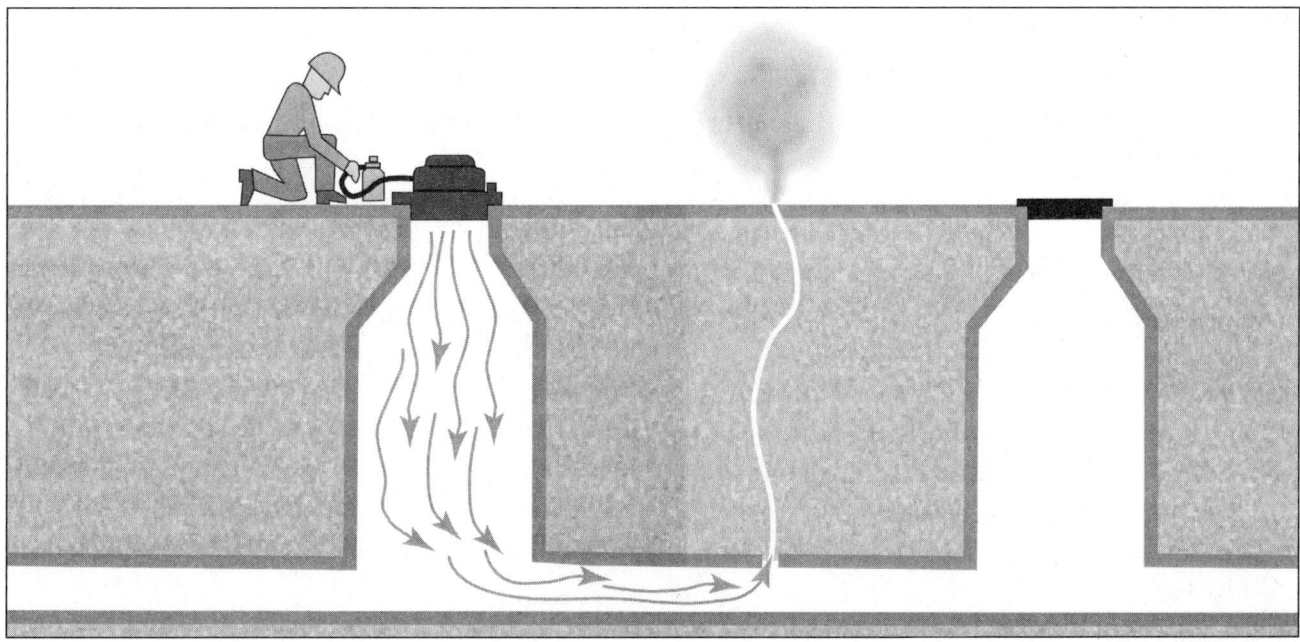

Figure 2.5 Smoke Testing a Sewer System

Air pressure tests of 50 psi are permitted on nonplastic supply systems. Plastic piping could be subject to catastrophic failure in an air test and individuals could be injured. Instead of simply rupturing in the way that metal pipes do, plastic materials can splinter and produce a kind of plastic shrapnel. As we all know, liquids are not compressible. When a water-filled pipe breaks, the pressure is instantly relieved. The property of **adhesion** will tend to keep the broken pieces of pipe connected to the water. If a plastic pipe filled with pressurized air breaks, the energy of the expanding air may carry broken pieces a great distance with considerable force.

Chapter Review

Note: Answers may vary based on whether the IPC or UPC is applicable.

1. New drainage systems constructed of cast-iron pipe can be tested with air pressure.
    a. true
    b. false

2. Storm drainage systems are tested with a different procedure than sanitary drainage systems.
    a. true
    b. false

3. For a test pressure of 50 psig, a gauge with increments of 2 psig may be used.
    a. true
    b. false

4. Piping in a trench must always be continuously supported on unexcavated earth.
    a. true
    b. false

5. If a sink will only have intermittent use, it is not necessary to connect it to the building's water supply system.
    a. true
    b. false

6. When testing a water supply system with air, the minimum test pressure must be 50 psi.
    a. true
    b. false

7. As long as a door can be opened at least 65 degrees, it is allowable for the door to come into contact with a plumbing fixture.
    a. true
    b. false

8. Toilets are required for construction sites with no permanent facilities.
    a. true
    b. false

9. Outdoor water supplies must be buried a minimum of 12 in. below grade.
    a. true
    b. false

10. If a building drain is made of cast iron it may pass through the building foundation without a sleeve.
    a. true
    b. false

11. The smallest openings in strainer plates on drain inlets must be no larger than _____.
    a. $\frac{1}{4}$ in.
    b. $\frac{3}{8}$ in.
    c. $\frac{1}{2}$ in.
    d. 1 in.

GENERAL REGULATIONS

12. Which item cannot be put into a building drainage system?
   a. food waste
   b. soap suds
   c. naphtha
   d. 130°F water

13. A large rock can be included in backfill above a pipe in a trench
   a. when it is 6 in. or more above the pipe.
   b. when it is 12 in. or more above the pipe.
   c. never.
   d. if it is no more than 4 in. in diameter.

14. Outdoor water supply pipes must be buried at least _____ inches below the frost line.

15. Horizontal copper tubing that is $1\frac{1}{4}$ in. or less must be supported at least every _____.

16. Which individual is responsible for testing a new plumbing system?
   a. building owner
   b. permit holder
   c. inspector
   d. plumber

17. Water supply systems made of any approved material can be tested with compressed air.
   a. true
   b. false

18. How would you bring a trench that has a bottom that is too soft to support the pipe up to code?

   _____
   _____
   _____

19. In testing a drainage system, the uppermost 10 ft. of piping needs to be subjected to 10 feet of head pressure.
   a. true
   b. false

20. Plumbing systems cannot be located in
   a. enclosed ceilings.
   b. elevator shafts.
   c. penitentiaries.
   d. cabinets.

21. Pipe and fitting sizes referred to in the code are all in ID sizes.
   a. true
   b. false

22. Pipes and fittings with the manufacturer's identification on them are the only ones that can be used.
   a. true
   b. false

23. Concrete is not considered corrosive to pipes.
   a. true
   b. false

24. What has to be done to industrial waste before it is sent down the building drain?

   _____
   _____
   _____
   _____

GENERAL REGULATIONS

25. The UPC requires plumbing systems in buildings installed before the current code was adopted to be brought to meet current code standards.
   a. true
   b. false

26. Townhouses can share water and sewer services.
   a. true
   b. false

27. Drilling and tapping hub and spigot–type cast-iron soil pipe is approved, as long as the drilling and tapping is completely in the hub area.
   a. true
   b. false

28. Sewers must be installed to a minimum depth of _____.

29. Piping installed through framing members that can be damaged by nails or screws must be 1 in. from the nearest edge or be protected with metal shields.
   a. true
   b. false

30. How many toilet facilities are required for 25 construction workers when they are serviced weekly?
   a. 6
   b. 1
   c. 2
   d. 3

31. A drainage system is pressurized to 5 psi when conducting a smoke test.
   a. true
   b. false

32. A trench for a soil pipe can be dug next to a footing to a depth of no more than 12 in. below the base of the footing.
   a. true
   b. false

33. When the testing of a new drainage system is under way, the inspector provides
   a. no testing equipment.
   b. gauges.
   c. test plugs.
   d. smoke.

34. Backflow preventers must be inspected and tested at least once a year.
   a. true
   b. false

35. All medical waste is approved for direct disposal in the sanitary sewer.
   a. true
   b. false

CHAPTER 3: MATH FOR PLUMBERS

CHAPTER SUMMARY

Math is a bit like vegetables. It may not be our favorite thing, but we know we need it. Keep an open mind and don't tense up. It's not that bad. We'll just keep it simple.

Trial and error is an inefficient way to run piping. Quite often, those pipes connect to various types and sizes of tanks. If an old tank with no markings on it must be replaced, the plumber must know how to calculate what size the replacement should be so that it will run properly. At times the plumber must know the water content of a system of piping. These are all fairly simple calculations when you know the right formula and constants to use. This chapter introduces the basic mathematical concepts and formulas that plumbers need to know to ensure that jobs go smoothly.

MATH FOR PLUMBERS

Learning Objectives

- Strengthen basic math skills
- Gain confidence in math skills related to plumbing
- Learn which calculations are appropriate for certain situations
- Memorize key constants

Key Terms

constant
offset

Basic Facts

Pipes are solid objects that need to fit in certain places in distinct ways. A few measurements and simple calculations can make any job easier.

Tape Measure Basics

When measuring, we here in the United States use the U.S. Customary System, also known as the foot-pound second system. We use inches and feet for our basic measurements of length in the plumbing trade. We all know that there are 12 inches in a foot. The harder part is dealing with fractions of inches, which you must do on a daily basis when you are installing piping.

Say that you want to find out if you can keep a drainage pipe up between the joists when you have a 32-foot run. You will want to maintain the normal $\frac{1}{4}$-inch per foot pitch. The pipe is 2-inch PVC, with an outside diameter of approximately $2\frac{1}{2}$ inches. A 2 × 12 inch joist is in reality about $1\frac{1}{2} \times 11\frac{1}{2}$ inches. Now, how much will the pipe drop in 32 feet to maintain the quarter-inches in an inch? Each foot in length of our 32-foot run will use up one of those quarters.

So, if we divide 32 by 4, we will get how many whole inches we will need. Go ahead, do the calculation in your head or on paper. You should get 8. That means the pipe must drop 8 inches from the beginning to the end of our 32-foot run in order to maintain a normal pitch.

Hints and Tips

All of the problems in this chapter are presented using a simple, step-by-step approach. Read slowly if math is difficult for you. Try the problems yourself, one step at a time, and avoid skimming the pages.

JOB CONNECTION

Julio, Evan, and Pat have their work cut out for them. They wonder if Carlos, the boss, knows what he has gotten them into. Running four parallel pipes, which are three different sizes, down through the ceiling of the hallway in this octagon-shaped hospital won't be easy. There are already two electrical conduits, a vacuum pipe, an oxygen pipe, two 2-inch diameter hot water heating pipes, and a couple of pipes about which they are clueless. There's not much room and it all has to look perfect. Julio and Evan think they are going to make Pat, the new whiz kid, do all of the math.

MATH FOR PLUMBERS

> **DID YOU KNOW?**
>
> Math can be fun!
> Try this:
>   Choose a number between 0 and 9.
>   Double that number.
>   Add 6 to the result.
>   Divide that number by 2.
>   Subtract the number you chose to begin with.
>   The answer is 3. It's always 3.
> One more math trick before we get to work:
>   Choose any number.
>   Double it.
>   Multiply that number by 5.
>   Drop the number on the right and you have the same number you started with.

So far we have only used up 8 inches of the $11\frac{1}{2}$ inches of joist depth we have to work with. But now we must consider the thickness of the pipe. If we start on the high end with the pipe tight to the underside of the floor above, the bottom of our 2-inch PVC pipe will be about $2\frac{1}{2}$ inches below the top of the joist, and the bottom of the pipe is the part we are trying to keep up within the joist space. So, we must add that $2\frac{1}{2}$ inches to the 8 inches in drop that will occur over our 32-foot distance. OK, so let's do the calculation: $2\frac{1}{2}$ plus 8 equals $10\frac{1}{2}$ inches. Does it fit? Will it work? Yes, but be careful to keep the starting end up tight to the floor and follow the $\frac{1}{4}$-inch per foot pitch, or you will end with the bottom of the pipe below the bottom of the joist. You only have how much extra to play with? That's right, only 1 inch.

Now, let's do another one. You are installing a $\frac{3}{4}$-inch copper pipe. The end of the pipe is $32\frac{1}{2}$ inches away from a wall. You must end your pipe $6\frac{1}{2}$ inches away from the wall. Go ahead and clean the end of the pipe and the coupling while you are thinking about this. Swab the flux on your pipe and inside the coupling and slide it on the pipe. OK, that's enough thinking time. This is pretty easy subtraction: $\frac{1}{2}$ minus $\frac{1}{2}$ is zero. Now, we know our answer will be a whole number. That doesn't happen too often, does it? We are left with 32 minus 6. You can probably do this calculation in your head, but you could also use your tape measure as a number line. Start at 32 inches and count down 6 inches: 1 inch (31), 2 inches (30), 3 inches (29), 4 inches (28), 5 inches (27), 6 inches (26). You must cut the pipe 26 inches long.

Let's try a calculation that's just a little harder. The end of the pipe still needs to end $6\frac{1}{2}$ inches from the wall, but this time the pipe you must put the coupling on is $21\frac{1}{2}$ inches from the wall. Last time we were able to get rid of the fraction at the start. This time, we will have a fraction left over, so let's deal with that first. We have to subtract $\frac{1}{2}$ from $\frac{1}{8}$. That means we have to make the $\frac{1}{8}$ "bigger" so we can take the $\frac{1}{2}$ away from it. Borrow a whole 1 from the 21, and add it to the $\frac{1}{8}$. We do this by making both denominators (that's the bottom number) the same. How many eighths are in one whole? That's right, 8:

$$\frac{8}{8} + \frac{1}{8} = \frac{9}{8}$$

That's big enough to subtract $\frac{1}{2}$ from, but we still need the denominators to be the same so we can finish the problem. We have to go to the larger denominator, which is 8. We already have the $\frac{9}{8}$ the way we want it; now, we must convert the $\frac{1}{2}$ into eighths. We know there are eight eighths in a whole, right? So how many eighths in a half? Half of a whole would be $\frac{4}{8}$. Now we can subtract, since our denominators are the same and we are subtracting a smaller number from a bigger number:

$$\frac{9}{8} - \frac{4}{8} = \frac{5}{8}$$

We now have the fraction out of the way, so we are only left with 20 minus 6. (Remember, we borrowed 1 from 21 to make our fraction bigger.) You can do the calculation in your head or use a tape measure: 1 inch (19), 2 inches (18), 3 inches (17), 4 inches (16), 5 inches (15), 6 inches (14). The answer is 14, and when we add the fraction, we know the next piece of pipe must be $14\frac{5}{8}$ inches long.

Fitting Allowance Basics

Since you have been working in the plumbing trade for a while now, you realize that when you assemble pipes with fittings, a certain amount of length is lost because the pipe and fitting must overlap to form a watertight joint. In the field, most of us hold a fitting in the air in the place we guess it will end up, and then measure the length of pipe needed. This works most of the time, but when you have restricted space and/or angles to traverse, guessing becomes more difficult. When you must install multiple pipes in a parallel configuration, doing the math becomes essential.

> **ALERT**
>
> Materials and labor are both very expensive. In order to win a bid, jobs are quoted with fairly thin profit margins, especially the big projects. Incorrectly installed piping and equipment can destroy any profit the plumbing company might have had. The bid is quoted intending the work to be done correctly on the first try. A few extra minutes spent with a measuring tape and a calculator can transform a marginal gain into more profit than you expected.

Take the example of the Job Connection box at the beginning of this chapter. The architect has drawn the pipes to be installed in the hallway ceiling as lines. It's easy to put those pipes in with a pencil, but it's the plumber's responsibility to turn those lines into pipes that function the way they were designed to function.

We will begin with the first 1-inch copper water supply pipe. On the straight run, the Ace Plumbing team must install a 1 × 1 × 1 inch copper sweat tee for a supply to another area of the hospital. The architect wants the pipe centered over a doorway to allow room for some ductwork. If Julio measures to the center of the doorway and cuts the pipe at that measurement, it will be too long and the tee branch will be off center. Pat tells Julio that he must allow for the fitting allowance. Pat has brought a pipe-fitting handbook along since Pat didn't expect Julio or Evan to be prepared properly.

The chart states that the fitting allowance for a 1-inch copper sweat tee is $\frac{3}{4}$ inch. That means the pipe must be cut $\frac{3}{4}$ of an inch shorter in order for the branch of the 1-inch tee to be centered over the door (Figure 3.1). The distance from the center of the branch (which is where the architect drew it) to the bottom of the fitting socket where the pipe will end is the fitting allowance. All other fitting patterns also have a fitting allowance. It is found in a similar way. You should have a plumbing math book or a pipe fitter's handbook so you have these dimensions handy.

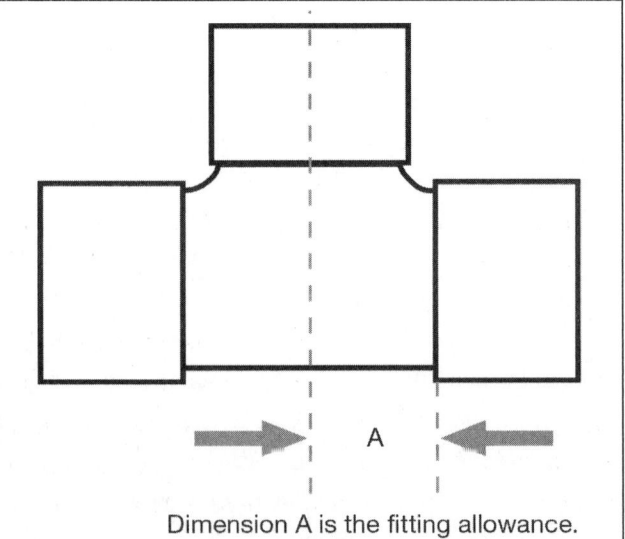

Dimension A is the fitting allowance.

Figure 3.1 Fitting Allowance for a Copper Tee

Sometimes a plumber must create an **offset** when there is an obstruction, such as a pillar that must be piped around (Figure 3.2). If the offset can be done with 90-degree elbows, it's a relatively simple task to figure the length of the offsetting pipe. You measure from the centerline of the existing pipe to the position where the centerline of the new pipe must be, deduct the length of the fitting allowance twice (once for each end of the pipe), and you have the length of the pipe needed.

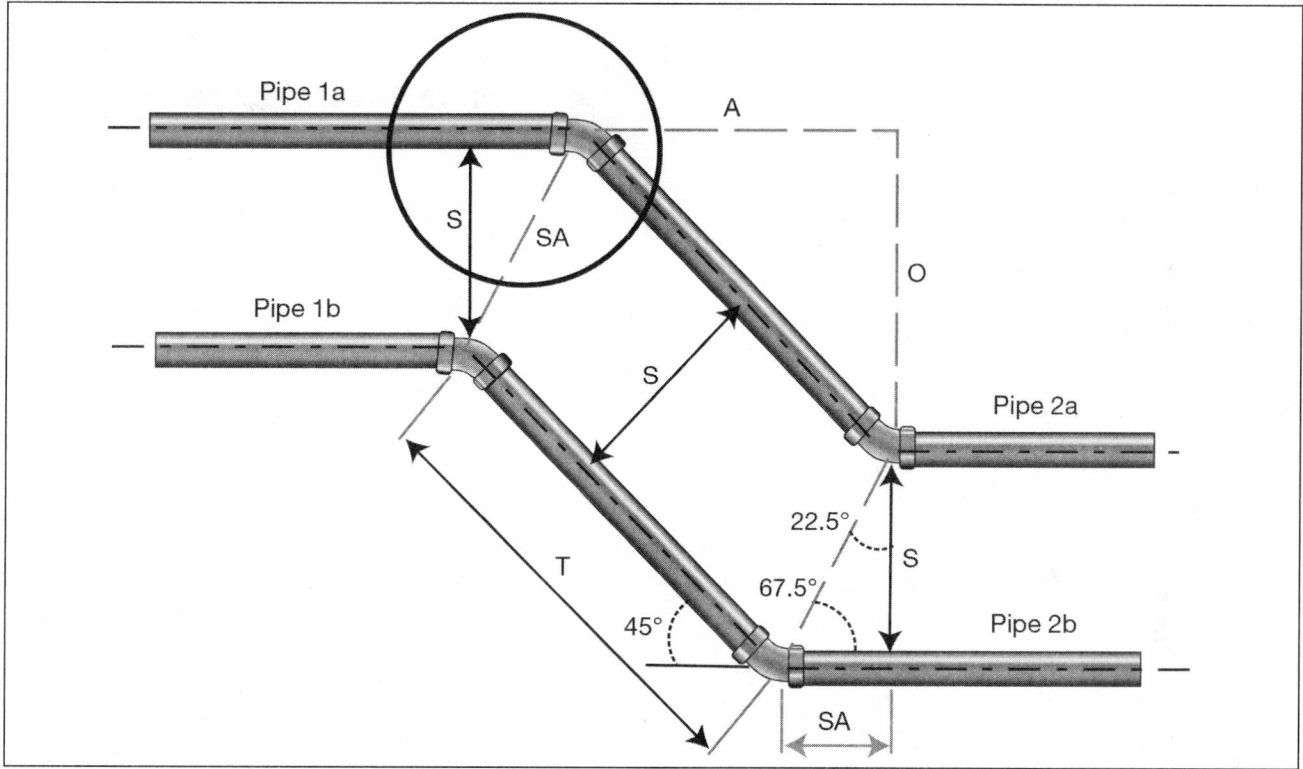

Figure 3.2 An Offset in Piping

The next most common fitting angle used for offsets is 45 degrees. A 45-degree angled fitting has much less resistance to flow than a 90-degree angled fitting. That explains its common usage as an offset. Figuring the length of the offsetting, or middle, pipe involves only one simple additional step. When using the 45-degree angled elbow, you will multiply the distance from the centerline of the original pipe to the centerline of the new pipe by 1.414. It will always be 1.414 when you are using 45-degree elbows. This number is called a **constant**. You then deduct the fitting allowance for a 45-degree elbow twice, once for each end again, and cut your pipe (Figure 3.3).

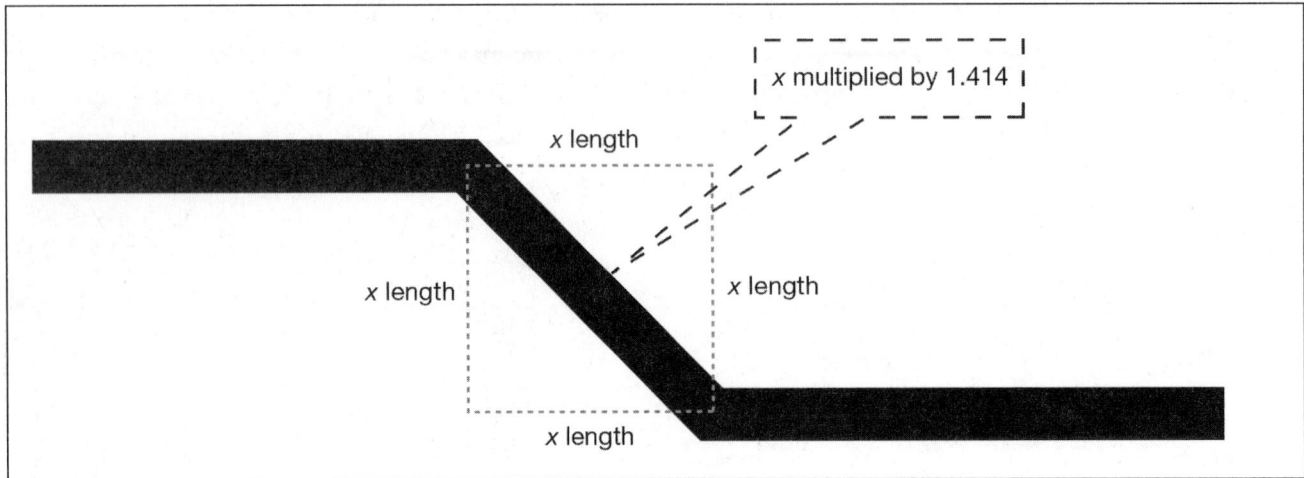

Figure 3.3 The Important Information Needed to Build an Offset

Occasionally, you know the length of the diagonal section of the piping. When you are using 45-degree elbows, the offset distance—that is, the measurement from the original pipe to the centerline of the continuation of the pipe—will be the length of the diagonal times 0.707 plus the fitting allowance. We will only touch on 45-degree offsets since they are the most common. It is unlikely that the licensing test would present you with any other angle for an offset.

Basic Calculations

After linear calculations, determining the volume of cylinders is the next most essential type of problem that must be solved by plumbers. In addition to the tank volume problem mentioned at the beginning of the chapter, two typical occasions that would demand volume calculations are figuring out the water content of a hydronic heating system to determine the amount of antifreeze that needs to be added for freeze protection and to determine the size of an expansion tank for the heating system.

Simple Geometry

Let's look at some basic shapes and their parts. Figure 3.4 illustrates a circle's radius, which is the distance from the center of a circle to the circle line. A diameter is the distance all the way across a circle (Figure 3.5). Radius and diameter are measurements we often use in plumbing since we deal with round pipes, round tanks, and round fittings. Circumference is the distance around a circle (Figure 3.6).

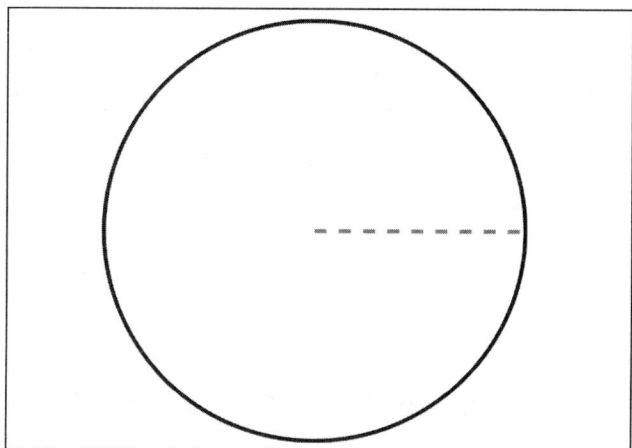

Figure 3.4 Circle Radius

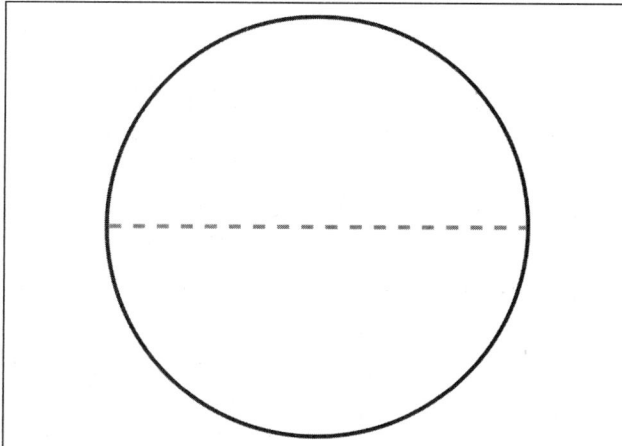

Figure 3.5 Circle Diameter

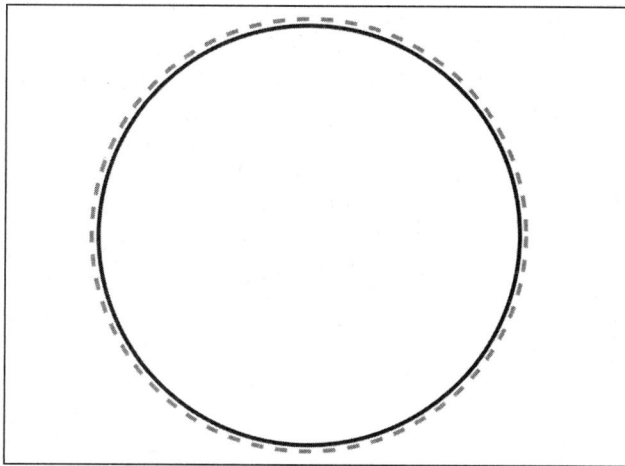

Figure 3.6 Circle Circumference—the Distance around a Circle

A cylinder is a circle with depth. We are confronted with cylinders often in the form of tanks (Figure 3.7).

Figure 3.7 Cylinder

We need to know the math involved for calculating the lengths, perimeters, areas, and volumes of these shapes. I know this is starting to sound harder, but it's not.

OK, we need to know that radius so we can figure out how much real estate the circle occupies. In other words, the area. The formula for the area of a circle is: $A = \pi r^2$. Read it out loud: Area equals pi (3.14) multiplied by the radius squared: $A = \pi r^2$.

Squared means the number is multiplied by itself. Pi is a Greek letter that stands for 3.14, which is a constant that is used for finding quantities in circles. The number really goes on indefinitely in decimals. (If you want to see how long it is, plug the formula $355 \div 113$ into a Cray supercomputer. That will keep it busy for a while.)

Rounding to two decimal places for the final answer is close enough for our purposes. So, what is the area of a circle that has a radius of $1\frac{1}{4}$ inches? I'm using a fraction because that's how we plumbers measure things. Before we can do any calculation, we have to convert the $\frac{1}{4}$ to a decimal. All you have to do is divide the top number by the bottom number:

$$1 \div 4 = 0.25$$

Let's plug these numbers into our area formula now: A = 3.14 × (1.25 × 1.25)². We always do the calculation in the parentheses first, so:

1.25 × 1.25 = 1.5625

Now, we have A = 3.14 × 1.56252. Of course, it's OK to use a calculator to do the squaring. You should now get something like A = 3.14 × 2.4414. Round it off to A = 3.14 × 2.44. I get A = 7.6616, which you can round off to A = 7.66. We always express area in square dimensions, like square feet, or square yards, or square miles. The original dimensions given to us for this problem were in inches, so we express the area of our circle with the $1\frac{1}{4}$-inch radius as 7.66 square inches.

> ### Hints and Tips
> Here are some of the constants commonly used in plumbing:
>
> **1.414** times the distance between the pipes (the offset) in a 45-degree offset = the length of the diagonal pipe
>
> **0.707** times the length of a diagonal pipe = the distance between the offset pipes
>
> **231** = cubic inches in a gallon
>
> 1 cubic foot = **1,728** cubic inches
>
> 1 cubic foot = **7.481** gallons
>
> 1 cubic foot of water weighs **62.426** pounds
>
> 1 gallon = **0.1339** cubic foot
>
> 1 gallon of water weighs **8.345** pounds

It's nice that we have an answer, but what in the world is 0.66 of an inch? We plumbers want our dimensions in feet, inches, and parts of inches. We need to convert 0.66 to inches, which is surprisingly easy. First, decide how accurate you need to be. If $\frac{1}{8}$ inch is accurate enough, you simply multiply the decimal number you have, in this case, it's 0.66, by the denominator, which is 8. That will tell you how many eighths the answer is. Let's do the calculation:

0.66 × 8 = 5.28

That means we have 5.28 eighths. Since the $\frac{1}{8}$-inch tolerance is good enough in this case, we will round off 5.28 to 5. Our answer is $\frac{5}{8}$ inch. So, if you need your measurement to the nearest $\frac{1}{8}$ inch, 0.66 equals $\frac{5}{8}$ inch.

Let's say you need a higher degree of accuracy. We'll work it to the nearest $\frac{1}{16}$ inch. All you have to do is multiply the decimal number, 0.66 in this case, by 16, which is the denominator of the level of accuracy you need for this measurement. Try it:

0.66 × 16 = 10.56

That means we have 10.56 sixteenths. Again, round it off to the nearest whole number, which is 11. Since we need the accuracy to the nearest sixteenth, we will now express our answer as $\frac{11}{16}$ inch.

This simple formula works no matter what degree of accuracy you need. It will work for quarter inches as well as sixty-fourths.

Take another look at Figure 3.5. It shows the diameter of a circle. We use the diameter to find the circumference of, or the distance around, a circle. The formula is C = πd. This formula means that the circumference of a circle is equal to pi (π) multiplied by the diameter of the circle. With a diameter of 8 inches, what is the circumference? Plug in the numbers and work it out:

C = 3.14 × 8
C = 25.12 inches

MATH FOR PLUMBERS

Here's another decimal we must convert to a fraction of an inch so we can measure it. This time we need to calculate to the nearest $\frac{1}{8}$ inch, so we multiply 0.12 by 8:

$0.12 \times 8 = 0.96$

That's how many eighths we have. Rounding 0.96 to the nearest whole number, we get 1. That gives us a circumference of $25\frac{1}{8}$ inches.

Knowing the volume of a cylinder is important in many plumbing situations. Once we find the area of one end of the cylinder by using the area formula, $A = \pi r^2$, all we have to do is multiply that area by the height of the cylinder. Take a look at the formula for the volume of a cylinder: $V = \pi r^2 h$. Volume equals pi (3.14) multiplied by the radius squared, and then multiplied by the height of the cylinder. Run through it once using 12 inches for the radius and 48 inches for the height:

$V = 3.14 \times 12^2 \times 48$
$V = 3.14 \times 144 \times 48$
$V = 21{,}703.68$ cubic inches

That's great, but you need to know how many *gallons* a tank holds so you can replace it with the same size tank. The supply house only stocks tanks measured in gallons, not cubic inches.

Here's another number to memorize: 1 gallon of water has 231 cubic inches in it. All you have to do is divide the cubic inches by 231 to get gallons. Using the cylinder example above, you get:

$21{,}703.68 \div 231 = 93.96$ gallons

All you have to do then is round 93.96 to 94 gallons.

Chapter Review

Note: Answers may vary based on whether the IPC or UPC is applicable.

1. What is the sum of $\frac{5}{8}$ in. and $\frac{7}{8}$ in.?

   _____
   _____
   _____
   _____

2. What is the sum of $\frac{3}{4}$ in. and $\frac{1}{8}$ in.?

   _____
   _____
   _____
   _____

3. What is $\frac{3}{4}$ in. minus $\frac{5}{8}$ in.?

   _____
   _____
   _____
   _____

4. What is the area of a hole drilled for a closet flange that has a radius of $2\frac{1}{2}$ in.?

   _____
   _____
   _____
   _____

5. The label is missing from a leaking water heater. What is the volume of the tank in gallons if the radius of the top is 9 in. and the height of the tank is 48 in.?

_____
_____
_____

6. Two parallel $\frac{1}{2}$-in. copper pipes must be connected on the end using two 90-degree elbows forming a U shape, which have a fitting allowance of $\frac{3}{8}$ in., and a piece of $\frac{1}{2}$-in. copper pipe. The two parallel pipes must be kept 8 in. apart, measured center to center. How long must the connecting piece of pipe be?

_____
_____
_____

7. A pipe that is 85 ft. long has an inside diameter of 2 in. How many gallons of water will it take to fill it?

_____
_____
_____

8. A 1-in. steel gas pipe must be offset 1ft. 6in., using 45-degree elbows. How long must the connecting piece be to the nearest $\frac{1}{16}$ in.? The fitting allowance for 1-in. steel pipe is 1 in.

_____
_____
_____

9. What is 4 ft. minus $6\frac{7}{8}$ in.?

_____
_____
_____

10. To the nearest $\frac{1}{16}$ in., how long must a piece of tape be to just go around a section of pipe insulation that has an outside diameter of $3\frac{1}{4}$ in?

_____
_____
_____

MATH FOR PLUMBERS

11. A pitch of $\frac{1}{4}$-in. per foot must be maintained on a drainpipe that is being hung under floor joists. On the upstream end of the pipe, it measures $12\frac{7}{8}$ in. from the underside of the floor above to the centerline of the pipe. What will the measurement be from the underside of the floor to the centerline of the pipe 18 ft. downstream?

12. Julio must offset a $1\frac{1}{2}$-in. copper pipe a distance of 12 in. using 45-degree elbows. How long must the diagonal pipe be if the fitting allowance for a $1\frac{1}{2}$-in. copper 45-degree elbow is $\frac{9}{16}$ in.?

13. What is 3 ft. $3\frac{1}{8}$ in. minus 2 ft. $\frac{3}{4}$ in.?

14. Antifreeze needs to be added to a heating system as a 50% antifreeze-water solution. How much straight antifreeze is needed to fill the two manifold pipes that are both 4 ft. 3 in. long and have a diameter of 2 in.?

15. What is the sum of 1 ft. 8 in. and 20 in.?

16. How many $\frac{1}{8}$-in. are there in 4 in.?

17. The volume of a section of pipeline is 182,822 cubic inches. How many gallons will it hold?

MATH FOR PLUMBERS

18. Evan must offset two parallel copper pipes 14 in. using 45-degree elbows. One of the pipes is $\frac{3}{4}$ in. in diameter and the other is 1 in. in diameter. The fitting allowance for $\frac{3}{4}$-in. pipe is $\frac{1}{4}$ in. and for 1-in. pipe, it's $\frac{5}{16}$ in. The two pipes must remain 3 in. apart, center to center, throughout the offset. How long must each diagonal pipe be?

_____

_____

_____

19. A $2\frac{3}{8}$-in. hole must be drilled through the center of a 2 × 6 stud. How much wood will be left on each side of the hole? Remember, a 2 × 6 is really $1\frac{1}{2} \times 5\frac{1}{2}$ in.

_____

_____

_____

20. A tank holds 6 gallons. How many cubic inches are in it?

_____

_____

_____

_____

21. A sheet of insulation blanket must be cut to cover the sides of a tank with a diameter of 10 in. and a height of 22 in. What must the dimensions of the blanket be to cover only the sides of the tank?

_____

_____

_____

22. 8 ft. $3\frac{1}{4}$ in. minus 1 ft. $4\frac{1}{2}$ in. equals what?

_____

_____

_____

23. Pat kicks over a bucket of antifreeze that is 12 in. in diameter and 19 in. tall. The level of the antifreeze was 10 in. down from the top before Pat kicked it over. How much antifreeze, in gallons, did Pat have to clean up?

_____

_____

_____

24. A 2-in. diameter pipe that is 3 ft. long has 1 ft. 6 in. added to its length. What is the volume of the lengthened pipe in gallons?

_____

_____

_____

_____

MATH FOR PLUMBERS

25. Julio drilled a hole through a wall for a pipe to pass through. The outside diameter of the pipe is $2\frac{1}{4}$ in. There is $\frac{1}{4}$ in. of clearance on all sides of the pipe to the edge of the hole. What is the circumference of the hole?

_____

_____

_____

26. A plumbing distribution system has 330 ft. of $\frac{1}{2}$-in. copper pipe. What size bucket, in gallons, would you need to drain the whole system?

_____

_____

_____

_____

27. The pitch of a large sewer is $\frac{1}{8}$-in. per foot. What is the difference in elevation from the sewage treatment plant to the end of the sewer main, which is 1,575 ft. away?

_____

_____

_____

28. What is the sum of $\frac{3}{4}$ in., $\frac{1}{8}$ in. and $\frac{5}{16}$ in.?

_____

_____

_____

29. Evan cut the 5-in. diameter hole for a water closet flange in the wrong place. He must cut a new hole $4\frac{1}{8}$ in., measured center to center, to the left of the original hole. Will the holes overlap?

_____

_____

_____

30. How far is it from the edge of a $2\frac{3}{8}$-in. diameter hole to the edge of another $2\frac{3}{8}$-in. diameter hole drilled 8 in. apart, center to center?

_____

_____

_____

_____

31. How long must a piece of 1-in. copper pipe be to act as an offset using 90-degree elbows when the two pipes are 12 in. apart? The fitting allowance for a 1-in. copper 90-degree elbow is $\frac{3}{4}$ in.

_____

_____

_____

_____

32. What is the area in square inches of a manhole cover that is 2 ft. in diameter?

33. If a 50-ft. length of a 24-in. diameter storm drain is running half full, how many gallons of water are in the pipe?

34. A hot water expansion tank holds 3 gallons and is 16 in. high. What is the diameter?

35. What is 35 in. minus 2 ft.?

CHAPTER 4  FIXTURES

CHAPTER SUMMARY

This is one of the areas of the plumbing trade where your work will be on display to the public. The requirements for the number of fixtures provided in public restrooms for each of the sexes has been debated exhaustively over the years. The IPC and the UPC have differing numbers throughout each of their charts, which demonstrates the intensity of the debates. Both codes have workable numbers, but you must be familiar with where to locate them in your respective codebook before you enter the testing room. Chances are you will need to pull some information from these charts.

The UPC explains some general requirements for fixtures, while the IPC goes into detail on a number of specific fixtures. The IPC includes as fixtures receptacles and devices that do not require connection to a water supply, such as waterless urinals and floor drains. (Refer to IPC 202) The IPC also clarifies the distinction between plumbing fixtures and appliances. Appliances are defined as having their operation or control dependent upon one or more energized components, such as motors, controls, and heating elements, manually adjusted or controlled by the owner-operator or operated automatically through one or more of the following actions: a time cycle, a temperature range, a measured volume or weight. (Refer to IPC 202, 2012) The results of both codes are approximately the same. The differences illustrate "design-by-committee" approach to code writing. We really wouldn't want it any other way. The folks involved in writing these plumbing codes are experts who bring a variety of experience and disciplines to the table.

FIXTURES

Learning Objectives

- Understand fixture material selection
- Explain the importance of access to facilities for all people
- Make appropriate connections of fixtures to plumbing systems

Key Terms

ADA
air break
closet carrier
concealed fouling surface
GPF

Fixture Material

Both the IPC and the UPC agree that plumbing fixtures need to have acid-resistant, waterproof surfaces that are smooth and have no defects. There is a slight difference between the IPC and the UPC regarding the issue of hidden fouling surfaces in fixtures. The IPC states that **concealed fouling surfaces** are prohibited. The UPC says that *unnecessary* concealed fouling surfaces are prohibited. The UPC leaves the discussion open for who decides what is unnecessary. Be aware of the concealed fouling surface issue with respect to the code used in your area, but realize that it's something you will probably never be involved in. Those issues are settled between code officials and the fixture manufacturers long before you handle the finished product. (Refer to IPC 402; UPC 401)

The IPC goes on to state that any specialty fixture must be made of materials suitable for its use. That keeps manufacturers from making, for instance, leather lavatories; that might be stylish, but certainly not appropriate for the function they are intended to perform.

JOB CONNECTION

Ace Plumbing Company has been invited to bid on the renovation of two separate, permanent bathroom facilities on the County Fairgrounds. The fair committee wants the contractors to carefully remove all of the old fixtures, replace all of the dividers, install new tile walls and floors, replace all of the supply and drain piping, install all new electrical wiring, and then reinstall all of the plumbing fixtures. The existing fixtures include wall-hung lavatories, water closets, and trough urinals. The lavatories and water closets in all the women's and men's rooms are in compliance with the Americans with Disabilities Act (**ADA**). Do you think Sophia, the Ace Plumbing Company estimator, will want to bid on this project?

FIXTURES

Fixture Requirements in Various Buildings

There is certainly a need for the requirements displayed in Table 4.1, which is Table 403.1 of the IPC, and Table 4.2, which is the UPC's Table 4–1. If the numbers of appropriate fixtures were not a requirement, imagine the lines you could face at a ball field or movie theater. Minimum numbers of plumbing fixtures must be provided according to the type of building occupancy as shown in UPC Table 422.1 and UPC 412.1.1, 422.2. The tables are included in this chapter for your reference and comparison. The table you need will also be found in your plumbing codebook. (Refer to IPC 403.1–403.4; UPC 412.1–413)

TABLE 4.1 MINIMUM NUMBER OF REQUIRED PLUMBING FIXTURES (IPC TABLE 403.1)

No.	Classification	Occupancy	Description	Water Closets (Urinals See Section 419.2)		Lavatories		Bathtubs/ Showers	Drinking Fountain (See Section 410.1)	Other
				Male	Female	Male	Female			
1	Assembly (see Sections 403.2, 403.4 and 403.4.1)	A-1[d]	Theaters and other buildings for the performing arts and motion pictures	1 per 125	1 per 65	1 per 200		—	1 per 500	1 service sink
		A-2[d]	Nightclubs, bars, taverns, dance halls and buildings for similar purposes	1 per 40	1 per 40	1 per 75		—	1 per 500	1 service sink
			Restaurants, banquet halls and food courts	1 per 75	1 per 75	1 per 200		—	1 per 500	1 service sink
		A-3[d]	Auditoriums without permanent seating, art galleries, exhibition halls, museums, lecture halls, libraries, arcades and gymnasiums	1 per 125	1 per 65	1 per 200		—	1 per 500	1 service sink
			Passenger terminals and transportation facilities	1 per 500	1 per 500	1 per 750		—	1 per 1,000	1 service sink
			Places of worship and other religious services	1 per 150	1 per 75	1 per 200		—	1 per 1,000	1 service sink
		A-4	Coliseums, arenas, skating rinks, pools and tennis courts for indoor sporting events and activities	1 per 75 for the first 1,500 and 1 per 120 for the remainder exceeding 1,500	1 per 40 for the first 1,500 and 1 per 60 for the remainder exceeding 1,500	1 per 200	1 per 150	—	1 per 1,000	1 service sink
		A-5	Stadiums, amusement parks, bleachers and grandstands for out-door sporting events and activities	1 per 75 for the first 1,500 and 1 per 120 for the remainder exceeding 1,500	1 per 40 for the first 1,500 and 1 per 60 for the remainder exceeding 1,500	1 per 200	1 per 150	—	1 per 1,000	1 service sink

(Continued)

TABLE 4.1 (Continued)

No.	Classification	Occupancy	Description	Water Closets (Urinals See Section 419.2)		Lavatories		Bathtubs/ Showers	Drinking Fountain (See Section 410.1)	Other
				Male	Female	Male	Female			
2	Business (see Sections 403.2, 403.4 and 403.4.1)	B	Buildings for the transaction of business, professional services, other services involving merchandise, office buildings, banks, light industrial and similar uses	1 per 25 for the first 50 and 1 per 50 for the remainder exceeding 50		1 per 40 for the first 80 and 1 per 80 for the remainder exceeding 80		—	1 per 100	1 service sink
3	Educational	E	Educational facilities	1 per 50		1 per 50		—	1 per 100	1 service sink
4	Factory and industrial	F-1 and F-2	Structures in which occupants are engaged in work fabricating, assembling, or processing of products or materials	1 per 100		1 per 100		(see Section 411)	1 per 400	1 service sink
5	Industrial	I-1	Residential care	1 per 10		1 per 10		1 per 8	1 per 100	1 service sink
		I-2	Hospitals, ambulatory nursing home patients[b]	1 per room[c]		1 per room[c]		1 per 15	1 per 100	1 service sink per floor
			Employees, other than residential care[b]	1 per 25		1 per 35		—	1 per 100	—
			Visitors, other than residential care	1 per 75		1 per 100		—	1 per 500	—
		I-3	Prisons[b]	1 per cell		1 per cell		1 per 15	1 per 100	1 service sink
		I-3	Reformitories, detention centers, and correctional centers[b]	1 per 15		1 per 15		1 per 15	1 per 100	1 service sink
		I-4	Adult day care and child care	1 per 15		—		—	1 per 100	1 service sink
6	Mercantile (see Sections 403.2, 403.4, 403.4.1 and 403.4.2)	M	Retail stores, service stations, shops, salesrooms, markets, and shopping centers	1 per 500		1 per 750		—	1 per 1,000	1 service sink

(Continued)

FIXTURES

TABLE 4.1 (Continued)

No.	Classification	Occupancy	Description	Water Closets (Urinals See Section 419.2)		Lavatories		Bathtubs/ Showers	Drinking Fountain (See Section 410.1)	Other
				Male	Female	Male	Female			
7	Residential	R-1	Hotels, motels, boarding houses (transient)	1 per sleeping unit	1 per sleeping unit	1 per sleeping unit		1 per sleeping unit	—	1 service sink
		R-2	Dormitories, fraternities, sororities and boarding houses (not transient)	1 per 10		1 per 10		1 per 8	1 per 100	1 service sink
		R-2	Apartment house	1 per dwelling unit		1 per dwelling unit		1 per dwelling unit	—	1 kitchen sink per dwelling unit; 1 automatic clothes washer connection per 20 dwelling units
		R-3	One- and two-family dwellings	1 per dwelling unit		1 per dwelling unit		1 per dwelling unit	—	1 kitchen sink per dwelling unit; 1 automatic clothes washer connection per dwelling unit
		R-4	Residential care/ assisted living facility	1 per 10		1 per 10		1 per 8	1 per 100	1 service sink
8	Storage (see Sections 403.2, 403.4 and 403.4.1)	S-1 S-2	Structures for the storage of goods, warehouses, storehouse and freight depots. Low and Moderate Hazard.	1 per 100		1 per 100		See Section 411	1 per 1,000	1 service sink

a. The fixtures shown are based on one fixture being the minimum required for the number of persons indicated or any fraction of the number of persons indicated. The number of occupants shall be determined by the *International Building Code*.
b. Toilet facilities for employees shall be separate from facilities for inmates or patients.
c. A single-occupant toilet room with one water closet and one lavatory serving not more than two adjacent patient sleeping units shall be permitted where such room is provided with direct access from each patient room and with provisions for privacy.
d. The occupant load for seasonal outdoor seating and entertainment areas shall be included when determining the minimum number of facilities required.
e. Service sinks are no longer required in Group B and M occupancies where the occupant load does not exceed 15. (Refer to IPC 403.1, Table 4.1)

FIXTURES

TABLE 4.2 MINIMUM PLUMBING FACILITIES (IPC TABLE 4-1) (Reprinted with the permission of the International Association of Plumbing and Mechanical Officials.)

Each building shall be provided with sanitary facilities, including provisions for persons with disabilities as prescribed by the Department Having Jurisdiction. For requirements for persons with disabilities, ICC/ANSI A117.1, Accessible and Usable Buildings and Facilities, may be used.

The minimum number of fixtures shall be calculated at fifty (50) percent male and fifty (50) percent female based on the total occupant load.

The occupant load and use of the building or space under consideration shall first be established using the Occupant Load Factor Table A. Once the occupant load and uses are determined, the requirements of Section 412.0 and Table 4-1 shall be applied to determine the minimum number of plumbing fixtures required.

This table applies to new buildings, additions to a building, changes of occupancy or type in an existing building resulting in increased occupant load (example: change an assembly room from fixed seating to open seating). Exception: New cafeterias for employee use are the only use exempted from this requirement.

Type of Building or Occupancy[2]	Water Closets[14] (Fixtures per Person)	Urinals[5,10] (Fixtures per Person)	Lavatories (Fixtures per Person)	Bathtubs or Showers (Fixtures per Person)	Drinking Fountains[3,12,19] (Fixtures per Person)
Assembly places—theatres, auditoriums, convention halls, etc.—for permanent employee use	Male / Female 1: 1–15 / 1: 1–15 2: 16–35 / 3: 16–35 3: 36–55 / 4: 36–55 Over 55, add 1 fixture for each additional 40 persons.	Male 0: 1–9 1: 10–50 Add one fixture for each additional 50 males.	Male / Female 1 per 40 / 1 per 40		
Assembly places—theatres, auditoriums, convention halls, etc.—for public use	Male / Female 1: 1–100 / 3: 1–50 2: 101–200 / 4: 51–100 3: 201–400 / 8: 101–200 / 11: 201–400 Over 400, add one fixture for each additional 500 males and 1 for each additional 125 females.	Male 1: 1–100 2: 101–200 3: 201–400 4: 401–600 Over 600, add 1 fixture for each additional 300 males.	Male / Female 1: 1–200 / 1: 1–200 2: 201–400 / 2: 201–400 3: 401–750 / 3: 401–750 Over 750, add one fixture for each additional 500 persons.		1: 1–150 2: 151–400 3: 401–750 Over 750, add one fixture for each additional 500 persons.
Dormitories[9]—School or labor[17]	Male / Female 1 per 10 / 1 per 8 Add 1 fixture for each additional 25 males (over 10) and 1 for each additional 20 females (over 8).	Male 1 per 25 Over 150, add 1 fixture for each additional 50 males.	Male / Female 1 per 12 / 1 per 12 Over 12, add one fixture for each additional 20 males and 1 for each 15 additional females.	1 per 8 For females, add 1 bathtub per 30. Over 150, add 1 bathtub per 20.	1 per 150[12]
Dormitories—for staff use[17]	Male / Female 1: 1–15 / 1: 1–15 2: 16–35 / 3: 16–35 3: 36–55 / 4: 36–55 Over 55, add 1 fixture for each additional 40 persons.	Male 1 per 50	Male / Female 1 per 40 / 1 per 40	1 per 8	
Dwellings[4] Single dwelling Multiple dwelling or apartment house[17]	1 per dwelling 1 per dwelling or apartment unit		1 per dwelling 1 per dwelling or apartment unit	1 per dwelling 1 per dwelling or apartment unit	
Hospital waiting rooms	1 per room		1 per room		1 per 150[12]
Hospitals—for employee use	Male / Female 1: 1–15 / 1: 1–15 2: 16–35 / 3: 16–35 3: 36–55 / 4: 36–55 Over 55, add 1 fixture for each additional 40 persons.	Male 0: 1–9 1: 10–50 Add one fixture for each additional 50 males.	Male / Female 1 per 40 / 1 per 40		

(Continued)

FIXTURES

TABLE 4.2 *(Continued)*

Type of Building or Occupancy[2]	Water Closets[14] (Fixtures per Person)	Urinals[5,10] (Fixtures per Person)	Lavatories (Fixtures per Person)	Bathtubs or Showers (Fixtures per Person)	Drinking Fountains[3,12,19] (Fixtures per Person)

Hospitals
  Individual room
  Ward room
	1 per room
	1 per 8 patients
			1 per room
			1 per 10 patients
				1 per room
				1 per 20 patients
					1 per 150[12]

Industrial[6] warehouses, workshops, foundries, and similar establishments—for employee use	Male Female 1: 1–10 1: 1–10 2: 11–25 2: 11–25 3: 26–50 3: 26–50 4: 51–75 4: 51–75 5: 76–100 5: 76–100 Over 100, add 1 fixture for each additional 30 persons.		Up to 100, 1 per 10 persons Over 100, 1 per 15 persons[7,8]	1 shower for each 15 persons exposed to excessive heat or to skin contamination with poisonous, infectious, or irritating material.	1 per 150[12]
Institutional—other than hospitals or penal institutions (on each occupied floor)	Male Female 1 per 25 1 per 20	Male 0: 1–9 1: 10–50 Add one fixture for each additional 50 males.	Male Female 1 per 10 1 per 10	1 per 8	1 per 150[12]
Institutional—other than hospitals or penal institutions (on each occupied floor)—for employee use	Male Female 1: 1–15 1: 1–15 2: 16–35 3: 16–35 3: 36–55 4: 36–55 Over 55, add 1 fixture for each additional 40 persons.	Male 0: 1–9 1: 10–50 Add one fixture for each additional 50 males.	Male Female 1 per 40 1 per 40	1 per 8	1 per 150[12]
Office or public buildings	Male Female 1: 1–100 3: 1–50 2: 101–200 4: 51–100 3: 201–400 8: 101–200 11: 201–400 Over 400, add one fixture for each additional 500 males and 1 for each additional 150 females.	Male 1: 1–100 2: 101–200 3: 201–400 4: 401–600 Over 600, add 1 fixture for each additional 300 males.	Male Female 1: 1–200 1: 1–200 2: 201–400 2: 201–400 3: 401–750 3: 401–750 Over 750, add one fixture for each additional 500 persons.		1 per 150[12]
Office or public buildings—for employee use	Male Female 1: 1–15 1: 1–15 2: 16–35 3: 16–35 3: 36–55 4: 36–55 Over 55, add 1 fixture for each additional 40 persons.	Male 0: 1–9 1: 10–50 Add one fixture for each additional 50 males.	Male Female 1 per 40 1 per 40		
Penal institutions—for employee use	Male Female 1: 1–15 1: 1–15 2: 16–35 3: 16–35 3: 36–55 4: 36–55 Over 55, add 1 fixture for each additional 40 persons.	Male 0: 1–9 1: 10–50 Add one fixture for each additional 50 males.	Male Female 1 per 40 1 per 40		1 per 150[12]

Penal institutions—for prison use
  Cell
  Exercise room
	1 per cell
	1 per exercise room
		Male
		1 per exercise room
			1 per cell
			1 per exercise room
					1 per cell block floor
					1 per exercise room

Public or professional offices[16]	Same as Office or Public Buildings for employee use[15]	Same as Office or Public Buildings for employee use[15]	Same as Office or Public Buildings for employee use[15]		Same as Office or Public Buildings for employee use[15]

(Continued)

TABLE 4.2 (Continued)

Type of Building or Occupancy[2]	Water Closets[14] (Fixtures per Person)	Urinals[5,10] (Fixtures per Person)	Lavatories (Fixtures per Person)	Bathtubs or Showers (Fixtures per Person)	Drinking Fountains[3,12,19] (Fixtures per Person)
Restaurants, pubs, and lounges[11,15,16]	Male / Female 1: 1–50 / 1: 1–50 2: 51–150 / 2: 51–150 3: 151–300 / 4: 151–300 Over 300, add 1 fixture for each additional 200 persons.	Male 1: 1–150 Over 150, add 1 fixture for each additional 150 males.	Male / Female 1: 1–150 / 1: 1–150 2: 151–200 / 2: 151–200 3: 201–400 / 3: 201–400 Over 400, add 1 fixture for each additional 400 persons.		

Retail or Wholesale Stores	Male / Female
	1: 1–100 / 1: 1–25
	2: 101–200 / 2: 26–100
	3: 201–400 / 4: 101–200
	               6: 201–300
	               8: 301–400
	Over 400, add one fixture for each additional 500 males and one for each 150 females.
		Male
		0: 0–25
		1: 26–100
		2: 101–200
		3: 201–400
		4: 401–600
		Over 600, add one fixture for each additional 300 males.
			One for each two water closets		0: 1–30[17]
					1: 31–150
					One additional drinking fountain for each 150 persons thereafter

Schools— for staff use All schools	Male / Female 1: 1–15 / 1: 1–15 2: 16–35 / 2: 16–35 3: 36–55 / 3: 36–55 Over 55, add 1 fixture for each additional 40 persons.	Male 1 per 50	Male / Female 1 per 40 / 1 per 40		
Schools— for student use Nursery	Male / Female 1: 1–20 / 1: 1–20 2: 21–50 / 2: 21–50 Over 50, add 1 fixture for each additional 50 persons.		Male / Female 1: 1–25 / 1: 1–25 2: 26–50 / 2: 26–50 Over 50, add 1 fixture for each additional 50 persons.		1 per 150[12]
Elementary	Male / Female 1 per 30 / 1 per 25	Male 1 per 75	Male / Female 1 per 35 / 1 per 35		1 per 150[12]
Secondary	Male / Female 1 per 40 / 1 per 30	Male 1 per 35	Male / Female 1 per 40 / 1 per 40		1 per 150[12]
Others (colleges, universities, adult centers, etc.)	Male / Female 1 per 40 / 1 per 30	Male 1 per 35	Male / Female 1 per 40 / 1 per 40		1 per 150[12]
Worship places educational and activities unit	Male / Female 1 per 150 / 1 per 75	Male 1 per 150	1 per 2 water closets		1 per 150[12]
Worship places principal assembly place	Male / Female 1 per 150 / 1 per 75	Male 1 per 150	1 per 2 water closets		1 per 150[12]

(Continued)

TABLE 4.2 (Continued)

[1] The figures shown are based upon one (1) fixture being the minimum required for the number of persons indicated or any fraction thereof.

[2] Building categories not shown on this table shall be considered separately by the Authority Having Jurisdiction.

[3] Drinking fountains shall not be installed in toilet rooms.

[4] Laundry trays. One (1) laundry tray or one (1) automatic washer standpipe for each dwelling unit or one (1) laundry tray or one (1) automatic washer standpipe, or combination thereof, for each twelve (12) apartments. Kitchen sinks, one (1) for each dwelling or apartment unit.

[5] For each urinal added in excess of the minimum required, one water closet may be deducted. The number of water closets shall not be reduced to less than two-thirds (2/3) of the minimum requirement.

[6] As required by ANSI Z4.1, Sanitation in Places of Employment.

[7] Where there is exposure to skin contamination with poisonous, infectious, or irritating materials, provide one (1) lavatory for each five (5) persons.

[8] Twenty-four (24) lineal inches (610 mm) of wash sink or eighteen (18) inches (457 mm) of a circular basin, when provided with water outlets for such space, shall be considered equivalent to one (1) lavatory.

[9] Laundry trays, one (1) for each fifty (50) persons. Service sinks, one (1) for each hundred (100) persons.

[10] General. In applying this schedule of facilities, consideration shall be given to the accessibility of the fixtures. Conformity purely on a numerical basis may not result in an installation suited to the needs of the individual establishment. For example, schools should be provided with toilet facilities on each floor having classrooms.
   a. Surrounding materials, wall, and floor space to a point two (2) feet (610 mm) in front of urinal lip and four (4) feet (1,219 mm) above the floor, and at least two (2) feet (610 mm) to each side of the urinal shall be lined with nonabsorbent materials.
   b. Trough urinals shall be prohibited.

[11] A restaurant is defined as a business that sells food to be consumed on the premises.
   a. The number of occupants for a drive-in restaurant shall be considered as equal to the number of parking stalls.
   b. Employee toilet facilities shall not be included in the above restaurant requirements. Hand-washing facilities shall be available in the kitchen for employees.

[12] Where food is consumed indoors, water stations may be substituted for drinking fountains. Offices or public buildings for use by more than six (6) persons shall have one (1) drinking fountain for the first one hundred fifty (150) persons and one (1) additional fountain for each three hundred (300) persons thereafter.

[13] There shall be a minimum of one (1) drinking fountain per occupied floor in schools, theatres, auditoriums, dormitories, offices, or public buildings.

[14] The total number of water closets for females shall be at least equal to the total number of water closets and urinals required for males. This requirement shall not apply to Retail or Wholesale Stores.

[15] For smaller-type Public and Professional Offices such as banks, dental offices, law offices, real estate offices, architectural offices, engineering offices, and similar uses. A public area in these offices shall use the requirements for Retail or Wholesale Stores.

[16] A unisex facility (one water closet and one lavatory may be used when the customer occupant load for the dining area, including outdoor seating area, is 10 or less and the total number of employees for the space is 4 or less.

[17] Recreation or community room in multiple dwellings or apartment buildings, regardless of their occupant load, shall be permitted to have separate single-accommodation facilities in common-use areas within tracts or multi-family residential occupancies where the use of these areas is limited exclusively to owners, residents, and their guests. Examples are community recreation or multi-purpose areas in apartments, condos, townhouses, or tracts.

[18] A drinking fountain shall not be required in occupancies of 30 or less. When a drinking fountain is not required, then footnotes 3, 12, and 13 are not applicable.

(Continued)

TABLE 4.2 (Continued)

Table A. Occupant Load Factor:

Occupancy*, **	Occupant Load Factor (square feet)
Group A	
1. Auditoriums, convention halls, dance floors, lodge rooms, stadiums, and casinos (use 1/2 "one-half" the number of fixed seating)	15 (where no fixed seating is provided)
2. Conference rooms, dining rooms, drinking establishments, exhibit rooms, gymnasiums, lounges, stages, and similar uses, including restaurants classified as Group B occupancies	30
3. Worship places; principal assembly area, educational and activity unit (use 1/2 "one-half" the number of fixed seating)	30 (where no fixed seating is provided)
Group B Office or public buildings (area accessible to the public)	200
Group E Schools for daycare, elementary, secondary	50
Educational Facilities Other than Group E Colleges, universities, adult centers, etc.	50
Group F Workshops, foundries, and similar establishments	2,000
Group H Hazardous materials fabrication and storage	2,000
Group I Hospital general use area, health care facilities	200
Group M Retail or wholesale stores	200
Group R Congregate residence, Group R-1	200
Group S Warehouse	5,000

* Any uses not specifically listed shall be based on similar uses listed in this table.

** For building or space with mixed occupancies, use appropriate occupancy group for each area (for example, a school may have an "A" occupancy for the gymnasium, a "B" occupancy for the office, an "E" occupancy for the classrooms, etc.). Accessory areas may be excluded (for example: hallway, restroom, stair enclosure).

FIXTURES

In the IPC code, when a unisex bathroom that is intended for use by people who need assistance is provided, the fixtures in that bathroom can be counted toward the total number of fixtures in the women's and the men's bathrooms in the building. An example would be a movie theater where perhaps six women's water closets and three men's water closets would be required. If there were a unisex bathroom provided, the one water closet in that bathroom would count as one women's water closet and as one men's water closet. Now the separate bathrooms would only be required to have five women's and two men's water closets. The total number of water closets in the building would now be eight, rather than the nine that would have been required if there was no unisex bathroom.

For family or assisted-use bathroom facilities serving as separate facilities, the IPC recently mandated the substitution of two-family/assisted-use bathrooms for each sex. It's more efficient to allow two-family or assisted-use bathrooms to serve as the required separate facilities because either bathroom can be used by either sex. (Refer to IPC 403.2.1)

The UPC does not have a provision similar to this. In general, the UPC requires a few more fixtures to be provided than the IPC. Each has their own scientific formulas for evaluating fixture needs and they both provide for toilet facilities that serve the public well.

Fixture Locations

Buildings that are not covered malls must have bathrooms for both employees and the public that are no more than 500 feet from any location in the building and the bathrooms must be no more than one floor away. If you are located on the third floor, for instance, the bathrooms would have to be no further away than a 500-foot trip, and they would have to be on the third floor where you are, or on the second or the fourth floor. This still seems like a long way, but at least a maximum distance has been established for the architects and engineers who design buildings. (Refer to IPC 403.4.1–403.5)

Fixtures for the use of the public must be accessible without going through any area of the building that is intended to be only for the employees. That means the public must be able to get to the bathrooms without going behind sales counters, through kitchens, through warehouse areas, or any other part of a building that is intrinsic to the operation of the business but is not a sales area. Also, bathrooms must not open directly into a room used to prepare food for public consumption. (Refer to IPC 403.3.2–403.31)

Covered malls are treated a little differently because malls can be very crowded at times and many people in a mall will not be familiar with the location of bathroom facilities. Therefore, the distance one must travel in a covered mall is reduced to 300 feet while the rule requiring no more than one story above or below is retained. This distance is measured from the front door of each store. For employees, the distance is measured from their individual work areas.

> **JOB CONNECTION**
>
> One of the ladies' rooms at the County Fairgrounds has eight water closets in a row installed along a wall that measures 19 feet 8 inches. These are all of the water closets in that bathroom. Can Ace Plumbing put them back in the same configuration and meet code? There is more than one issue here.

The IPC now requires that any pay toilets must be additional fixtures beyond the number required for a certain area. Regarding the locking of bathroom doors, the IPC now requires that the door at the entrance to a multiple-occupant bathroom not have a lock on the inside. This is intended to reduce the possibility of inappropriate activities that are more likely to occur when someone can lock others out of the bathroom. (Refer to IPC 403.3.6)

Bathroom Design Relative to Fixtures

Bathrooms in public areas must comply with ADA (Americans with Disabilities Act) standards. These rules make it possible for individuals with a variety of physical disabilities to use public restrooms unrestricted by the structure. This includes the allowance of room for wheelchairs. Before the ADA, people with disabilities had a very difficult, if not impossible, time simply using a bathroom; the ADA gives these individuals far more independence than they previously had. The content of the ADA that directly relates to plumbing is included in the appendix of this book. (Refer to IPC 404.1–405.3.2; UPC 413.0)

Section 405.1 of the IPC stresses the importance of protecting all water supplies to fixtures from backflow. There are no specifics explained in Section 405.1. If you are answering questions on your licensing test, and you are unsure of an answer, try to remember that the protection of all water supplies from any sort of contamination is one of the most important aspects of both codes. This concept may help you pick a correct answer.

Ease of cleaning any fixture and the area around it is necessary for proper sanitation. The IPC makes it a requirement in Section 405.2.

The specification of space between and around fixtures helps a bathroom look neat and allows most people to have normal access to all the fixtures. On your test, you will probably be expected to know that according to the IPC, water closets and urinals need to be a minimum of 15 inches from a wall to the left or right of the fixture, and there must be 30 inches from the center of one water closet to the center of the next water closet when there are multiple fixtures in a bathroom. The IPC requires a minimum of 21 inches clear area in front of a water closet, urinal, bidet, or lavatory. This means a completely open area with no door, wall, or fixture encroaching on that area (Figure 4.1). A wall-hung, nonaccessible water closet compartment is permitted to be 56 inches in depth, which is 4 inches less than is required for a compartment containing a floor-mounted water closet.

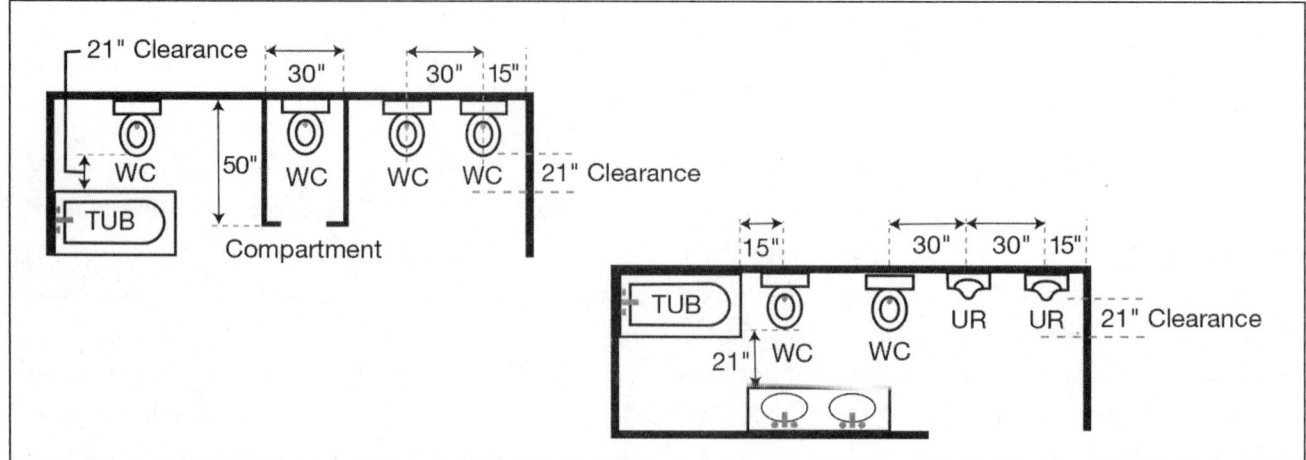

Figure 4.1 Fixture Clearances (IPC Figure 405.3.1)

FIXTURES

> **Hints and Tips**
> Always remember that the spacing dimensions for fixtures required by the codes are the bare minimum allowed. If space permits, and the building owner approves, spread your fixtures out a little more to give people more room than they may need or prefer. People may need to carry packages, luggage, or heavy winter coats into a stall with them. They will appreciate some extra room.

The UPC measurement requirements are a little different. The UPC agrees with the IPC on water closet spacing, but for urinals the UPC only demands 12 inches from the center of a urinal to any wall or partition next to it, and 24 inches from the centerline of one urinal to the next. The UPC allows a little more room in front of all of these fixtures than the IPC: 24 inches is the space required by the UPC.

Section 405.3.2 of the IPC states that a required lavatory must be in the same room with the required water closet. It doesn't seem necessary that a statement like that would have to be part of the regulations, but it could be inconvenient or embarrassing for the lavatory to be installed outside the bathroom or in some other area of the building.

Approved Fixture Connections

This seems almost too obvious to require a code section, but is it really? All of us who have worked on plumbing fixtures in the field have our own fond memories of rocking toilets and leaking urinal flanges (Figure 4.2). (Refer to IPC 405.4–405.9; UPC 404.2–404.4, 406.1)

Figure 4.2 Leakage from a Poorly Set Water Closet

Let's get specific. Part of IPC 405.4 states, "the flange shall be attached to the drain and anchored to the structure." Generally, we find the flange anchored to the drain, but all too often there has been a tendency to overlook or ignore the "anchored to the structure" part. Perhaps the installer cut the hole a little too big and was only able to get a couple of screws to hold to the floor. A little leakage and some corrosion, combined with a little wood rot, and the floor flange under the water closet no longer mates to the floor. The floor needs to be rebuilt when the repair is made for the leakage problem, or the source of the problem will persist. Being an attentive installer will eliminate most of those issues. Remember, those closet bolts must be brass.

Wall-hung water closets have to be mounted on a **closet carrier**, which is a cast-iron and steel structure concealed in the wall. The carrier supports the entire weight of the closet and provides a secure means for making a leakproof seal for the outlet of the water closet. The carrier is essential to make a secure and consistent mount (Figure 4.3).

FIXTURES

Figure 4.3 Wall-Hung Water Closet Carrier
(Photo courtesy of Watts)

> **ALERT**
>
> Closet carriers and lavatory carriers must be securely installed. Anticipate that unreasonable loads will be placed on them. Expect the fixtures to be broken by extreme force, but install the carriers so that a new fixture can easily reuse the existing, secure, and solid carrier that was installed.

Wherever a fixture makes contact with a wall or floor surface, it must be sealed with a material that will minimize movement and keep out water and bugs. So seal those edges with silicone caulk, plaster of paris, grout, or some other suitable material that will help prevent the creation of a concealed fouling surface.

Mental health facilities have special needs with respect to plumbing fixtures. In order to eliminate the plumbing as a means by which clients could harm themselves or others, all the piping must be concealed and the fixtures must be bolted through the wall from the back.

The overflows for fixtures must be able to handle the capacity of the filler to prevent overflow and it must drain out completely after use to eliminate trapped wastewater. The overflow must also drain into the drainage system on the fixture side of the trap. That way, the overflow is also protected from sewer gas by the fixture trap. In the case of flush tanks for water closets or urinals, the overflow drains directly into the fixture to which it is connected.

Slip-joint connections can be used on drains from the fixture to the trap. After the trap, only one slip-joint connection is allowed. This makes good sense, since slip-joint connections are the most susceptible to dislocation and leaking. No slip-joint connection can be in any inaccessible location. Most slip-joint connections are found under a sink or lavatory. They are easily handled in that location. If a slip joint happens to be located behind a wall, it must be made accessible with an access panel.

Regulations for Specific Fixtures

Up to this point, the fixture regulations have been worded in generalities. Following are regulations that control the design and installation of specific fixtures.

Automatic Clothes Washers

The most important aspect of the code dealing with automatic clothes washers is the prevention of backflow. All new residential washers are shipped with an integral air gap or backflow preventer. No extra backflow prevention is needed. Some commercial washers are not supplied with a backflow preventer. If that is the case, a backflow preventer must be added. The drain of an automatic clothes washer must have its outlet discharge through an **air break** into a standpipe that is at least 2 inches in diameter and no more than 30 inches high above the trap weir. (Refer to IPC 406.1–406.3)

FIXTURES

Bathtubs
All bathtubs must be manufactured with a slip-resistant surface on the bottom. The waste outlet must be a minimum of $1\frac{1}{2}$ inches in diameter. An overflow must be installed on all bathtubs. The waste outlet should be equipped with a water-tight stopper. (Refer to IPC 407.2 and UPC 409.0) If shower doors or solid glass panels surround a bathtub or shower, that glass must be safety glass. The hot water must not exceed 120°F and be controlled by a device that is in accordance with ASSE 1070 or CSA B 125.3, not by the thermostat or water heater. The filler valve must be protected by an air gap in order to protect against backflow. (Refer to IPC 407.1–407.4; UPC 414.5)

Bidets
The water outlet in a bidet is located below the flood-level rim of the fixture. Therefore, the bidet presents a high-risk possibility of backflow and contamination of the supply system. A bidet must be equipped with independent backflow prevention. The water supply to a bidet must be protected by an air gap. (Refer to UPC 410.1, Section 704.3) The water temperature of a bidet must be limited to 110°F by an approved tempering valve. The thermostat on the water heater is not considered to be the tempering valve. (Refer to IPC 408.1–408.3)

Dishwashers
The water supply of a dishwasher must be protected against backflow. New residential dishwashers have integral backflow prevention so no further backflow device is needed. Commercial dishwashers need an air gap or a backflow preventer installed on the inlet pipe. The waste from a residential dishwasher can be connected to a Y-tailpiece under a sink on the inlet side of the trap as long as the discharge hose is connected with a strap at a point as high as possible under the cabinet. Commercial dishwashers must discharge indirectly through an air gap or by a direct connection with floor drain protection. (Refer to IPC 409.1–409.3; UPC 414.0, Section 704.3)

Drinking Fountains
In any building where water fountains are required, no fewer than two drinking fountains shall be installed, with at least one being in compliance with ADA standards. A restaurant that provides drinking water is exempt from this rule. (Refer to IPC 410) The outlet of a drinking fountain must be above the flood-level rim of the fixture and a guard must be installed on the outlet to reduce direct human contact with the outlet. Thankfully, it is prohibited to install a drinking fountain in any bathroom. With respect to the location of drinking fountains, the IPC requires that the travel distance in multiple tenant facilities should not be more than 300 feet and not more than one floor above and below. (Refer to IPC 403.5, 410.1–410.2)

Floor Drains
The strainer on every floor drain must be removable to allow access to the drain for cleaning. The nominal pipe size for a floor drain must be a minimum of 2 inches. When a floor drain is in a public laundry or a central laundry for a multifamily residential unit such as an apartment complex or dormitory, the floor drain must be a minimum of 3 inches in diameter. (Refer to IPC 412.1–412.4; UPC 411.1)

> **Hints and Tips**
> If a floor drain is in an area that is unlikely to receive a regular flow of water, the water in the trap seal will evaporate and sewer gas will enter the building. It will be difficult for a building owner to find the source of the smell. A floor drain such as this should be equipped with a trap primer. This is a fitting that sends a small amount of water into the floor drain whenever a nearby fixture is used. Thus, a trap seal is maintained.

FIXTURES

Food Waste Grinders

In a residence, the food waste grinder is connected to the kitchen sink. That installation satisfies the need for a water supply to carry the waste down the drainpipe and the requirement of connection to a $1\frac{1}{2}$-inch drain. Commercial food waste grinders are connected to a 2-inch drain. The commercial grinder has a sink attached on top of it with a cold water supply equipped with a vacuum breaker to prevent contamination of the water supply. The commercial food waste grinder must have its own independent trap. (Refer to IPC 413.1–413.4)

Garbage Can Washers

The lowly garbage can washer is a fixture with no third-party standards established for it. The basic design is a modified floor drain with a washing spray nozzle aiming up through the center of it (Figure 4.4). This creates a high-risk possibility for a backflow situation. To guard against backflow, the code calls for a backflow preventer to be installed on the water supply to the nozzle. The drainage part of the garbage can washer is a floor drain that must be equipped with a basket to collect the solids before they go down the drain. The basket must be removable to make it easy to lift out and clean. (Refer to IPC 414.1–414.2)

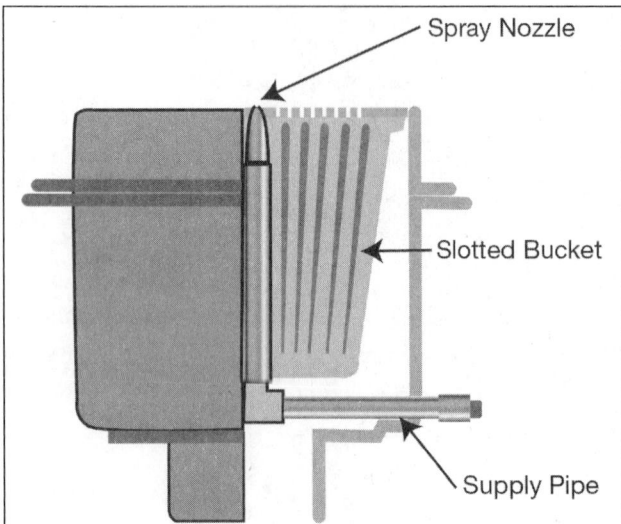

Figure 4.4 Combination Garbage Can Washer and Drain

Laundry Trays

As with other fixtures, laundry trays must have a smooth, impervious surface. The drain must be a minimum of $1\frac{1}{2}$ inches in diameter. Laundry trays have many uses that can range from washing out socks to cleaning paintbrushes to potting plants. Large debris is practically guaranteed to get into the drain. The outlet of a laundry tray must have a strainer or at least a crossbar to prevent large items from going down the drain. (Refer to IPC 415.1–415.2)

Lavatories

Surprisingly, lavatories are no longer required to have an overflow. The reason given is that bacteria can grow easily in this area where the surface is rarely washed. That reason has merit, but some believe an additional reason was pressure from the manufacturers who want to produce stylish products, such as above-the-counter vessels without the encumbrance of an overflow (that is strictly my opinion).

The required size of the lavatory waste is $1\frac{1}{4}$ inches. Once again, as in the laundry tray, a strainer or crossbar is required. Public lavatories must provide tempered water through a temperature-limiting device. (Refer to IPC 416.1–416.5)

Showers

A practical provision of this section is the requirement that the riser pipe to the shower head must be fastened to the building. The waste outlet and trap must be a minimum of $1\frac{1}{2}$ inches in diameter according to the IPC. The UPC requires a minimum trap size of 2 inches. I strongly recommend always using 2 inches on a shower unless structural problems make the use of a $1\frac{1}{2}$-inch drain more practical for the situation.

The walls in a shower stall have to be smooth, impervious, and noncorrosive to a height of 6 feet above the floor of the room and at least 70 inches above the floor of the shower. The floor of the shower also needs to be an impervious and noncorrosive material. No surprise there, but one surprise in the IPC is that a nonslip floor surface is not required. The

UPC does require a nonskid surface on the floor. The IPC has recently recognized, for the on-site construction of shower floors, another form of waterproofing. It is a self-curing liquid rubber polymer and a reinforcing fabric that is quickly applied to form a flexible, seamless waterproofing membrane that bonds to a wide variety of surfaces. (Refer to IPC 417, 417.5.2.6; UPC 411.5–411.9)

Sinks

Sinks come in many forms, like kitchen sinks, service sinks, mop sinks, bar sinks, and wash sinks. Sinks of all types have the usual impervious material provision. The outlet must be $1\frac{1}{2}$ inches at a minimum. (Refer to IPC 418.1–418.3)

Urinals

Urinals can have either integral or external traps. The water usage per flush is limited to 1 **GPF**, or gallon per flush. The newer waterless urinals have been assigned a drainage fixture unit, or DFU, of 0.5 since they use no water for flushing (Figure 4.5). Urine passes through a sealing liquid that is less dense than water, passes through a traplike cartridge, and into the drainage system. Standard 1 GPF urinals have a DFU of 2. The surrounding wall areas must be covered with an impervious, easily cleaned surface to help prevent bacteria growth. (Refer to IPC 419.1–419.3; UPC 409.0)

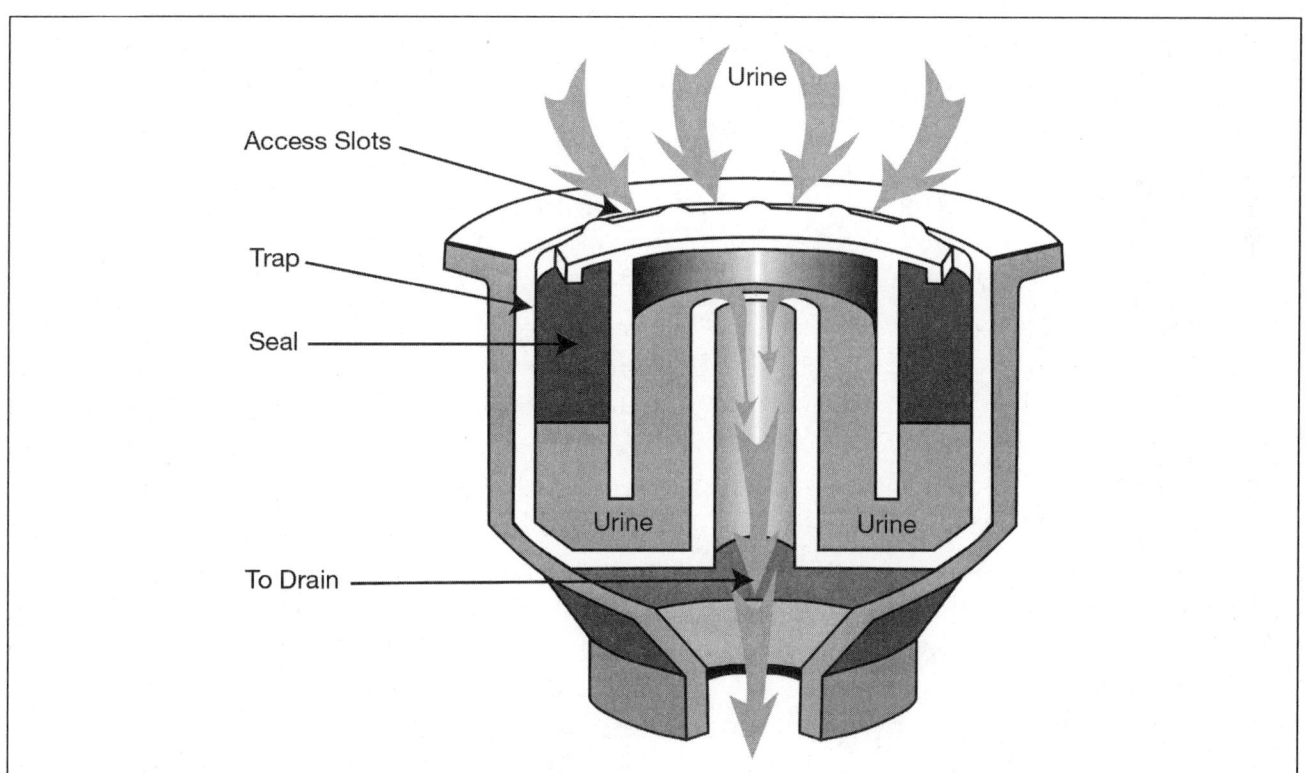

Figure 4.5 Representative Cartridge for a Waterless Urinal

FIXTURES

Water Closets

Water closets must flush with an average water usage of 1.6 GPF. In the IPC, one exception is the blowout toilet, which may use up to 3.5 GPF. The UPC allows no such exception. Any public-use water closet must be elongated. The bowl is made longer by 2 inches in the front, helping to protect the user from contact with pathogens. Public-use seats must be open front for the same reason. Water closets and other personal hygiene devices must conform to ASME A112.4.2 standards that ensure the protection of the public by setting temperature limits and minimum acceptable backflow protection requirements. (Refer to IPC 424.9)

Although the plumbing codes do not allow any reduction in the size of a drainage pipe, the use of a 4 × 3 inch closet flange or a 4 × 3 inch closet elbow is not considered a reduction in size. No reason is given for this, but one possible explanation is that the closet flange or closet elbow is being viewed as a type of receiver. The outlet of most water closets ranges from $1\frac{7}{8}$ inches to $2\frac{1}{2}$ inches. Therefore, the waste is actually discharging into a larger pipe than the outlet. (Refer to IPC 420.1– 420.4; UPC 408.0)

> **DID YOU KNOW?**
>
> The term "water closet" began with early indoor plumbing when an enclosed wooden chair would allow waste to drop through the floor into a cistern located below. "Water closet" gained popularity as the reference of choice since the room where the commode was located was generally about the size of a closet. The term has evolved into the present technical term for what some would consider to be the most important fixture in a building.

Whirlpool Baths

The installation instructions are considered a part of the plumbing code for whirlpool baths. The great variety in designs accounts for this. The circulation pump must be at a level above the weir of the trap to ensure that the pump drains out whenever the tub is drained. Once again, this regulation displays the primary purpose of plumbing codes, which is to protect people's health. If the pump had water remaining in it all the time, it would become the most popular vacation home for bacteria. People immerse themselves in these things, so it's important to keep them sanitary and safe. The codes require that the whirlpool's pump be accessible for servicing. The hot water must not exceed 120°F and be controlled by a device that is in accordance with ASSE 1070 or CSA B 125.3, not by the thermostat or water heater. The filler valve must be protected by an air gap in order to protect against backflow. The waste outlet should be equipped with a water-tight stopper. (Refer to IPC 407.2, 421.1–421.6; UPC 409.1–409.6)

Specialized Healthcare Fixtures

Some general rules govern healthcare fixtures. All the fixtures and devices that are connected to a water supply or drainage system must have backflow protection for the water supply system and from the drainage system. Any fixture that needs to be cleaned with hot water in excess of 180°F must be designed to withstand those temperatures with no damage. All working parts and parts that may need to be serviced must be accessible.

A clinical sink, or bedpan washer, is a primary fixture in healthcare facilities. It has an integral trap and rim flush like a water closet, and allows for complete elimination of the contents by a siphon action like a water closet. The only real difference between the clinical sink and a water closet is that a clinical sink has a flexible rinser available to clean out bedpans and the like. Sometimes water closets double as clinical sinks, which reduces the number of fixtures needed.

Steam supplied to sterilizers must have the condensate drained to prevent it from entering the sterilizer. Condensate could contaminate a sterilizer due to its lower temperature than the steam. (Refer to IPC 422.1–422.10)

Specialty Fixtures

There are many fixtures that are of a special design in nature, which makes it difficult to meet many provisions of plumbing codes. Some of these are baptisteries, fountains, aquariums, ornamental pools, and fountains (Figure 4.6).

The overriding principle concerning this type of fixture is protection of the water supply. Any such fixture requires the installation of a backflow preventer. Any special fixture with a water and/or a drain connection must be offered for approval to the authority having jurisdiction. (Refer to IPC 423.1–423.2)

Emergency Eye Wash and Shower

An emergency eye wash and shower must be of the type approved by American National Standard and the International Safety Equipment Association (ISEA Z 358.1). It is to be located in a workplace where people may be exposed to fire, chemicals, and products that could cause harm to the body and eyes. The water flow rates, discharge pattern, and temperature of the flushing fluid must be in accordance with ISEA Z 358.1 based on the type of hazardous material. The manufacturer's installation instructions must be adhered to. Its location must be at the same level as the potential hazard and immediately accessible. A drain is not required. If a drain is provided, it must be in accordance with Section 811.0. (Refer to UPC 416.0–416.5)

Figure 4.6 A Baptistery

FIXTURES

Chapter Review

Note: Answers may vary based on whether the IPC or UPC is applicable.

1. The minimum drainage pipe size for an automatic washing machine is _____.

2. A slip-joint connection can be concealed as long as there is access to it.
   a. true
   b. false

3. A 4 × 3 inch closet flange is considered a reduction in pipe size and is therefore considered a code violation.
   a. true
   b. false

4. Water closet seats in public restrooms must be closed front.
   a. true
   b. false

5. Piping in mental health facilities must be _____.
   a. plastic
   b. concealed
   c. kept overhead
   d. ADA approved

6. Siphon jet water closets are limited to _____ GPF.
   a. 1
   b. 3.5
   c. 2.0
   d. 1.6

7. How many lavatories are required in the women's room in a movie theater that seats 200 people?
   a. 1
   b. 2
   c. 3
   d. 4

8. How many water closets are required for 350 men in a factory?
   a. 2
   b. 4
   c. 8
   d. 14

9. If a unisex bathroom in a theater has two water closets, the required number of water closets in the men's room and the women's room can each be reduced by two.
   a. true
   b. false

10. Separate men's and women's restrooms are required if there are more than _____ employees in a location.
    a. 5
    b. 10
    c. 15
    d. 20

11. The condensate from the steam piped to a sterilizer must be drained directly through the sterilizer.
    a. true
    b. false

12. A water closet can be accessorized to be used as a clinical sink.
    a. true
    b. false

FIXTURES

13. Service sinks can be substituted for clinical sinks.
    a. true
    b. false

14. To quiet a whirlpool bath pump, it is acceptable to install it one story below the whirlpool bath.
    a. true
    b. false

15. The drain of an ice machine must be protected from backflow.
    a. true
    b. false

16. The circulation piping on a whirlpool bath must
    a. be larger than 1 in. in diameter.
    b. drain completely.
    c. be made of PVC.
    d. be flexible.

17. Fixtures used for therapy must be able to withstand water temperatures above which of the following temperatures?
    a. 180°F.
    b. 160°F.
    c. 140°F.
    d. 120°F.

18. A lavatory can be installed outside a restroom as long as it is within 6 feet of the door.
    a. true
    b. false

19. A closet flange must be attached only to the water closet base.
    a. true
    b. false

20. Multiple water closets must be set no closer than _____ inches on center.
    a. 32
    b. 30
    c. 15
    d. 24

21. Trough urinals are allowed _____.
    a. for temporary uses.
    b. never.
    c. in prisons.
    d. for commercial use.

22. Sheet copper cannot weigh less than _____ ounces per square foot.
    a. 8
    b. 10
    c. 12
    d. 16

23. How many water closets are required for a dormitory with 18 female residents?
    a. 1
    b. 2
    c. 3
    d. 4

24. In a covered mall, toilet facilities must be within _____ feet of the front door of any store.
    a. 200
    b. 300
    c. 400
    d. 500

25. It is not necessary to seal the base of a water closet to the floor.
    a. true
    b. false

FIXTURES

26. Overflows are _____ for lavatories.
   a. required
   b. optional
   c. forbidden
   d. required in pairs

27. Sinks need a minimum outlet size of
   a. $\frac{1}{4}$ in.
   b. $1\frac{1}{2}$ in.
   c. 1 in.
   d. 2 in.

28. Water closet seats
   a. must be closed-front in a public restroom.
   b. must have a cover.
   c. must all be elongated.
   d. must fit the closet bowl.

29. A whirlpool bath pump must be accessible.
   a. true
   b. false

30. An automatic washing machine drain must be at least _____ inches in diameter.
   a. $1\frac{1}{4}$
   b. $1\frac{1}{2}$
   c. 2
   d. $2\frac{1}{2}$

CHAPTER

FAUCETS AND FIXTURE FITTINGS

CHAPTER SUMMARY
When people turn on a faucet, they expect the delivery of safe water at a volume and temperature suitable for their use. They aren't concerned with low volume, scalding, or an unsafe supply due to a backflow condition at the faucet. That's how it should be. Faucets are meant to be taken for granted. The plumbing codes ensure this.

FAUCETS AND FIXTURE FITTINGS

Learning Objectives

- Gain familiarity with faucet requirements
- Identify various types of flushing devices
- Explain water conservation requirements
- Understand the importance of water temperature control

Key Terms

flush tank
flushometer tank
flushometer valve

From the mundane to the exotic, faucets are some of the most taken-for-granted items in buildings. Figures 5.1 and 5.2 show examples of a plain, utilitarian faucet and a highly decorative faucet.

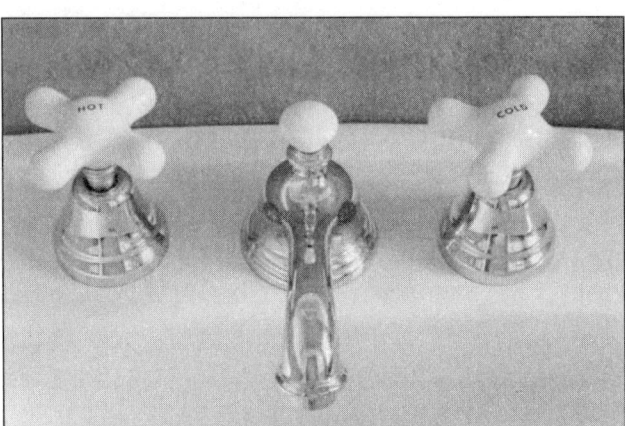

Figure 5.1 Basic Lavatory Faucet

Figure 5.2 Fancy Lavatory Faucet

FAUCETS AND FIXTURE FITTINGS

> ## JOB CONNECTION
> Ace Plumbing Company is bidding on another renovation. This time it involves the one common bathroom on the second floor of a college dormitory dating from the 1930s. All the walls are large tile. The old floor-mount urinals are all flushed together by a tank behind the wall. Tiny dental lavatories line the interior dividing walls. The water closets have washout bowls, and their flush tanks are equipped with flush ells. All of the showers have two-handle faucets that are original. This equipment was great stuff to have been in service this long, but can and should it be salvaged? Give Sophia some help deciding what to do here. Does any of this stuff still meet code requirements? Remember, we're dealing with an institution here—budget, budget, budget.

UPC. Section 415.0 clearly states that single-handle mixing faucets need to have markings on the faucet trim to indicate hot and cold to the user. (Refer to IPC 425.1.1; UPC 404.0, 415.0, 603.4.5)

TABLE 5.1 MAXIMUM FLOW RATES OF CERTAIN FAUCETS

Private lavatory faucet	2.2 GPM at 60 psi
Public metered lavatory faucet	0.25 gallon per cycle
Public nonmetered lavatory faucets	0.5 GPM at 60 psi
Shower head	2.5 GPM at 80 psi
Sink faucet	2.2 GPM at 60 psi

Handheld showers create a strong potential for a backflow condition, because the business end could be left hanging in a filled bathtub. Therefore, all hand showers are required to have either an integral backflow preventer or they must be connected to an external backflow preventer. Without this protection, bathtub water could be sucked into the potable water distribution system of the building—definitely not a good thing.

Water Consumption

Heightened awareness of, and in many cases the necessity for, water conservation drove the authorities having jurisdiction to limit maximum flows from faucets. We are only touching on faucet flows at this point in the book in keeping with the subject of this chapter. Table 5.1 provides a quick look at the maximum water flows specified by the IPC.

Design

We all assume that with any two-handle faucet the hot will be on the left and the cold will be on the right. That's tradition, right? It is also part of the

> ### Hints and Tips
> Some faucet manufacturers make tub and shower faucets and lavatory faucets in which the hot and cold side can be reversed by making an internal adjustment. This is useful when it is more convenient to rough-in the hot and cold pipes on the "wrong" side. This situation can occur when bathtubs or lavatories are back to back on the same wall, as can be found in apartment complexes. One hot supply pipe and one cold supply pipe can supply both fixtures.

FAUCETS AND FIXTURE FITTINGS

> **DID YOU KNOW?**
>
> The first modern plumbing code was the English Public Health Code of 1848. The first American plumbing code was the Metropolitan Health Law adopted in New York City in 1870. The connection between bacteria and disease was made after the Civil War and the importance of sanitation increased the focus on sanitary plumbing systems.

Faucet outlets must be designed to have an air gap of at least 1 inch above the flood-level rim of the fixture it is serving. In the event that the design of a fixture prevents this, the outlet must be protected with a vacuum breaker mounted at least 6 inches above the flood-level rim of the fixture. Much of the body of both plumbing codes is aimed at the prevention of backflow conditions. That's the whole point of having a plumbing code: protecting the health of all of those who use the plumbing systems we create. The advancement of modern plumbing codes has increased the health of our nation many times over.

All plumbing fixtures must have strainers in their drains to catch debris and minimize the size of solids passing down the drain. This helps prevent blockages. The exception to this rule is water closets and siphon-action or blowout urinals. Even field-built fixtures like garbage can washers and trough drains need a strainer.

Every water closet is required to be equipped with some kind of device to flush it. (Thank goodness!) That goes for any fixture that drains with a siphon, like urinals and bedpan washers. Either a **flush tank**, **flushometer tank**, or a **flushometer valve** is required. *Every* fixture must have its own individual flushing device. The days of the common flush tank and automatic common flushing valves are gone. This rule helps with water conservation. All flushometers must have a vacuum breaker integral with the valve. The fill valve in the gravity flush tanks must prevent back-siphonage.

Temperature Control

According to the UPC, 120°F is the highest temperature water allowed to flow from a lavatory faucet installed in a public facility. For the IPC, the maximum temperature for public lavatories is 110°F. Table 424.4 of the IPC shows this requirement (Table 5.2). Jump ahead to Chapter 6 and take a look at Table 6.1 to see how severe burns can be, even at 120°F. (Refer to IPC Table 424.4; UPC 413.1)

TABLE 5.2 MAXIMUM TEMPERATURES ALLOWED AT USE LOCATIONS		
PLUMBING FIXTURE	MAXIMUM DISCHARGE TEMPERATURE	REFERENCED STANDARD
1. Bidet	≤110°F	ASSE 1070
2. Public hand-washing facilities	>85°F and <110°F	ASSE 1070
3. Individual shower valves	≤120°F	ASSE 1016 OR CSA B125
4. Multiple (gang) showers	≤120°F	ASSE 1016 OR CSA B125
5. Bathtub/whirlpool bathtubs	≤120°F	ASSE 1070, Optional Combination tub/shower valve conforming to ASSE 1016 or CSA B 125

> **ALERT**
>
> An average of 300 children a day are taken to emergency rooms in the United States for burns from hot water. According to the National Safety Council, the United States has the highest rate of burns in the industrialized world. In the United States, after congenital malformations, disorders related to short gestation and low birth weight, and sudden infant death syndrome, burns are the leading cause of death for children from birth age through five years. The thinner skin of a child burns much faster than an adult's skin.

It is important to note that the temperature control for public lavatories cannot be the thermostat on the water heater. This rule obligates the installer to control the temperature with an approved tempering valve or some other type of secondary temperature control. This is a smart rule since water heater thermostats are good for general temperature control, but are not designed for highly accurate temperature control. This same temperature (120°F) applies to all bathtubs and showers, whether they are public or private.

Both the UPC and the IPC require pressure-balancing faucets for tubs and shower valves. Pressure-balancing faucets react to a sudden drop in pressure of either the cold or the hot water supply to the faucet. If the hot water pressure drops, such as when a washing machine starts to fill with hot water, a mechanism inside the body of the faucet reduces the water delivery from the cold side of the faucet, which keeps the outlet water at a relatively stable temperature. The same is true when someone flushes a water closet and the cold water supply pressure suddenly drops. The internal mechanism, usually a sliding piston, moves to reduce the hot water flow to the shower head or tub spout to prevent scalding of the bather. When multiple showers are installed in a locker-room type of facility, hot water temperature control can be achieved with one main mixing valve rather than a temperature control at each shower fixture.

FAUCETS AND FIXTURE FITTINGS

Chapter Review

Note: Answers may vary based on whether the IPC or UPC is applicable.

1. Flushometer valves
    a. close with water pressure.
    b. open with water pressure.
    c. always have a flush handle.
    d. are common in homes.

2. When you open the lid of a flushometer tank you see
    a. another tank.
    b. a lever actuator.
    c. no water.
    d. all of the above.

3. Water closets flush through a siphon action.
    a. true
    b. false

4. A group of urinals in a large public restroom can have a common flushing device.
    a. true
    b. false

5. Transfer valves must have
    a. a rounded handle.
    b. a vacuum breaker.
    c. two vents.
    d. three shower heads.

6. The highest hot water temperature allowed in a shower is
    a. 110°F.
    b. 100°F.
    c. 115°F.
    d. 120°F.

7. All water closet fill valves must be antisiphon.
    a. true
    b. false

8. According to the UPC, an exposed tubular trap must be constructed of 20-gauge drawn brass.
    a. true
    b. false

9. A public nonmetered lavatory can flow no more than
    a. 2 GPM.
    b. 0.5 GPM.
    c. 2.2 GPM.
    d. 1 GPM.

10. The thermostat on a water heater for a public restroom must be carefully set at no more than 120°F since it is the only temperature control.
    a. true
    b. false

11. Residential showers need no temperature control.
    a. true
    b. false

12. "Hot on the left, cold on the right" is no more than a tradition.
    a. true
    b. false

13. A continuous waste can be made of PVC.
    a. true
    b. false

14. The drains of what fixtures require strainers?
    a. laundry tray
    b. tank-type water closet
    c. blowout urinal
    d. none of the above

FAUCETS AND FIXTURE FITTINGS

15. The open area of a strainer must be equal to the area of the tailpiece.
    a. true
    b. false

16. A vacuum breaker on a flushometer must be _____ inches above the top of a urinal.
    a. 4
    b. 5
    c. 6
    d. 7

17. The water filler for a baptistery requires no vacuum breaker if the termination of the filler is below the flood level rim of the baptistery.
    a. true
    b. false

18. A handheld shower is considered an extension of the shower head and has no further regulations than the shower head.
    a. true
    b. false

19. Bathtubs can be supplied with higher water temperature than showers.
    a. true
    b. false

20. A handheld shower must be controlled by a transfer valve.
    a. true
    b. false

21. The amount of water delivered to a water closet by a flushometer valve is
    a. 1 GPF.
    b. 1.6 GPF.
    c. 2.1 GPF.
    d. 3.5 GPF.

22. A wye branch under a sink that receives the drainage from a dishwasher must connect
    a. at the trap weir.
    b. before the trap.
    c. after the trap.
    d. horizontally.

23. How many urinals are permitted to be flushed by the same flushing device?
    a. 1
    b. 2
    c. 3
    d. any number as long as adequate flow to all fixtures is maintained

24. A deck-mounted transfer valve on a bathtub must be equipped with
    a. a metal handle.
    b. an indication arrow.
    c. a vacuum breaker.
    d. a silicone gasket.

25. If the filler outlet for a baptistery is located below the water line of the baptistery, it must be equipped with
    a. a vacuum breaker.
    b. a ball valve.
    c. a tempering valve.
    d. a soft end piece.

26. The maximum hot water temperature permitted for a public lavatory is
    a. 110°F.
    b. 115°F.
    c. 120°F.
    d. 140°F.

FAUCETS AND FIXTURE FITTINGS

27. The waste from an emergency eyewash station
   a. must be connected directly to the sanitary drainage system.
   b. must be indirectly connected to the sanitary drainage system.
   c. requires no connection.
   d. must drain to a floor drain.

28. The drain that a domestic food waste grinder connects to must have a minimum diameter of
   a. 1 in.
   b. $1\frac{1}{4}$ in.
   c. $1\frac{1}{2}$ in.
   d. 2 in.

29. The water supply nozzle of a bidet is
   a. below the flood-level rim of the fixture.
   b. above the flood-level rim of the fixture.
   c. pointed downward.
   d. controlled remotely.

30. A shower head is limited to a flow rate of
   a. 2.5 GPM.
   b. 1.6 GPM.
   c. 3.5 GPM.
   d. 1 GPM.

CHAPTER 6 ▶ WATER HEATERS

CHAPTER SUMMARY
The purpose of this chapter may seem straightforward and you might even think you can skip it. After all, you install and service water heaters all the time. However, don't be too quick to take this topic for granted. Water heater installation and safety guidelines are very strict, so these rules are absolutely worth reviewing.

WATER HEATERS

Learning Objectives

- Identify proper locations to install water heaters
- Explain water temperature control, how it's done, and the ranges allowed
- State general installation requirements, including piping, valves, and drain pans
- Discuss safe installations, including siphon prevention, shutdown requirements, and relief valves

Key Terms

aquastat
BTU
dielectric unions
domestic hot water
grade
indirect waste
natural gas
plenum
propane gas
thermostat
vacuum relief valve

JOB CONNECTION

Ace Plumbing Company received a service call at 3:30 P.M. The 30-gallon natural gas water heater at the Springfield Little League concession stand is leaking. The concession operator has turned off the water supply and the gas supply to the heater. You have the job ticket to replace this water heater tomorrow morning. The heater is in a closet behind a louvered door in the same room where the volunteers work. The vent is B-vent. The permit has been paid for and posted. You must be sure the new heater is properly installed. What do you have to look for?

Quite often, small volunteer organizations will take advantage of the skills, or at least the willingness to help, of their members. Every dollar saved is, well, a dollar saved. Or is it? Harry, the guy who mowed the Springfield baseball field enough years to go through two John Deere riding mowers and about 97 weed whackers, installed the water heater in the concession stand with a little help from his friend Freddie. Freddie said he "knew about" this kind of stuff since he worked at the hardware store.

Now it's your turn to replace the water heater, and you are a professional. Your license, reputation, and, most importantly, your ego are at stake.

Location

Both plumbing codes are quite specific about issues related to where water heaters can be installed. In this section, we explore this aspect of the codes.

Accessibility

Every water heater must be installed so that *all parts that could need inspection or service are accessible.* Accessibility means that you do not need to remove any part of the permanent structure to get to all parts of the water heater. For example, there might be a panel or door that covers access to the water heater; it would need to be removable without doing any damage to the building and easily replaced after inspection. These instructions seem a little vague, and leave a whole lot of room for interpretation. Thankfully, you will have the installation instructions that come with the new water heater. *Read them!* (Refer to IPC 501.4, 502.3; UPC 509.4)

The installation manual is considered to be a part of the plumbing code. You must strictly follow the instructions included with the water heater. The good thing about the instructions is that they take care of just about any code issue you might have. Consequently, you should have the instructions available when the inspector examines your installation, especially if the heater is unusual in some way or if it involves new technology. Vent terminals that terminate through an outside wall of a building shall be located not less than 10 feet horizontally from an operable opening (e.g., window) in an adjacent building. This shall not apply to operable openings that are not less than 2 feet below or 25 feet above the elevation of the vent terminals.

Sometimes a water heater will be installed in an attic because of lack of space in the structure. Access takes on a whole new meaning in attics. There now must be a passageway large enough to allow the removal of the water heater, and that space cannot be smaller than 30 inches high and 22 inches wide, and no longer than 20 feet (Figure 6.1). Where the water heater stands, there must be a platform in front of the heater that is at least 30 inches by 30 inches, enough room to allow a serviceperson to work safely on the heater. And these are really only minimum requirements. Imagine dragging or pushing a water heater along in a passageway for 20 feet that is only 30 inches high and 22 inches wide! At least it is also required that the passageway have continuous solid flooring so you won't have to worry about falling through. The UPC has added additional enforcement requirements to the manufacturer's requirements to provide clearances and accessibility to serviceable components within gas utilization appliances.

Figure 6.1 A Narrow Passageway for Water Heater Access

Room as a Plenum

In large commercial buildings, such as warehouses, a room with large air-handling equipment can be used as a plenum. If a water heater is fired by any solid, liquid, or gas type of fuel, it cannot be installed in this room. If a heater of this type were installed in a room used as a **plenum**, the products of combustion could be distributed throughout the building if a negative pressure is created in the combustion chamber. That pretty much leaves only electric water heaters as options for a room used as a plenum. (Refer to IPC 502.2)

Both the installer and the inspector are now required to ensure makeup air is provided for fire-burning appliances in order that the products of combustion are not distributed through the building.

Garages

Any part of a water heater that can be a source of ignition must be installed at least 18 inches above floor level. The water heater stand in Figure 6.2 is 18 inches high and is designed to support a water heater at a height that will always meet this provision of the code. This is to prevent starting a fire or explosion due to flammable gasses or fumes from products used in a garage. Gasoline fumes are a major concern in garages. (Refer to UPC 508.14)

Figure 6.2 Water Heater Support Stand (*Photo courtesy of Watts*)

Temperature Control

Let's start with some really basic information. Some of this may seem obvious, but we don't want to overlook anything. *Every* hot water supply system must have some kind of temperature control, which is adjustable from a low setting to whatever maximum temperature is allowed for the purpose of the system. Therefore, an on-off switch, a hand-controlled valve to mix cold water with hot water, or any other kind of manual control *is not allowed*. The temperature control must be *fully automatic* and require no human supervision. (Refer to IPC 501.8)

Type of Control

A typical **natural gas** or **propane gas** water heater will have a gas control **thermostat**. An automatic valve that controls the gas flow is controlled by a probe that monitors the temperature of the water in the tank.

An electric water heater is controlled by one or two thermostats that sense the tank temperature through conduction, by being in contact with the surface of the tank.

The control for an oil-fired water heater is another probe-type control called an **aquastat**.

Control Range

All tankless water heaters must be restricted to an output temperature of 140°F. A practice that is becoming more and more common is to use water heaters both to heat a building and also to provide domestic hot water. Often the space-heating portion of the system is radiant tubing. All water heaters used for both **domestic hot water** and for space heating must have an approved tempering valve that limits the domestic hot water temperature to 140°F.

The potential for scalding must be taken seriously. Every year people are injured and some die from scalding injuries. All people are at risk, but extra consideration must be given to children, the handicapped, and anyone who may have difficulty controlling the temperature of a bath or shower. People who would have difficulty getting out of a hot shower or bathtub face the same danger (Table 6.1).

TABLE 6.1 APPROXIMATE TIMES AND TEMPERATURES CAUSING A THIRD-DEGREE BURN

TEMPERATURE	ADULTS SKIN THICKNESS OF 2.5 MM	CHILDREN 0–5 YEARS SKIN THICKNESS OF 0.56 MM
160°F	1 second	0.5 second
155°F	1.5 seconds	1 second
150°F	2 seconds	1.5 seconds
145°F	3 seconds	2 seconds
140°F	5 seconds	3 seconds
135°F	15 seconds	4 seconds
130°F	35 seconds	10 seconds
125°F	3 minutes	2 minutes
120°F	10 minutes	5 minutes
100°F	Safest Water Temperature for Bathing	

Source: Adapted from the following articles:
Hot water burn and scalding graph. (2004). Retrieved June 20, 2005, from Hot Water Burn Prevention and Consumer Safety Web site: http://www.accuratebuilding.com/services/legal/charts/hot_water_burn_scalding_graph.html.
Christophersen, Ph.D., E. G. (2004). Burn safety: hot water temperature. Retrieved June 20, 2005, from Pediatric Advisor Health Library
Web site: http://www.fairview.org/healthlibrary/content/pa_hotwatr_hhg.htm.
Scalds: a burning issue. (n.d.). Retrieved Aug. 10, 2005, from http://www.ameriburn.org.

Installation Requirements

Primarily, this section addresses piping configurations that need to be adhered to when installing water heaters. Drain pan requirements are also discussed.

Piping Requirements

A drain valve needs to be installed at the bottom of every tank-type water heater and every hot water storage tank (Figure 6.3). Drain valves are required to facilitate servicing or replacement of a water heater. (Refer to IPC 501.3, 503.1)

You might wonder why there needs to be a code rule requiring a valve on the cold water inlet. It's only logical that you would want a valve on the cold inlet of a water heater, and one that's in a handy place. But think, if there was a house that had a problem with the basement flooding occasionally, it would be better to have the water heater up on the first floor rather than in the basement where it is susceptible to water damage. If the cold water supply valve were left in the basement, it would be very inconvenient to work on this water heater; if there were a reason that the building occupant needed to shut the water off in some kind of emergency situation, he or she wouldn't be able to tell which valve was the right one to close.

DID YOU KNOW?

The cold water inlet to the water heater must have a valve on it, which is only for the water heater. It cannot influence the flow of cold water in any other part of the cold water supply system. The valve must be located near the water heater and on the same floor as the water heater.

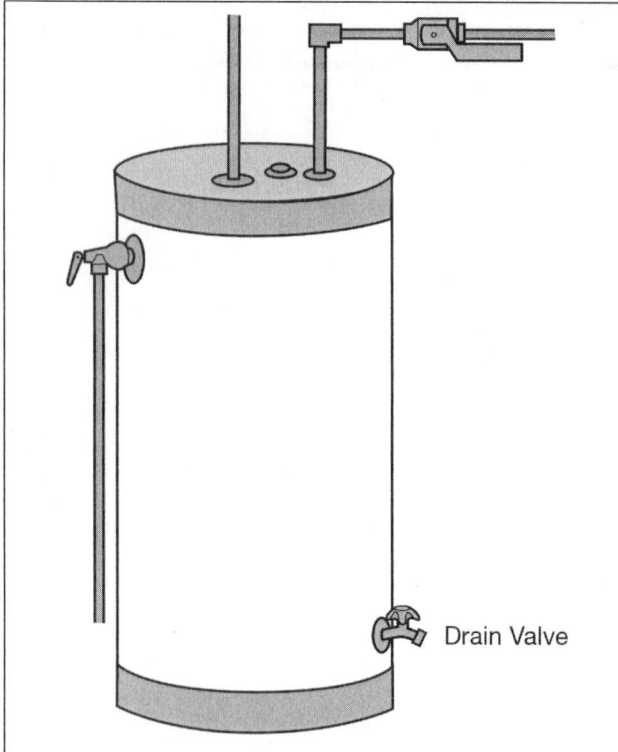

Figure 6.3 Water Heater Drain Valve Location

Drain Pans

Any water heater that is installed in a location where leaking could cause damage to the property must have a drain pan installed under it (Figure 6.4). (Refer to IPC 504.7; UPC 508.4)

Figure 6.4 Drain Pan to Capture Leaks from a Water Heater

When tank-type water heaters leak, the problem usually starts slowly. Because the leak is small but steady, a lot of damage to the structure can occur. Evaporation hides the leak from the building inhabitants for some time while the slowly leaking water begins to soften, swell, and rot the structure. Flooring, drywall, plaster, and ceilings get ruined, making for a very expensive repair that goes well beyond the cost of the water heater replacement.

The drainpipe from a water heater pan must be a minimum of $\frac{3}{4}$ inch in diameter and must be extended to drain into an **indirect waste** receptor. If the drain terminates outside the building, it must be extended to a distance not less than 6 inches and not greater than 24 inches above **grade**.

Most of the time, the pipe from the drain pan is run to a floor drain. Other places to lead the pipe to could be a laundry tray, service sink, or a floor sink. Do not connect the drainpipe directly to the sanitary drainage system. The pipe cannot be reduced in size over its entire length. If the most effective way to safely end this pipe is outdoors, it must be brought to a reasonable height above the ground as stated above to prevent snow, leaves, and plant growth from plugging it; at the same time, it must be low enough to not be a dripping nuisance.

Safety

Incidents involving unsafe water heaters are rare. For that reason, it might be easy to overlook safety regulations when installing a water heater. To the credit of code writers and code enforcers, safety is taken seriously and approached on a no-exception basis. All new water heaters have a new temperature and pressure relief valve shipped with them now. This was not the case even just a few years ago. Safe installations must be ensured by the plumber. The building occupants' lives could literally be on the line: it's up to the plumber to make sure that water heater mishaps due to unsafe installations continue to be rare.

WATER HEATERS

Antisiphon Strategies

An approved means of preventing siphoning of water from a tank-type water heater or hot water storage tank must be installed. A siphon anywhere in a water distribution system could draw contaminants into the piping system. (Refer to IPC 504.1)

> ### JOB CONNECTION
>
> OK, so far so good. The location of the water heater in the concession stand is fine. Let's move on. The old water heater has no drain pan, but it really doesn't need it. It is sitting on a concrete floor with a floor drain about 2 feet away. Harry and Freddie put a $\frac{3}{4}$-inch gate valve on the cold supply pipe about 14 inches up the pipe. Looks good enough to reuse on the new installation. They even installed **dielectric unions** to prevent corrosion.

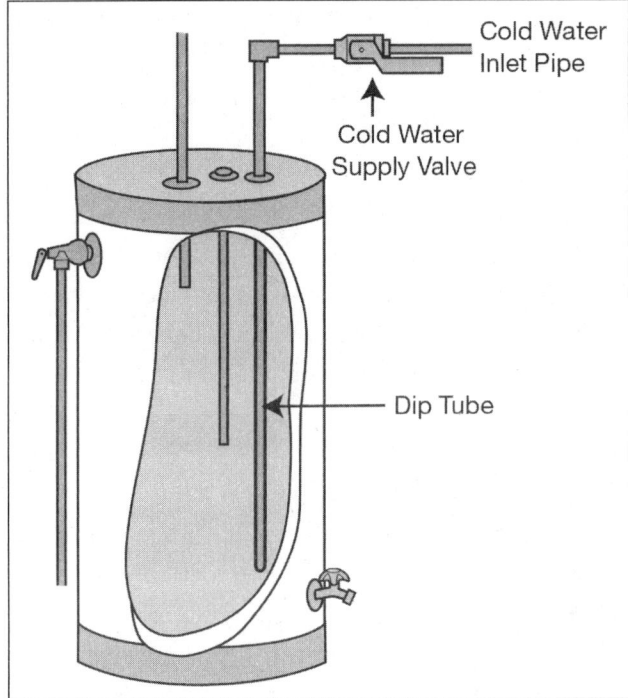

Figure 6.5 Dip Tube in Tank on Cold Water Inlet

Tank-type water heaters that have the cold water connection at the top of the tank are equipped with a dip tube, which is a tube extending inside the tank from the cold water inlet at the top, down to a point near the bottom of the tank (Figure 6.5). The purpose of the dip tube is to take the incoming cold water to the bottom of the tank where it can be heated, while the heated water rises to the top of the tank. Hot water tends to stay at the top to be drawn off at the hot water outlet and the cold at the bottom because cold water is denser, and therefore heavier than hot water per cubic volume. If conditions arise that could cause a siphon, the dip tube has a small hole drilled in the side of it near the top to break any siphon that may occur.

> ### DID YOU KNOW?
>
> All tank-type water heaters must have a pressure relief valve installed directly into the body of the tank and the relief valve must have a pressure rating of no more than 150 psi or the rated working pressure of the tank, whichever is lower. This valve must be self-closing and located in the top 6 inches of the tank.

Water heaters that have the cold water inlet at the bottom must have a **vacuum relief valve** installed in the cold water inlet pipe. If the drain valve is opened with both the cold water inlet and all the hot water outlets closed, a vacuum will occur that might cause some water heater tanks to collapse. The vacuum relief valve allows air to enter the tank through the cold water inlet and prevent the vacuum condition.

Shutdown

Every electric water heater must have a dedicated electrical disconnect switch in the same room with the water heater to allow for complete shutdown. All water heaters using other fuels must have a dedicated valve to shut off the fuel supply. (Refer to IPC 504.3)

Relief Valves

Quite possibly the most important item in the water heater section of the plumbing code is the section on relief valves. First, the nuts and bolts of the code. (Refer to IPC 504.4, 504.5, 504.6; UPC 505.4, 505.5)

As you most likely already know, a water heater is usually equipped with a combination temperature and pressure relief valve (T & P). The plumbing code allows for them to be separate, so they are treated separately. The IPC has now changed the code with respect to water heaters with separate storage tanks. It now specifies that both water heaters with separate storage tanks and storage water heaters must be installed with approved, self-closing (levered) pressure relief and temperature combination valve.

> **DID YOU KNOW?**
>
> Tank-type water heaters must be equipped with a temperature relief valve installed directly into the body of the tank. The opening temperature cannot be set at more than 210°F. The temperature relief valve must be able to relieve at or above the **BTU** input rating of the water heater. This relief valve must also be self-closing.

> **ALERT**
>
> *Never* reuse the old T & P valve when you change a water heater. Never plug the outlet end of a relief valve, no matter how much it is dripping. Replace it. *Always* check the T & P valve on a water heater to make sure it is installed correctly. Lives could be at risk.

If there were no T & P relief valve on a water heater, there could be catastrophic consequences. If the pressure in a water heater exceeded the pressure the tank could tolerate, obviously the tank would fail, expelling the water under great pressure, for a moment anyway. Personal injury is certainly a possibility, along with extensive water damage to the building.

The greater risk is if the temperature control on a water heater fails and the temperature of the water continues to climb above the boiling point. Initially, there is a very real danger of severe scalding if anyone opens a hot water faucet. A catastrophe can occur if the temperature exceeds 212°F and the system pressure climbs above the bursting pressure of the tank through expansion of the heating water. This can occur if there is a backflow preventer on the water main, giving the expanding water no place to go. Now, you have a tank of superheated water, which is water remaining in the liquid state above the boiling point because it is under pressure. Let's say the rated working pressure of the tank is 150 psi. That means the burst pressure is at least $1\frac{1}{2}$ times the working pressure. That would be 225 psi.

At that pressure, the water has been able to remain a liquid water, *but it's now at a temperature of 397°F!* Remember, water boils in your teapot at 212°F, and you can't say that's not hot.

At this point, the tank just can't take it anymore, and a dangerous explosion is the likely result. Forty gallons (almost 300 cubic feet) of superheated water just turned into 480,000 cubic feet of very hot steam! This will send the water heater on the next Mars mission. There are only 200,000 cubic feet of limestone in the Empire State Building. The results can be devastating. Incredible damage and loss of life have been attributed to exploding water heaters. They have been known to go up through multiple stories in large buildings. You can check this out for yourself on the Internet. There are lots of stories and some of them can send chills up your spine.

Relief Valve Outlet

The outlet of the relief valve must be a plain-end pipe, which remains at the full diameter of the relief valve outlet. It must drain by gravity, indirectly to an indirect waste receptor, to the floor or to the outdoors.

If a relief valve were dripping outdoors in freezing weather it would soon be plugged solid with ice. Therefore, if an outlet pipe is drained to the outdoors, it must first drain into an indirect waste receptor in a heated area, and then the waste pipe can be directed outdoors. When the pipe simply drains to the floor, it must end within 6 inches of the floor.

The outlet pipe cannot be directly connected with any other relief valve piping. The relief valve outlet is never connected directly to any drainage system.

Hints and Tips

Premade water heater discharge pipes are available for purchase wherever you buy water heaters. They are inexpensive, quick and easy to install, and they meet code regulations.

JOB CONNECTION

There is a $\frac{1}{2}$-inch gas cock 8 inches away from the water heater gas connection, which serves only the water heater. There is a $\frac{3}{4}$-inch T & P relief valve installed directly into the side of the tank about 4 inches down from the top. It has a BTU rating label that is far greater than the BTU input rating of the water heater. For the drop pipe on the outlet of the relief valve, Harry used a piece of $\frac{3}{4}$-inch galvanized pipe, threaded on both ends and long enough to end 5 inches off the concrete floor. So how did old Harry do? Pretty good job for an amateur, but with one thing wrong—it shouldn't have passed inspection.

Chapter Review

Note: Answers may vary based on whether the IPC or UPC is applicable.

1. The end of the outlet pipe for a temperature and pressure relief valve is permitted to
   a. connect directly to a stack vent.
   b. end within 6 in. of the floor.
   c. end with a pipe nipple, threaded on both ends.
   d. connect with another relief valve outlet before ending as an indirect waste.

2. Water heater installation instructions from the manufacturer
   a. must be followed, unless the plumbing code is more stringent.
   b. are considered a part of the plumbing code.
   c. are supplied with every new water heater.
   d. all of the above.

3. A passageway to a water heater in an attic
   a. can be 22 ft. long.
   b. must, at the lowest part, be at least 5 ft. tall.
   c. needs a platform in front of the water heater at least 30 in. wide by 22 in. deep.
   d. must have continuous flooring.

4. A water heater pan
   a. is essential.
   b. must be drained indirectly.
   c. is installed to create a noncombustible floor under the water heater.
   d. must have a drainpipe with at least a 1-in. inside diameter.

5. The highest pressure rating allowed for a relief valve on a water heater is
   a. 150 psi.
   b. 125 psi.
   c. the rated working pressure of the tank if it is lower than 150 psi.
   d. the rated working pressure of the tank.

6. The outlet pipe from a T & P relief valve must
   a. end outdoors.
   b. connect directly to the sanitary sewer.
   c. have pipe threads so it can be extended.
   d. be at least the same diameter as the T & P valve outlet.

7. When superheated water flashes to steam, it expands
   a. approximately 1,600 times in volume.
   b. approximately 300 times in volume.
   c. slowly.
   d. only when the T & P relief valve has a plug screwed into it.

8. A tank-type water heater
   a. must have a full-port valve installed on the hot water outlet pipe.
   b. must have the relief valve located within 12 in. of the top of the tank.
   c. installed in a garage must have all parts that are a source of ignition at least 18 in. above the floor.
   d. can have the temperature controlled by a manual valve as long as it is supervised by qualified maintenance personnel at all times.

WATER HEATERS

9. Tank-type water heaters must have a temperature and pressure relief valve installed directly into the _____ of the water heater.
   a. bottom
   b. top
   c. side
   d. body

10. A temperature and pressure relief valve
    a. must be manually closed after a release so someone is aware of the situation.
    b. must have a BTU rating of at least $\frac{1}{2}$ of the BTU input rating of the water heater.
    c. must automatically close.
    d. is only required when the water heater is in a public building.

11. The maximum temperature water allowed in a potable hot water distribution system is
    a. 120°F.
    b. 140°F.
    c. 160°F.
    d. 110°F.

12. The inlet male threads on a water heater drain valve must be what pipe size?
    a. $\frac{1}{2}$ in.
    b. $\frac{3}{4}$ in.
    c. 1 in.
    d. $\frac{3}{8}$ in.

13. The maximum water outlet temperature at a tub or shower is 115°F.
    a. true
    b. false

14. All water heaters must be installed 18 in. off the floor.
    a. true
    b. false

15. All water heater supply systems must have automatic temperature control.
    a. true
    b. false

16. The outlet pipe of a T & P relief valve must be metal.
    a. true
    b. false

17. A water heater cannot be installed in any attic area.
    a. true
    b. false

18. The valve on the inlet pipe for a water heater must be a full-port design.
    a. true
    b. false

19. A valve is required on both the cold inlet and the hot outlet pipes of the water heater.
    a. true
    b. false

20. If a water supply is equipped with a backflow preventer, the T & P relief valve may serve as control for thermal expansion.
    a. true
    b. false

21. The relief valve outlet pipe must end within _____ inches of the floor.

22. A water heater drain pan must be installed when a water heater leak could cause _____.

23. A water heater drain pan must be at least _____ inches deep.

WATER HEATERS

24. Hot water storage tanks must be insulated to lose no more than _____ BTUs per hour.

25. A water heater relief valve may discharge into the water heater drain pan.
   a. true
   b. false

26. Seismic supports are always optional if a homeowner desires the added security.
   a. true
   b. false

27. A water heater must be provided with a way to prevent siphon emptying of the tank.
   a. true
   b. false

28. The relief valve must be installed in the hot outlet pipe near the water heater.
   a. true
   b. false

29. Since water will release from a relief valve forcefully, the outlet pipe of the relief valve may be piped upward to a drain point.
   a. true
   b. false

30. A relief valve outlet must be piped directly into the sanitary sewer.
   a. true
   b. false

CHAPTER 7: GAS PIPING INSTALLATIONS

CHAPTER SUMMARY

The proper and safe installation of natural gas to the population is essential. To ensure that this is done, interested persons combined their experiences and knowledge to develop what we now know as the UPC. The principal hazards of improper and inadequate fuel gas installation include poor equipment operation, fire, explosion, and asphyxiation.

GAS PIPING INSTALLATIONS

Learning Objectives

- Identify materials that are permitted by the codes for the installation of fuel gas supply and distribution systems.
- Calculate the proper size of a gas main and gas distribution system.
- Develop familiarity with the potential hazards present in a gas distribution system.

General Rules

Coverage of gas piping systems must extend from the point of delivery to the appliance connection. The point of delivery is the outlet of the service meter assembly or the service shutoff valve if no meter is provided (UPC 1201.1). Upon installation of a fuel gas system, it is required that the systems be inspected by the Authority Having Jurisdiction. There are two inspections: the rough piping inspection, which is done before the installed pipes are covered up in walls; and the final inspection, when the portions of gas piping that are to be concealed have been concealed and fixtures and appliances have not yet been attached.

The rough inspection is done to determine that the gas piping size, material, and installation meet the requirements of the code.

The final inspection is an air test, and the gauges used in conducting the test shall be in accordance with Section 318.0 of the UPC. This inspection shall include an air, CO_2, or nitrogen pressure test, and the piping must withstand a pressure of 10 psi gauge pressure. The length of time that this test should last will be determined by the Authority Having Jurisdiction, but in no case less than 15 minutes, and there must not be a drop in pressure. For welded piping and for piping carrying gas pressure that is in excess of 14 inches water column pressure, the test pressure shall not be less than 60 psi for a period of time satisfactory to the Authority Having Jurisdiction. The permit holder will supply all the necessary apparatuses to perform the test (UPC 1213.13).

Gas connection and appliances can be disconnected by the Authority Having Jurisdiction or the serving gas supplier if it is found that the gas piping is not in accordance with the requirements of the UPC 1201, or is defective and in such condition as to endanger life or property. Upon the completion of the final inspection and the finding that the gas piping installation is in accordance with the provisions of the relevant codes, the Authority Having Jurisdiction shall then issue a certificate of inspection.

Before a gas piping system can be installed, a plan or piping sketch shall be prepared for approval of the Authority Having Jurisdiction authorizing the installation of the gas system. This plan must show the proposed location of piping, the size of the different branches, the various load demands, and the location of the point of delivery.

Materials

When installing gas piping systems, the UPC states that pipes, fitting, and valves must be free of any foreign material and ascertained as approved for the system by NFPA 54:5.6.1.2. Cast-iron pipe shall not be used—only steel and wrought-iron pipe, which shall not be less than standard weight (schedule 40). It shall comply with the following standards: (1) ASME B36.10, (2) ASTM A53, and (3) ASTM A106 (NFPA 54:5.6.2.2).

Copper and brass pipes shall not be used where the gas contains more than an average 0.3 grains of hydrogen sulfide per 100 standard cubic feet (scf) of gas (NFAP 54:5.6.2.3). The method used to join this piping will be either with threads or the pipe should be welded together (UPC 1208).

GAS PIPING INSTALLATIONS

Pressure

Correct gas pressure must be supplied to an appliance, or else it will not perform correctly and may become harmful to life and property. As a result, a line pressure regulator must be installed where the gas supply pressure exceeds that at which the appliance is designed to operate (UPC 1208.7, NFPA 54–12:5.8.1).

In light of this code, the installer needs to verify the appliance design pressure and ensure that the gas supply pressure does not exceed that design pressure. Inspectors need to verify that overpressure protection is installed to code and manufacturer's instructions, when appropriate (UPC 1208.7.1). When gas lines are installed under grade or slab it must be done in a conduit. This conduit, if it terminates inside a building, must not be sealed. This will allow for any gas buildup to escape and not accumulate inside of the conduit (UPC 1210.1.6.2). This termination point must also be made accessible.

All gas piping systems for industrial, commercial, and public facilities must be purged to the outdoors only. The purging of these systems shall be in accordance with the provisions of the UPC Section 1213.6.1.1 through Section 1213.6.1.4, where the piping systems meet either of the following: (1) the design operation gas pressure exceeds 2 psig, or (2) the piping being purged contains one or more sections of pipe or tubing meeting the size and length criteria of UPC Table 1213.6.1.

Sizing

When installing a gas piping system, you must ensure that it is sized correctly or else you may end up with insufficient gas to supply the demand that is needed. As a result, when having to size piping for gas systems, it is necessary to refer to UPC Table 1216.2(1) through Table 1216.2(36) (UPC 1215.0 and 1216.0 references).

Chapter Review

1. If gas system piping is not installed correctly, what can occur?
   a. failed inspection
   b. appliance malfunction
   c. explosion
   d. all of the above

2. Where does coverage of gas piping systems begin?
   a. the gas main in the street
   b. the point of delivery
   c. at the appliance
   d. the point of entry to the gas meter

3. At completion of installation of a gas piping system, what must be done?
   a. paint the pipes
   b. conceal the pipes
   c. inspect the pipes
   d. turn on the gas supply

4. Who conducts the gas piping inspection?
   a. the permit holder
   b. the Authority Having Jurisdiction
   c. the plumber
   d. the engineer

5. How many inspections are there for a gas piping system?
   a. 1
   b. 2
   c. 3
   d. 4

6. Gas piping installed in the walls is called
   a. concealed piping.
   b. rough piping.
   c. DWV.
   d. none of the above.

GAS PIPING INSTALLATIONS

7. What is determined at the first inspection?
   a. correct sizing of piping
   b. use of correct material
   c. the code requirements are met
   d. all of the above

8. If gas piping is not installed in accordance with the relevant codes, the Authority Having Jurisdiction can
   a. revoke the license of the permit holder.
   b. shut down the job.
   c. disconnect the gas supply.
   d. all of the above

9. What must be prepared for approval by the Authority Having Jurisdiction before a gas system can be installed?
   a. an invoice
   b. a contract
   c. a sketch or plan
   d. a payment schedule

10. What schedule of metallic pipe must be used on gas piping?
    a. 10
    b. 20
    c. 40
    d. 80

11. What must pipes, valves, and fittings be free of before they are installed in gas piping systems?
    a. threads
    b. bends
    c. foreign material
    d. paint

12. Copper or brass cannot be used for gas piping if the gas contains more than how many grains of hydrogen sulfide per 100 standard cubic feet of gas?
    a. 0.1
    b. 0.2
    c. 0.3
    d. 0.4

13. What is an acceptable way of joining gas piping for installations?
    a. solvent weld
    b. threads
    c. poured lead joints
    d. with no hub couplings

14. Not enough gas pressure in a system can cause
    a. backflow.
    b. overcharging for gas.
    c. appliance malfunction.
    d. none of the above

15. When do you install a gas pressure regulator on a gas piping system?
    a. when there is too much gas pressure
    b. when there is too little gas pressure
    c. when a meter is installed
    d. for a gas inspection

16. Who needs to verify that overpressure protection is installed?
    a. the architect
    b. the plumber
    c. the general contractor
    d. the inspector

GAS PIPING INSTALLATIONS

17. All gas conduits terminating in buildings must not be _____ in order to prevent gas buildup.
   a. pointed
   b. sealed
   c. lighted
   d. none of the above

18. The termination of gas conduits in buildings must be
   a. 20 ft. above grade.
   b. 5 ft. above grade.
   c. capped.
   d. accessible.

19. What type of gas piping systems must be purged?
   a. industrial
   b. commercial
   c. public facilities
   d. all of the above

20. When gas lines are installed under grade or slab, they must be
   a. copper.
   b. in a conduit.
   c. solvent welded.
   d. painted.

CHAPTER 8 ▶ WATER SUPPLY AND DISTRIBUTION

CHAPTER SUMMARY

Providing a safe and clean water supply to the population is the cornerstone of modern plumbing. The occurrence of multiple diseases has been minimized, and some diseases have been virtually eliminated through the implementation of modern plumbing techniques. The fact that clean water is often taken for granted in the United States is testament to the success of American plumbing. The plumbing codes of our country will continue to develop, guide, and enforce the provisions necessary to continue this tradition.

WATER SUPPLY AND DISTRIBUTION

Learning Objectives

- Identify materials that are permitted for water supply and distribution
- Calculate the proper size of a water main and water distribution system
- Develop familiarity with the potential contamination hazards present in a water distribution system
- Recognize how electrical grounding can be affected with different piping materials
- State how to maintain sufficient water pressure throughout a water distribution system

Key Terms

flow pressure
individual water supply

General Rules

The water supply and distribution sections of both the IPC and the UPC cover all the aspects of water supply piping, including the materials used, the design of water supply systems, and the proper manner of installation for water supply systems. The IPC takes special note of the possible effects that a solar water heating system can have on a water distribution system. The IPC starts the chapter on water supply and distribution by clearly stating that any form of solar water heating must be installed in compliance with all of the rules of the plumbing code. This was more than likely done due to the high potential for cross-connections with systems containing antifreeze liquids. In reality, the code regulations contain enough rules to protect water distribution systems from cross-connections—one particular application should not be singled out as a potential cross-connection offender when the problem can occur almost anywhere. However, with any unfamiliar system, it is a good idea to be aware of the possible dangers that can occur with the application. (Refer to IPC 601.0–603.0; UPC 601.1–601.4)

JOB CONNECTION

Sophia, the estimator at Ace Plumbing Company, is earning her pay again this week. Converting the old Williams Hardware building into apartments is going well, but the size of this job requires thinking beyond the usual residential standards. With six apartments, two large washing machines, and one laundry tray, water supply pipe sizing becomes an important issue. Is a 1-inch water service big enough? At what point in the supply train can we reduce to $\frac{3}{4}$ inch? These questions must be answered before the actual water supply piping can begin.

Section 601.3 of the IPC requires that metallic water piping used for electrical grounding must be replaced with nonmetallic piping, unless a continuity of grounding is installed that meets with the code official's approval. This is referred to as bonding the system. Many older homes relied on the plumbing system for electrical grounding, so always be careful! It makes the electrical system not only potentially dangerous for the residents, but also for you while you are working on the system. Imagine cutting an old, leaking section of galvanized pipe. The water drains on the floor where you are standing, and, as the continuity is broken when the flow of water stops, the electricity that used to go to ground through the pipe now needs to find a new path to ground—down your wet arm to the puddle you are standing in seems like a good possibility. Not a healthy way to last long enough to collect your Social Security.

Both the IPC and the UPC require that all buildings equipped with plumbing fixtures that are intended for people to live in must be supplied with potable water at pressures specified later in the chapter. Potable water must be supplied to all fixtures except where gray-water systems are installed in an approved manner. Gray-water systems may become more common as a way to conserve clean water supplies around our country and the world. The UPC prohibits any connection whatsoever between a potable water system and a reclaimed water system. Even an approved backflow preventer is not enough of a guarantee for prevention of a cross-connection. Reclaimed water is only allowed to be used to flush water closets and urinals, and as a water supply for trap primers on floor drains and floor sinks.

The spirit of the opening sections of the water supply and distribution chapter of both the IPC and the UPC demonstrates the concern with the safety of water supplied to the fixtures in all buildings. It is a basic premise of modern plumbing that safe, clean water is supplied to the population. It is your responsibility to carry out that mandate.

Quantity

All the fixtures in a building intended for human occupancy must be supplied with an adequate water supply. The source can include any storage capacity on the site as well as the supply entering through the water service from the public system or from the **individual water supply** system. Before a new supply system or a repaired supply system is placed in service, it must first be disinfected. (Refer to IPC 602.3.2–602.3.5.1; UPC 608.1)

The individual water supply system has some additional rules. The pump must be designed to safely handle potable water. It cannot allow any kind of contamination into the water system. The pump has to be designed to keep a prime, and all parts of the pump need to be accessible. The pump and the immediate piping have to be protected from freezing and any pump installed in a basement must be mounted on a surface that is at least 18 inches above the basement floor. This helps prevent flooding of the pump. A well cannot be located in a pit due to the potential for contamination of the water supply if the well pit floods.

Water Service

In parallel water distribution systems, the hot and cold water piping may now be grouped in the same bundle.

No matter how well designed a water distribution system is, it cannot function well without a properly designed and sized water service. The whole distribution system is dependent on the quality, quantity, and pressure of the water supply through the building's water service. (Refer to IPC 603.1–603.2.1; UPC 609.2– 609.10.1)

WATER SUPPLY AND DISTRIBUTION

JOB CONNECTION

The building that Sophia needs to size the water service for will contain six apartments when it is completed. All of them have one full bathroom and a kitchen sink. Using Table 8.1 we can find what the water service fixture unit (WSFU) load is for the building.

The six full bathrooms can each be sized as a bathroom group:

BATHROOM GROUP, PRIVATE, FLUSH TANK 3.6 WSFU

3.6 °— 6 = **21.6 WSFU**

Remember that every apartment has a kitchen sink:

KITCHEN SINK, PRIVATE, FAUCET 1.4 WSFU

1.4 °— 6 = **8.4 WSFU**

We still have the laundry room with two washing machines and a laundry tray. Let's include those figures:

WASHING MACHINE (15 LB.), PUBLIC, AUTOMATIC 4.0 WSFU

4.0 °— 2 = **8 WSFU**

LAUNDRY TRAY (1 TO 3), PRIVATE, FAUCET 1.4 WSFU

1.4 °— 1 = **1.4 WSFU**

Now, add up all the water service fixture units in the building:

21.6
8.4
8.0
+ 1.4
39.4 WSFU

We now know that the total number of water service fixture units in the building is 39.4. The next step is to convert the WSFU total into GPM (gallons per minute). Use Table 8.2 to make the conversion. We must round up our 39.4 WSFU to 40 WSFU to fit the chart. On the chart, 40 WSFU converts to 26.3 GPM.

The next step is to use Table 8.3 to determine what size pipe will carry 26.3 GPM at a velocity no greater than 10 FPS (feet per second). Beyond that velocity, noise and pipe erosion increase dramatically. We can see that a 1-inch pipe will handle the volume and remain under 10 FPS.

Sizing

The smallest diameter water service pipe allowed is $\frac{3}{4}$ inch. This refers to the nominal, or inside, pipe diameter as stated earlier in the code. Remember, all pipe sizes referred to in the codes are nominal unless stated otherwise.

The smallest diameter water service pipe allowed is $\frac{3}{4}$ inch, but what if you have a building with lots of plumbing fixtures and you know the main will have to be larger than $\frac{3}{4}$ inch? How do we size it properly? Follow this next example to find out. Use Tables 8.1, 8.2, and 8.3 to help you.

TABLE 8.1 LOAD VALUES ASSIGNED TO FIXTURES

FIXTURE	OCCUPANCY	TYPE OF SUPPLY CONTROL	LOAD VALUES, IN WATER SUPPLY FIXTURE UNITS (WSFU)		
			COLD	HOT	TOTAL
Bathroom group	Private	Flush tank	2.7	1.5	3.6
Bathroom group	Private	Flush valve	6.0	3.0	8.0
Bathtub	Private	Faucet	1.0	1.0	1.4
Bathtub	Public	Faucet	3.0	3.0	4.0
Bidet	Private	Faucet	1.5	1.5	2.0
Combination fixture	Private	Faucet	2.25	2.25	3.0
Dishwashing machine	Private	Automatic	—	1.4	1.4
Drinking fountain	Offices, etc.	$\frac{1}{8}$" valve	0.25	—	0.25
Kitchen sink	Private	Faucet	1.0	1.0	1.4
Kitchen sink	Hotel, restaurant	Faucet	3.0	3.0	4.0
Laundry trays (1 to 3)	Private	Faucet	1.0	1.0	1.4
Lavatory	Private	Faucet	0.5	0.5	0.7
Lavatory	Public	Faucet	1.5	1.5	2.0
Service sink	Offices, etc.	Faucet	2.25	2.25	3.0
Shower head	Public	Mixing valve	3.0	3.0	4.0
Shower head	Private	Mixing valve	1.0	1.0	1.4
Urinal	Public	1" flush valve	10.0	—	10.0
Urinal	Public	$\frac{1}{4}$" flush valve	5.0	—	5.0
Urinal	Public	Flush tank	3.0	—	3.0
Washing machine (8 lb)	Private	Automatic	1.0	1.0	1.4
Washing machine (8 lb)	Public	Automatic	2.25	2.25	3.0
Washing machine (15 lb)	Public	Automatic	3.0	3.0	4.0
Water closet	Private	Flush valve	6.0	—	6.0

(Continued)

WATER SUPPLY AND DISTRIBUTION

TABLE 8.1 (Continued)

FIXTURE	OCCUPANCY	TYPE OF SUPPLY CONTROL	LOAD VALUES, IN WATER SUPPLY FIXTURE UNITS (WSFU)		
			COLD	HOT	TOTAL
Water closet	Private	Flush valve	2.2	—	2.2
Water closet	Public	Flush valve	10.0	—	10.0
Water closet	Public	Flush valve	5.0	—	5.0
Water closet	Public or private	Flushometer tank	2.0	—	2.0

For SI: 1 inch = 25.4 mm. 1 pound = 0.454 kg.

Note: For fixtures not listed, loads should be assumed by comparing the fixture to one listed using water in similar quantities and at similar rates. The assigned loads for fixtures with both hot and cold water supplies are given for separate hot and cold water loads and for total load. The separate hot and cold water loads being three-fourths of the total load for the fixture in each use.

TABLE 8.2 TABLE FOR ESTIMATING DEMAND (IPC TABLE 603.1)

SUPPLY SYSTEMS PREDOMINANTLY FOR FLUSH TANKS

LOAD	DEMAND	
(WATER SUPPLY FIXTURE UNITS)	(GALLONS PER MINUTE)	(CUBIC FEET PER MINUTE)
1	3.0	0.4104
2	5.0	0.6684
3	6.5	0.86892
4	8.0	1.06944
5	9.4	1.2566
6	10.7	1.4308
7	11.8	1.57743
8	12.8	1.71111
9	13.7	1.83142
10	14.6	1.95174
11	15.4	2.05868
12	16.0	2.13889
13	16.5	2.20573
14	17.0	2.27257
15	17.5	2.33941
16	18.0	2.40624
17	18.4	2.45972
18	18.8	2.51319
19	19.2	2.56667
20	19.6	2.62014
25	21.5	2.87413
30	23.3	3.11476
35	24.9	3.32865
40	26.3	3.5158
45	27.7	3.70295

For SI: 1 gallon per minute = 3.785 L/m, 1 cubic foot per minute = 0.0004719 m^3/s.

TABLE 8.3 FRICTION LOST IN SMOOTH PIPE

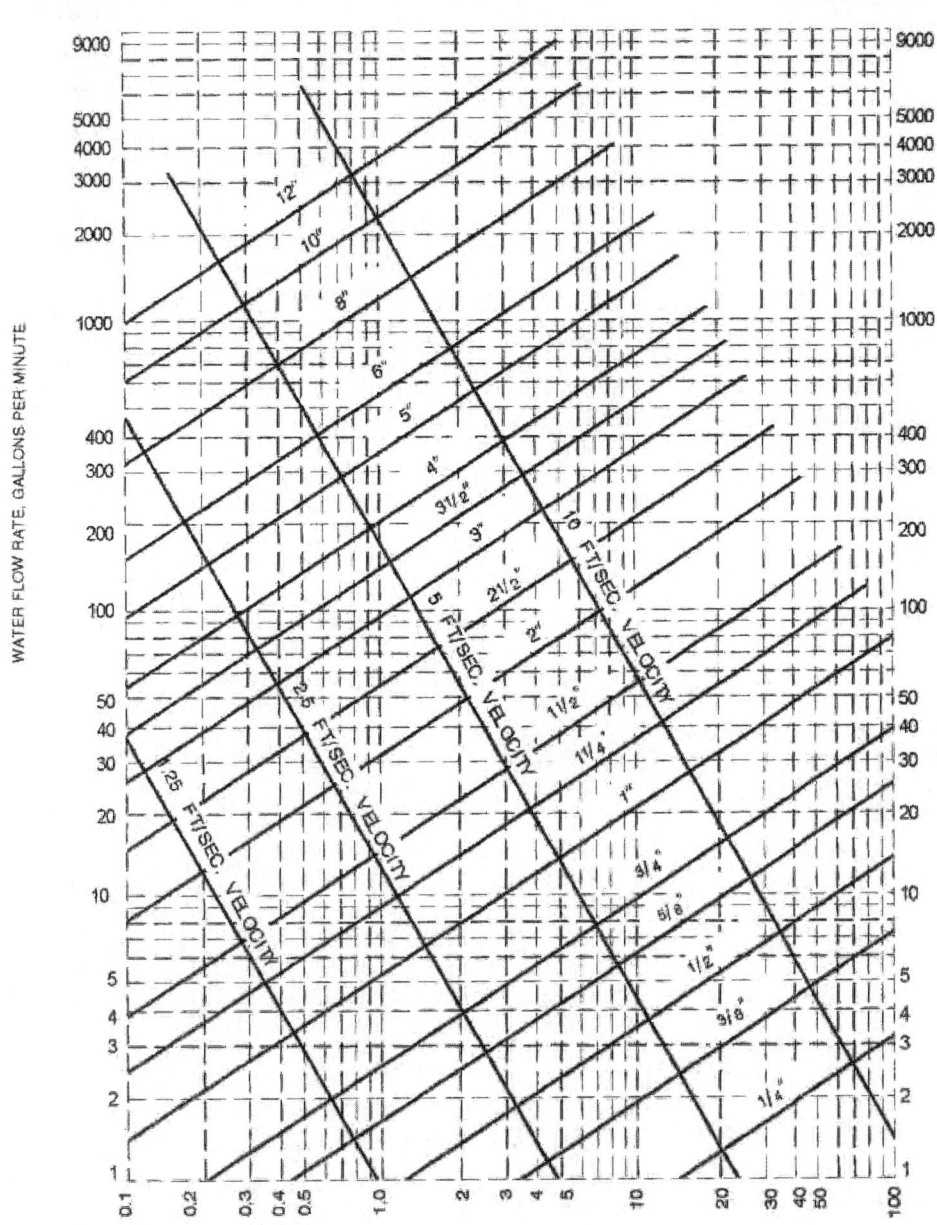

PRESSURE DROP PER 100 FEET OF TUBE, POUNDS PER SQUARE INCH

Note: Fluid velocities in excess of 5 to 8 feet/second are not usually recommended.

**FIGURE E103.3(3)
FRICTION LOSS IN SMOOTH PIPE[a] (TYPE L, ASTM B 88 COPPER TUBING)**

For SI: 1 inch = 25.4 mm, 1 foot = 304.8 mm, 1 gpm = 3.785 L/m, 1 psi = 6.895 kPa, 1 foot per second = 0.305 m/s.

a. This chart applies to smooth new copper tubing with recessed (streamline) soldered joints and to the actual sizes of types indicated on the diagram.

WATER SUPPLY AND DISTRIBUTION

Installation

When installing the water service and the sewer main for a building, keep in mind that they can be no closer than 5 feet apart if they are buried at the same depth. You are allowed to put both pipes in the same trench if the water service is installed at least 12 inches above the sewer main. This isn't always possible due to frost depth considerations. If a water main is sleeved at least 5 feet in each direction, it can cross over a sewer main without having to meet the 12 inches above the sewer rule. A further protection of the potable water supply pipe is the prohibition against installing any water service pipe above, below, or through a cesspool, septic tank, or leach field. Table 8.4 gives the distances the water service pipe must be from various components of a sewage system. (Refer to IPC 603.2–603.2.1)

TABLE 8.4 MINIMUM HORIZONTAL SEPARATION DISTANCES (IPC TABLE 603.2.1)

TABLE 603.2.1
MINIMUM HORIZONTAL SEPARATION DISTANCES BETWEEN WATER SOURCE ELEMENTS AND SOIL ABSORPTION SYSTEM COMPONENTS

Element	Absorption Field	Septic or Treatment Tank
	Distance (feet)	
Cistern	50	25
Lot line	5	2
Spring	100	50
Water main	50	—[a]
Water service	10	5
Water well	50	25

For SI: 1 foot = 304.8 mm

[a] No limitation is listed in Table 802.8 of the IPSDC.

Distribution System

We can now move inside the building and concentrate on the water distribution system to the fixtures within the building. (Refer to IPC 604.1–604.11)

Sizing

The IPC simply states that the water distribution system must be designed according to accepted engineering practices and the method used must be approved. A particular sizing system is not mandated because it is not possible to predict the demands that will be placed on a particular distribution system in any particular part of it at any particular place. (Refer to IPC 604.1)

Pressure

When a system is designed, considerations must be made for the pressures required to operate the various fixtures in a system (Table 8.5). The most demanding fixture is the blowout water closet with a flushometer valve. These require a **flow pressure** of 25 psig at a flow rate of 35 GPM. That's why you don't find these fixtures in residential installations. The water supply pipe must be 1 inch in diameter from the main to the flushometer valve. (Refer to IPC 604.4–604.7)

While we are talking about pressure, we should also mention velocity concerns. Table 8.6, which is IPC Table 604.1, shows the maximum velocity allowed in various materials and sizes. If the stated velocities are exceeded, significant erosion of the pipe material will occur, along with noise in the pipes.

TABLE 8.5 WATER DISTRIBUTION DESIGN CRITERIA (IPC TABLE 604.3)

TABLE 604.3
WATER DISTRIBUTION SYSTEM DESIGN CRITERIA REQUIRED CAPACITY AT FIXTURE SUPPLY PIPE OUTLETS

Fixture Supply Outlet Serving	Flow Rate[a] (gpm)	Flow Pressure (psi)
Bathtub	4	8
Bidet	2	4
Combination fixture	4	8
Dishwasher, residential	2.75	8
Drinking fountain	0.75	8
Laundry tray	4	8
Lavatory	2	8
Shower	3	8
Shower, temperature controlled	3	20
Sillcock, hose bibb	5	8
Sink, residential	2.5	8
Sink, service	3	8
Urinal, valve	15	15
Water closet, blow out, flushometer valve	35	25
Water closet, flushometer tank	1.6	15
Water closet, siphonic, flushometer valve	25	15
Water closet, tank, close coupled	3	8
Water closet, tank, one piece	6	20

For SI: 1 pound per square inch = 6.895 kPa, 1 gallon per minute = 3.785 L/m

[a]For additional requirements for flow rates and quantities, see Section 604.4

TABLE 8.6 MAXIMUM FLOW VELOCITY (IPC TABLE 604.1)

TABLE 604.1
MAXIMUM FLOW VELOCITY

Material	Pipe or Tube Size (inches)	Maximum Velocity (feet per second)
All pipe and tube material	$\frac{1}{2}$ and smaller	5
Brass pipe	$\frac{5}{8}$ and larger	8
Copper or copper-alloy pipe and tubing (hot water systems)	$\frac{5}{8}$ and larger	5
Copper or copper-alloy pipe and tubing (cold water systems)	$\frac{5}{8}$ and larger	8
Chlorinated polyvinyl (CPVC) plastic pipe	$\frac{5}{8}$–1	8
	$1\frac{1}{4}$ and larger	10
Cross-linked polyethylene (PEX) plastic tubing	$\frac{5}{8}$ and larger	8
Cross-linked polyethylene/aluminum/ cross-linked polyethylene (PEX-AL-PEX) pipe	$\frac{5}{8}$ and larger	8
Galvanized steel pipe	$\frac{5}{8}$–1	8
	2–4	10
	4 and larger	12
PE-AL-PE	$\frac{5}{8}$ and larger	8

For SI: 1 inch = 25.4 mm, 1 foot = 0.3048 m/s.

DID YOU KNOW?

Head pressure (the pressure created by the vertical height of water) can be expressed as psig by multiplying the height of the water in feet by 0.433.

$$psig = 0.433 \times height$$

And, if you know the gauge pressure (psig), you can multiply the psig by 2.31 to get the height of the water column.

$$psig \times 2.31 = height$$

Materials

Many materials are accepted for use as water service pipe by the IPC and the UPC. Some of the most common materials used today are copper, cross-linked polyethylene (PEX), polyethylene (PE), and galvanized steel pipe. Polyethylene of raised-temperature (PE-RT) plastic hot and cold water tubing and distribution systems are now recognized by the IPC. It is important to remember that it is prohibited to install any water service pipe in soil that is corrosive or detrimental to the pipe material. (Refer to IPC 605.1–605.3)

Some of the more common materials allowed for distribution systems within a building are copper, PEX, chlorinated polyvinyl chloride (CPVC), and brass pipe. All piping materials used in distribution systems must have a minimum working pressure of 100 psi at 180°F.

WATER SUPPLY AND DISTRIBUTION

> **ALERT**
>
> Translucent piping materials, such as some varieties of PEX and PE, should not be installed where sunlight will strike them. UV rays will break down the molecular structure of the tubing and the tubing can act like a miniature greenhouse and support algae and bacteria growth in the water within the tubing.

> **Hints and Tips**
>
> In many jurisdictions, shutoff valves will also be required on the water supplies to tubs and showers. Be sure you are referring to the local version of the plumbing code when you are preparing for your test.

Water Hammer Controls

The design of the water distribution system with respect to limiting flow velocities should control a water hammer in the system in general. Where quick-closing valves, such as solenoid valves in automatic washing machines, are used, a water hammer arrestor must be installed. The manufacturer's instructions for installation are considered part of the plumbing code. (Refer to IPC 604.9)

Installation Criteria

There are several places in a water distribution system that require a full-opening valve. The water service pipe at the curb needs one. The place at which the water service pipe enters the building is another. The outlet, or discharge, side of every water meter, and at the base of all water risers in multifamily buildings over two stories tall are two other targeted areas. In multifamily buildings, every down-feed pipe requires a full-opening valve at its top. The entrance to every individual dwelling needs the full-opening valve. The water supply pipe to a water tank and the cold water supply pipe into a water heater both require a full-opening valve. This is done to maintain control over the system without adding any significant restriction to the flow. All full-opening and shutoff valves must be accessible. (Refer to IPC 606)

In addition to the full-opening valves placed throughout the building, shutoff valves must be installed on the fixture supply to each plumbing fixture with the exception of showers and tubs. The water supply to every sillcock requires a shutoff valve. The water supply to every piece of mechanical equipment, such as the heating boiler, and every water-using appliance also must have a shutoff valve.

If there is a pressure tank in the water distribution system, it must be equipped with a pressure relief valve rated no higher than the rated working pressure of the tank. Pressure reducing valves are required to be installed wherever the water service pressure is above 80 psi.

Hot Water

The cold-water society ended decades ago. Now, all fixtures that are intended for bathing, washing, culinary, cleansing, laundry, or building maintenance must be supplied with hot water. Tempered water (maximum 110°F) must be supplied to baths and showers. A tempering control other than the thermostat of the water heater must be utilized. (Refer to IPC 607) The IPC prohibits a water heater thermostat from being used as the temperature-limiting device where the code requires a limit for hot or tempered water.

When the hot water must travel 50 feet or more to the farthest fixture, some method of maintaining the water temperature must be used. A common method would be recirculation of the hot water to the water heater either by convection or through the use of a pump, and ensuring that water piping in automatic temperature-maintenance systems is insulated. This is now a provision of the IPC.

Backflow Prevention

The potable water system must be protected against backflow through sound installation practices and the use of prevention devices such as air gaps, antisiphon fill valves, double check valve backflow preventers, barometric loops, and vacuum breakers. Connections to boilers must use a double check valve backflow preventer with a vent. (Refer to IPC 608.1) Potable water supply to beverage dispensers, carbonated beverage dispensers, or coffee machines must also be protected by an air gap or a vented backflow preventer in accordance with ASSE.1022.

Expansion Accommodation

If thermal expansion is likely to occur, some means of controlling it must be installed. Usually a pressure-reducing valve is installed on the water main to control the incoming water service pressure. Many pressure-reducing valves act as a check valve and will prevent water from expanding out through the water service pipe. Therefore, an expansion tank must be installed to absorb any expansion that might occur in the hot water system. (Refer to IPC 607.3)

Chapter Review

Note: Answers may vary based on whether the IPC or UPC is applicable.

1. The concern about cross-connections with solar water heating systems is
   a. that panels may be subject to freezing.
   b. possible antifreeze solutions in the solar panel piping system.
   c. that solar water heating systems are often connected to boilers.
   d. unwarranted.

2. Metallic piping used for grounding
   a. can be replaced with nonmetallic pipe if another approved ground is used.
   b. must be replaced with PEX tubing only.
   c. must be replaced with metallic pipe.
   d. must be left in place.

3. A potable water supply is required
   a. in every house.
   b. in every school.
   c. in every hotel.
   d. all of the above.

4. If a fixture is installed in a building intended for occupancy,
   a. it is only required to be connected to the drainage system.
   b. it is only required to be connected to the water supply system.
   c. it is required to be connected to the water supply and drainage system.
   d. none of the above.

WATER SUPPLY AND DISTRIBUTION

5. Water must be supplied to a building at
   a. 90% of the demand rate.
   b. 100% of the demand rate.
   c. 75% of the demand rate.
   d. 125% of the demand rate.

6. A water pump must be installed
   a. 10 in. above the basement floor.
   b. 16 in. above the basement floor.
   c. 12 in. above the basement floor.
   d. 18 in. above the basement floor.

7. The minimum diameter of a water service pipe is
   a. 1 in.
   b. $\frac{1}{2}$ in.
   c. $\frac{3}{4}$ in.
   d. $1\frac{1}{4}$ in.

8. A water service pipe can be installed in the same trench with a building sewer
   a. if they are at the same depth.
   b. if the water pipe is 12 in. above the entire sewer pipe.
   c. if the sewer pipe is sleeved.
   d. none of the above.

9. A water service pipe can be
   a. copper.
   b. 1 in. in diameter.
   c. 5 ft. from a septic tank.
   d. all of the above.

10. The water distribution system in a building
   a. must be installed according to the plumbing code sizing method.
   b. can have $\frac{1}{2}$-in. water mains.
   c. must deliver sufficient flow pressure for all the fixtures.
   d. must have a greater flow velocity than 12 FPS.

11. The water distribution system design criteria for a bathtub
   a. has a flow rate of 4 GPM.
   b. has a flow rate of 12 GPM.
   c. has a flow rate of 7 GPM.
   d. has a flow rate of 6 GPM.

12. A blowout water closet with a flushometer valve
   a. has the highest flow rate for fixtures.
   b. always cycles in 6 seconds.
   c. needs a $\frac{3}{4}$-in. supply pipe.
   d. all of the above.

13. The size of a fixture supply pipe
   a. cannot be reduced.
   b. can be reduced for the last 30 in.
   c. must always be $\frac{1}{2}$ in. in diameter.
   d. none of the above.

14. A water hammer arrestor
   a. must be installed by quick-closing valves.
   b. must be installed by all fixture shutoffs.
   c. is optional.
   d. none of the above.

15. If the water pressure from the street is inadequate to maintain fixture flow pressure
   a. a new water service must be installed.
   b. a booster pump may be installed.
   c. a booster pump must be installed.
   d. a booster pump is not allowed.

16. Approved water hammer arrestor types include
   a. piston, diaphragm, and compression element.
   b. air chamber, compression element, and piston.
   c. piston, air chamber, and diaphragm.
   d. air chamber and compression element.

WATER SUPPLY AND DISTRIBUTION

17. Water supply pipes and fittings can contain lead up to
   a. 0.03%.
   b. 18%.
   c. 8%.
   d. 3%.

18. Water service pipe can be made of
   a. PEX.
   b. asbestos-cement pipe.
   c. PE.
   d. all of the above.

19. Full-opening valves are required at
   a. the cold water inlet to a water heater.
   b. a water heater outlet pipe.
   c. a water closet supply.
   d. sillcocks.

20. A drainpipe with a $1\frac{1}{2}$-in. diameter is required on a water tank with a capacity of
   a. up to 750 gal.
   b. 500 to 1,000 gal.
   c. 751 to 1,500 gal.
   d. over 7,500 gal.

21. A pressure relief valve must be installed in a
   a. backflow preventer.
   b. booster pump.
   c. water tank.
   d. water closet.

22. Hot water temperature maintenance is required for pipes longer than
   a. 50 ft.
   b. 100 ft.
   c. 60 ft.
   d. 75 ft.

23. Hot water recirculation pipes must be
   a. $\frac{1}{2}$ in.
   b. insulated.
   c. flexible.
   d. copper.

24. Pressure-reducing valves can
   a. reduce pressure.
   b. be made of bronze.
   c. act as a check valve.
   d. all of the above.

25. Hot water temperature for showers must be limited to
   a. 3 GPM.
   b. 100°F.
   c. 110°F.
   d. 4.5 GPM.

26. Backflow preventers prevent vacuum.
   a. true
   b. false

27. Stop and waste valves are approved for underground water service.
   a. true
   b. false

28. If a pipe that had been used for another purpose has been sterilized, it can be used for potable water.
   a. true
   b. false

29. Hospitals must have at least two water service pipes.
   a. true
   b. false

WATER SUPPLY AND DISTRIBUTION

30. A water heater thermostat can be adjusted and used as the temperature control for hot water to showers.
   a. true
   b. false

31. The pressure relief valve on a water tank must be rated for 50 percent less than the working pressure of the tank.
   a. true
   b. false

32. An overflow for a 200-gal. water supply tank must be 3 in. in diameter.
   a. true
   b. false

33. Ten feet of head create 23 psig.
   a. true
   b. false

34. A water pump must be protected from freezing.
   a. true
   b. false

35. PE-AL-PEX pipe can be used for electrical grounding.
   a. true
   b. false

CHAPTER 9: SANITARY DRAINAGE

CHAPTER SUMMARY

More emphasis is placed on the proper design of a sanitary drainage system than almost any other part of a plumbing code, with the possible exception of ensuring a potable water supply. It could be argued that sanitary drainage systems have contributed more toward the elimination and/or reduction of disease than the medical profession. By reducing the population's exposure to pathogens, the health of the nation has been vastly improved.

SANITARY DRAINAGE

Learning Objectives

- Discuss the importance of a properly designed drainage system
- Explain the correct connection techniques for similar pipes and fittings, as well as the connection methods for dissimilar pipes and fittings
- Recognize the appropriate materials for the different parts of a drainage system
- Learn to properly size a sanitary drainage system
- Identify the proper places for cleanouts and manholes
- Detect when a backwater valve is necessary
- Pinpoint the proper piping and design for subdrains, sumps, and sewage ejectors

Key Terms

building drain
building sewer
DFU
hydraulic jump
local vent
Transite pipe

General Requirements

Take a look at the title of this chapter—"Sanitary Drainage." It seems like a contradiction in terms. We all know what really goes down those "sanitary" drains, but something more important is going on here. A couple of centuries ago, typhoid fever, dysentery, and cholera were common, life-threatening diseases. Modern sanitary drainage systems and clean water supplies accomplished more toward the suppression of these dangerous scourges than the science of medicine did. The system name arises from the job these drainpipes do. They help keep our personal environments sanitary. Therefore, the name is more appropriate than one would have thought. The whole idea of the plumbing codes is to ensure that sanitation is maximized through well-designed plumbing systems. (Refer to IPC 701.1–701.3; UPC 713.0–713.6

JOB CONNECTION

Manuel and Will have the job of roughing in the drainage piping for the new Springfield Old Elms Community, a retirement complex of apartments and townhouses. The Ace Plumbing Company estimator, Sophia, has done none of the sizing of the drainage piping. She quoted the job using rule-of-thumb averages. It is up to Manuel and Will to be certain they load the different sections of drainage piping with the proper **DFU** load or risk inspection failures and plugged drains.

All buildings that have plumbing fixtures installed in them must be connected to the public sewer if there is one. This rule creates some turmoil when a locality installs a sewer system where none existed before. Many individuals will want to cling to their own septic tank; not for sentimental reasons, but in order to save the expense of installing the building sewer. The rule exists to prevent individual sewage disposal systems from possibly polluting or contaminating nearby properties, water supplies, or waterways by leaking into storm sewers. The individual sewer systems are mainly uncontrolled after the original installation, and local governments prefer being able to provide a safe, sanitary system by ensuring that all sewage goes directly to a sewage treatment plant through the public sewer.

SANITARY DRAINAGE

Individual buildings under different ownership must have their own connection to the public sewer. If a complex of buildings is under one ownership, which is common with commercial properties, buildings can share a single connection to the public sewer. This approach strives to eliminate conflicts that can arise when trying to determine which owner (or owners) is responsible for repairs and maintenance of a sewer lateral being shared by different buildings with different owners.

Allowable Sewage

Sewage cannot be released into any part of the environment until it has been rendered harmless through some form of sewage treatment. This provision prevents any individual or public/private entity from dumping sewage directly into the soil or any waterway. This has gone a long way toward cleaning up many waterways and wetlands. Only methods approved by the local Authority Having Jurisdiction can be used to treat sewage. (Refer to IPC 701.4–701.9; UPC 714.0– 714.5)

No material that will hinder the function of a sewage treatment plant can be introduced into a sewer system. This includes chemical waste and any items that can clog a system. Nothing that can break down, attack, or damage the drainage piping is allowed. When an industrial plant, for example, has chemical waste, it must be neutralized before sending it through drains into the sewage system.

Another often-overlooked provision prevents water that is at a temperature higher than 140°F from being sent down the drainage piping. High-temperature waste can damage piping through stress and strain from expansion and, of course, plastic drainpipes can be warped by high temperatures. This regulation also keeps any steam exhausts from being connected to the sewer system.

Piping Materials

It is essential that quality materials are used in drainage systems. Both aboveground and especially belowground systems remain in service for many decades. Sewer piping must be able to withstand physical stress and be resistant to the liquids and gasses that flow through them. (Refer to IPC 702.1–702.6; UPC 701.0– 701.2.3)

Aboveground Piping

Table 9.1 displays all the materials that are permitted for aboveground drainage piping. Brass pipe, copper or copper alloy pipe, galvanized steel pipe, and glass pipe are only allowed in aboveground applications. The other materials shown in the table are also allowed to be used in underground building drain applications.

Belowground Piping

Table 9.2, which lists the approved materials for underground **building drain** and vent piping, contains most of the same materials as Table 9.1 for aboveground drainage piping, with a few exceptions. Asbestos-cement pipe is allowed for use as the underground building drain, but not for use above ground.

SANITARY DRAINAGE

TABLE 9.1 ABOVEGROUND DRAINAGE AND VENT PIPE

MATERIAL	STANDARD
Acrylonitrile butadiene styrene (ABS) plastic pipe	ASTM D 2661: ASTM F 628; CSA B 43
Brass pipe	ASTM B 43
Cast-iron pipe	ASTM A 74; ASTM A 888; CISPI 301
Coextruded composite ABS DWV schedule 40 IPS pipe solid	ASTM F 1488
Coextruded composite ABS DWV schedule 40 IPS pipe (cellular core)	ASTM F 1488
Coextruded composite PVC DWV schedule 40 IPS pipe solid	ASTM F 1488
Coextruded composite PVC DWV schedule 40 IPS pipe (cellular core)	ASTM F 891; ASTM F 1488
Coextruded composite PVC IPS-DR, PSI140, PS200 DWV	ASTM F 1488
Copper or copper-alloy pipe	ASTM B 42; ASTM B 302
Copper or copper-alloy tubing (Type K, L, M or DWV)	ASTM B 75; ASTM B88; ASTM B 251; ASTM B 306
Galvanized steel pipe	ASTM A 53
Glass pipe	ASTM C 1053
Polyolefin pipe	CSA B 181.3
Polyvinyl chloride (PVC) plastic pipe (Type DWV)	ASTM D 2665; ASTM D 2949; ASTM F 1488; CSA B 181.2
Stainless steel drainage systems, Types 304 and 316L	ASME A 112.3.1

TABLE 9.2 UNDERGROUND BUILDING DRAINAGE AND VENT PIPE

MATERIAL	STANDARD
Acrylonitrile butadiene styrene (ABS) plastic pipe	ASTM D 2661: ASTM F 628; CSA B 43
Asbestos-cement pipe	ASTM C 428
Cast-iron pipe	ASTM A 74; ASTM A 888; CISPI 301
Coextruded composite ABS DWV schedule 40 IPS pipe solid	ASTM F 1488
Coextruded composite ABS DWV schedule 40 IPS pipe (cellular core)	ASTM F 1488
Coextruded composite PVC DWV schedule 40 IPS pipe solid	ASTM F 1488
Coextruded composite PVC DWV schedule 40 IPS pipe (cellular core)	ASTM F 891; ASTM F 1488
Coextruded composite PVC IPS-DR, PSI140, PS200 DWV	ASTM F 1488
Copper or copper-alloy tubing (Type K, L, M or DWV)	ASTM B 75; ASTM B88; ASTM B 251; ASTM B 306
Polyolefin pipe	CSA B 181.3
Polyvinyl chloride (PVC) plastic pipe (Type DWV)	ASTM D 2665; ASTM D 2949; ASTM F 1488; CSA B 181.2
Stainless steel drainage systems, Types 304 and 316L	ASME A 112.3.1

SANITARY DRAINAGE

For the underground **building sewer** pipe, the list of accepted materials once again changes. Some additional configurations of PVC and ABS pipe are allowed; concrete pipe is allowed; only types K and L copper pipe are allowed; polyethylene (PE), which is one form of polyolefin pipe, is allowed; and finally, that old (very old) drainage standby, vitrified clay tile pipe, aka terra-cotta pipe, is allowed for use as building sewer pipe (Table 9.3).

TABLE 9.3 BUILDING SEWER PIPE

MATERIAL	STANDARD
Acrylonitrile butadiene styrene (ABS) plastic pipe	ASTM D 2661: ASTM F 628; CSA B 43
Asbestos-cement pipe	ASTM C 428
Cast-iron pipe	ASTM A 74; ASTM A 888; CISPI 301
Coextruded composite ABS DWV schedule 40 IPS pipe (solid)	ASTM F 1488
Coextruded composite ABS DWV schedule 40 IPS pipe (cellular core)	ASTM F 1488
Coextruded composite PVC DWV schedule 40 IPS pipe (solid)	ASTM F 1488
Coextruded composite PVC DWV schedule 40 IPS pipe (cellular core)	ASTM F 891; ASTM F 1488
Coextruded composite PVC IPS-DR, PSI140, PS200 DWV	ASTM F 1488
Coextruded composite ABS sewer and drain DR-PS in PS35, PS50, PS 100, PS140, PS200	ASTM F 1488
Coextruded PVC sewer and drain PS25, PS50, PS100, PS140, PS200	ASTM F 1488
Coextruded PVC sewer and drain PS25, PS50, PS100 (cellular core)	ASTM F 891
Concrete pipe	ASTM C14; ASTM C 76; CSA A257.1M; CSA a257.2M
Copper or copper-alloy tubing (type K or L)	ASTM B 75; ASTM B88; ASTM B 251
Polyethylene (PE) plastic pipe (SDR-PR)	ASTM F 714
Polyvinyl chloride (PVC) plastic pipe (type DWV, SDR26, SDR35, SDR41, PS50 or PS100)	ASTM D 2665; ASTM D 2949; ASTM D3034; CSA B 181.2; CSA B182.4
Stainless steel drainage systems (types 304 and 316L)	ASME A 112.3.1
Vitrified clay pipe	ASTM C 4; ASTM C 700

SANITARY DRAINAGE

An important fact about building sewers that may very well be on your licensing test is that it is allowable to lay a storm sewer side by side with the sanitary sewer. There is far less risk of a sanitary sewer leaking into a storm sewer than a sanitary sewer leaking into a potable water system.

The grade, or pitch, of a drainage system is essential to ensuring that waste flows properly though the sewage system. A velocity of approximately 2 feet per second is the best speed for wastewater to flow in order to keep suspended solids moving along with the water. A minimum pitch of $\frac{1}{2}$-inch per foot for drains $2\frac{1}{2}$ inches in diameter reaches that critical velocity. For drains of 3 inches to 6 inches in diameter, $\frac{1}{8}$-inch per foot is the minimum slope allowed, and $\frac{1}{16}$-inch per foot is the minimum slope for pipes that are 8 inches or larger. One should note that no maximum pitch is stated in the codes; however, for good piping practice, a pitch of no more than $\frac{1}{2}$-inch per foot should be maintained.

Never reduce the diameter of a drainage pipe in the direction of flow. Never reduce the diameter of a drainage pipe in the direction of flow. Yes, I repeated that. It's important, and no, a 4 × 3 inch water closet connection is not considered to be a reduction in size by both codes. Why? you ask. Take a look at the outlet size of a water closet bowl. It will range somewhere between $1\frac{7}{8}$ inches up to about $2\frac{1}{4}$ inches. The closet flange is made larger in order to create a larger surface area and steadier mounting surface for the sealing ring and the closet bolts.

ALERT

Hydraulic jump can be observed in the water at the base of a dam or even in a flat-bottomed sink when a stream of water from the faucet lands in the sink. Water "piles up" at the point where the velocity of the flow can no longer keep the water moving at the same pace and the body of water slows to the pace determined by flow in a horizontal plane. When wastewater piles up just past the base of a stack where the flow transitions from a rapid vertical descent to a slower horizontal flow, wastewater would flow backward up a branch that is connected within 10 pipe diameters of the stack. The consequence is the depositing of solids in the branch, which creates the potential for blockage of the branch.

Connection of horizontal branches to the horizontal drain at the base of a stack is prohibited until a distance equal to 10 times the diameter of the stack is reached. If a connection is closer than that to the base of the stack, waste from the stack will be pushed up the connected branch by the turbulence at the base of the stack known as **hydraulic jump**.

Dead ends in drains are prohibited. A dead end is most often created when capping drainpipes that are taken out of service. A discontinued drain must be cut and capped within 24 inches of an active drain in order to meet code. In new work, extension of a cleanout beyond 24 inches is permitted and drainage pipes for future plumbing are permitted to go beyond 24 inches.

The IPC (Section 702.6) establishes the minimum thickness for lead bends and traps at 0.125 inch. Nowadays, lead pipes and fittings are rarely installed, generally in a chemical piping system.

SANITARY DRAINAGE

Piping Joints

Just as the weakest link determines the strength of the chain, proper joints maintain the integrity of a piping system. Correct joining methods must be used to ensure not only a sturdy system, but also that the piping system will pass inspection. (Refer to IPC 705.1– 705.22; UPC 704.0–705.3.3)

Connections between Similar Fittings and Pipe

ABS plastic—allowable joining methods:

- Elastomeric seals
  Elastomeric seals can be used to connect ABS to ABS or to other piping materials.

- Solvent cement
  The only approved solvent cementing is ABS solvent cement to connect ABS to ABS.

- Threaded joints
  Threaded joints can be used to connect ABS to ABS or to any other "threadable" material.

Asbestos cement pipe—commonly referred to as **Transite pipe**—can only be joined using the asbestos cement sleeve coupling with elastomeric rings. When connecting asbestos cement pipe to other materials, mechanical joints utilizing elastomeric couplings are commonly used, although the plumbing code does not recognize this coupling method.

Brass pipe—can be connected to other brass pipe by brazing; mechanical joints, which include elastomeric seals and brass ferules or compression rings; and by welding.

Cast iron—caulked oakum and lead joints, compression gaskets, and mechanical joints are the approved methods for joining cast-iron pipe.

Concrete pipe—an elastomeric ring is the only approved joining method.

PVC plastic—mechanical joints, solvent cementing using purple primer and clear cement, and threaded connections are permitted.

Copper pipe—brazing, mechanical connections, soldering, and threading are the approved connections for copper pipe.

Glass pipe—bolted compression couplings, or caulked joints when connecting glass pipe to a cast-iron hub are permitted.

Steel pipe—threading and mechanical joints are permitted.

Lead pipe—wiped joints and burned joints are approved for lead pipe.

Vitrified clay tile pipe—only elastomeric seals are to be used.

PE pipe—heat-fusion and mechanical joints are approved.

Polyolefin pipe—heat-fusion and mechanical joints are the permitted joining methods.

Connections between Dissimilar Fittings and Pipe

Copper to cast iron—use a ferrule, or soil pipe adapter, to connect copper to cast iron. The adapter can then be caulked or installed with a compression gasket.

Copper to galvanized steel—threaded fittings with dielectric properties must be used.

Cast iron to galvanized steel—a caulked joint or mechanical elastomeric coupling may be used.

Plastic pipe to other materials—threaded fittings can be joined, or in the case of cast iron, a caulked joint is permitted.

Lead pipe to other materials—a wiped joint to a caulking ferrule or approved threaded fitting.

SANITARY DRAINAGE

Glass pipe to other materials—adapter couplings containing a TFE seal.

Stainless steel drainage pipe—approved mechanical couplings may be used.

Drainage slip joints—elastomeric rings are approved.

Use of Fittings

Table 9.4 shows the allowable positions in which fittings can be used. In general, one can expect that if there is a possibility of waste or wastewater being thrown in an upstream direction in a fitting, the position for that fitting will not be allowed. The point is to ensure a relatively smooth flow of waste from the fixture to the sewer with the least turbulence and opportunity for blockage. (Refer to IPC 706.1–706.4; UPC 706.1–706.4)

TABLE 9.4 FITTINGS FOR CHANGE IN DIRECTION

TYPE OF FITTING PATTERN	CHANGE IN DIRECTION		
	HORIZONTAL TO VERTICAL	VERTICAL TO VERTICAL	VERTICAL TO HORIZONTAL
Sixteenth bend	X	X	X
Eighth bend	X	X	X
Sixth bend	X	X	X
Quarter bend	X	X[a]	X
Short sweep	X	X[a,b]	X[a]
Long sweep	X	X	X
Sanitary tee	X[c]	—	—
Wye	X	X	X
Combination wye and eighth bend	X	X	X

[a]The fittings shall only be permitted for a 2-inch or smaller fixture drain.

[b]Three inches or larger.

[c]For a limitation on double sanitary tees, see Section 706.3 of IPC.

A heel-inlet quarter bend can be used unless it serves a water closet. Only a high heel fitting can be used for a wet venting application. Side-inlet quarter bends are permitted.

Hints and Tips
Tee wyes that sit on their back with the branch in a vertical position are not permitted for use when waste is flowing down the vertical pipe. This position is, however, allowed if the vertical pipe is used exclusively as a vent.

SANITARY DRAINAGE

Some *prohibited connections* include cement, concrete, mastic, hot bituminous joints, nonapproved fittings, elastomeric rolling O-rings (when used to connect different diameter pipes), solvent cementing different types of plastic pipes, and saddle fittings.

Cleanouts

Blockages in a drainage system are virtually unavoidable. Cleanouts provide the necessary access to the interior of the drainage pipes through threaded plugs, without the need to cut pipes to gain access. (Refer to IPC 708.1–708.3.2; UPC 707.0–707.10)

Cleanouts must be placed in the following locations:

- In all horizontal drains not more than 100 feet apart.
- In building sewers less than 8 inches in diameter, and not more than 100 feet apart.
- When a horizontal drain, a building drain, or a building sewer with a diameter less than 8 inches has a change of direction of more than 45 degrees, a cleanout is required. No more than one cleanout is required in any 40-foot length of pipe, even if there is more than one change in direction greater than 45 degrees.
- The base of every soil stack or waste stack must have a cleanout.
- A cleanout must be located near the connection between the building drain and the building sewer. If there is a cleanout on a 3-inch or larger soil stack within 10 feet of the connection between the building drain and the building sewer, no other cleanout is necessary.

If a cleanout is in a crawl space or other space with less than 24 inches of clearance, the cleanout must be brought up to an accessible grade. In addition, a cleanout cannot be used for a fixture connection unless it is approved and there is another cleanout available to service the area.

All cleanouts must be the same size as the pipe served, up to 4 inches. Pipes larger than 4 inches can have 4-inch cleanouts.

Manholes

When a manhole is part of a building drain, in other words, inside the building, it must have a gas-tight seal and the cover must be secured. This is done to prevent the escape of sewer gasses inside the building and to help prevent incidental access to the manhole. For building sewers greater than or equal to 8 inches in diameter, manholes must be installed within 200 feet of the connection with the building drain, and after that, no more than 400 feet apart. Building sewers greater than or equal to 8 inches in diameter must also have a manhole installed at every change in direction of the pipe.

> **DID YOU KNOW?**
>
> Water flowing down a vertical stack adheres to the walls of the pipe, leaving an open central column of air. The water also flows downward in a clockwise spiral, which is caused by the rotation of the earth. In the southern hemisphere, the rotation is counterclockwise.

Sizing

Sizing a drainage system is an exercise in chart reading and some basic math, assembled here in this chapter. I have assembled all the charts you need together in one place (Tables 9.5 through 9.8). Using these charts, you can determine the required size of any portion of a

SANITARY DRAINAGE

drainage system. Table 9.5 gives you the drainage fixture unit (DFU) value of the individual fixtures. Add up the fixtures on a particular branch and you can easily find the pipe size needed by looking at Table 9.8. Table 9.8 will also help you figure the maximum DFU load that can be put on stacks. Table 9.6 gives you the DFU values for fixture drains and traps. (Refer to IPC 709.1–710.2; UPC 702.0–703)

TABLE 9.5 DRAINAGE FIXTURE UNITS FOR FIXTURES AND GROUPS

FIXTURE TYPE	DRAINAGE FIXTURE UNIT VALUE AS LOAD FACTORS	MINIMUM SIZE OF TRAP (INCHES)
Automatic clothes washers, commercial[a,g]	3	2
Automatic clothes washers, residential[g]	2	2
Bathroom group as defined in Section 202 (1.6 gpf water closet)[f]	5	—
Bathroom group as defined in Section 202 (water closet flushing greater that 1.6 gpf)[f]	6	—
Bathtub (with or without overhead shower or whirlpool attachments)	2	$1\frac{1}{2}$
Bidet	1	$1\frac{1}{4}$
Combination sink and tray	2	$1\frac{1}{2}$
Dental lavatory	1	$1\frac{1}{4}$
Dental unit or cuspidor	1	$1\frac{1}{4}$
Dishwashing machine,[c] domestic	2	$1\frac{1}{2}$
Drinking fountain	0.5	$1\frac{1}{4}$
Emergency floor drain	0	2
Floor drains	2	2
Kitchen sink, domestic	2	$1\frac{1}{2}$
Kitchen sink, domestic with food waste grinder and/or dishwasher	2	$1\frac{1}{2}$
Laundry tray (1 or 2 compartments)	2	$1\frac{1}{2}$
Lavatory	1	$1\frac{1}{4}$
Shower	2	$1\frac{1}{2}$
Service sink	2	$1\frac{1}{2}$
Sink	2	$1\frac{1}{2}$
Urinal	4	Note d
Urinal, 1 gallon per flush or less	2[e]	Note d
Urinal, nonwater supplied	0.5	Note d
Wash sink (circular or multiple), each set of faucets	2[e]	$1\frac{1}{2}$
Water closet, flushometer tank, public or private	4[e]	Note d
Water closet, private (1.6 gpf)	3[e]	Note d

(Continued)

SANITARY DRAINAGE

TABLE 9.5 (Continued)

FIXTURE TYPE	DRAINAGE FIXTURE UNIT VALUE AS LOAD FACTORS	MINIMUM SIZE OF TRAP (INCHES)
Water closet, private (flushing greater than 1.6 gpf)	4[e]	Note d
Water closet, public (1.6 gpf)	4[e]	Note d
Water closet, public (flushing greater than 1.6 gpf)	6[e]	Note d

For SI: 1 inch = 25.4 mm, 1 gallon = 3.785 L (gpf = gallon per flushing cycle).

[a]For traps larger than 3 inches, use Table 709.2.

[b]A showerhead over a bathtub or whirlpool bathtub attachment does not increase the drainage fixture unit value.

[c]See Sections 709.2 through 709.4 for methods of computing unit value of fixtures not listed in this table or for rating of devices with intermittent flows.

[d]Trap size shall be consistent with the fixture outlet size.

[e]For the purpose of computing loads on building drains and sewers, water closets and urinals shall not be rated at a lower drainage fixture unit unless the lower values are confirmed by testing.

[f]For fixtures added to a dwelling unit bathroom group, add the DFU value of those additional fixtures to the bathroom group fixture count.

[g]See Section 406.3 for sizing requirements for fixture drain, branch drain, and drainage stack for an automatic clothes washer standpipe.

TABLE 9.6 DRAINAGE FIXTURE UNITS FOR FIXTURE DRAINS OR TRAPS

FIXTURE DRAIN OR TRAP SIZE (INCHES)	DRAINAGE FIXTURE UNIT VALUE
$1\frac{1}{4}$	1
$1\frac{1}{2}$	2
2	3
$2\frac{1}{2}$	4
3	5
4	6

For SI: 1 inch = 25.4 mm

Let's look at an example. A two-story building has two offices on the first floor, each with one water closet and one lavatory. In the custodian's closet, there is a service sink. On the second floor are four apartments, each with a kitchen sink and a full bathroom consisting of a water closet, a bathtub, and a lavatory. Start on the second floor. All of the apartments will drain down a single soil stack to the building drain. Each bathroom counts as 5 DFUs, even though the water closet is 3, the tub is 2, and the lavatory is 1. That adds up to 6, but the bathroom group counts as 5 since all fixtures are not expected to be used at the same time. The kitchen sink accounts for 2 DFUs.

Bathroom group	5
Kitchen sink	+2
	7 DFU per apartment
All apartments	×4
	28 DFU total for second floor

If you look at Table 9.8, you will see that a $2\frac{1}{2}$-inch pipe would be enough for an apartment drainpipe. However, when you look at Table 9.7 you can see that the minimum pipe size allowed for a water closet is 3 inches, so each apartment must drain

SANITARY DRAINAGE

with a 3-inch pipe. If the branch from each apartment ties together before entering the soil stack, the horizontal drain carrying all four of the apartment drains must be 4 inches since the combined DFU load of the four apartments is 28 DFUs. A 3-inch horizontal branch can only carry up to 20 DFUs, but the 4-inch branch can carry 90 DFUs as the total load from one branch interval, being the second floor in this case. As we enter the vertical soil stack, it would appear that we can reduce our stack size to 3 inches since a 3-inch stack can handle 48 DFUs and we are only sending it 28 DFUs. We know we cannot do this though, because we can never reduce the diameter of a drainage pipe in the direction of flow. That leaves us with continuing on with a 4-inch stack. Obviously, there are other considerations that go into drain sizing than simply the sizes on the charts.

TABLE 9.7 BUILDING DRAINS AND SEWERS

MAXIMUM NUMBER OF DRAINAGE FIXTURE UNITS CONNECTED TO ANY PORTION OF THE BUILDING DRAIN OR THE BUILDING SEWER, INCLUDING BRANCHES OF THE BUILDING DRAIN[a]

DIAMETER OF PIPE (INCHES)	SLOPE PER FOOT			
	$\frac{1}{16}$ INCH	$\frac{1}{8}$ INCH	$\frac{1}{4}$ INCH	$\frac{1}{2}$ INCH
$1\frac{1}{4}$	—	—	1	1
$1\frac{1}{2}$	—	—	3	3
2	—	—	21	26
$2\frac{1}{2}$	—	—	24	31
3	—	36	42	20
4	—	180	216	250
5	—	390	480	575
6	—	700	840	1,000
8	1,400	1,600	1,920	2,300
10	2,500	2,900	3,500	4,200
12	3,900	4,600	5,600	6,700
15	7,000	8,300	10,000	12,000

For SI: 1 inch = 25.4 mm, 1 inch per foot = 83.3 mm/m.

[a]The minimum size of any building drain serving a water closet shall be 3 inches.

SANITARY DRAINAGE

TABLE 9.8 HORIZONTAL FIXTURE BRANCHES AND STACKS
MAXIMUM NUMBER OF DRAINAGE FIXTURE UNITS (DFU)

DIAMETER OF PIPE (INCHES)	TOTAL FOR HORIZONTAL BRANCH	STACKS[b] TOTAL DISCHARGE INTO ONE BRANCH INTERVAL	TOTAL FOR STACK OF THREE BRANCH INTERVALS OR LESS	TOTAL FOR STACK GREATER THAN THREE BRANCH INTERVALS
$1\frac{1}{2}$	3	2	4	8
2	6	6	10	24
$2\frac{1}{2}$	12	9	20	42
3	20	20	48	72
4	160	90	240	500
5	360	200	540	1,100
6	620	350	960	1,900
8	1,400	600	2,200	3,600
10	2,500	1,000	3,800	5,600
12	2,900	1,500	6,000	8,400
15	7,000	Note c	Note c	Note c

For SI: 1 inch = 25.4 mm

[a]Does not include branches of the building drain. Refer to Table 710.0(1).

[b]Stacks shall be sized based on the total accumulated connected load at each story or branch interval. As the total accumulated connected load decreases, stacks are permitted to be reduced in size. Stack diameters shall not be reduced to less that one-half the diameter of the largest stack size required.

How many DFUs do the first floor fixtures add up to? For the two water closets: 3 + 3 = 6 DFUs. For the two lavatories: 1 + 1 = 2 DFUs. One service sink equals 2 DFUs. So the total DFU load for the first floor is 10. The first floor requires a 4-inch drain.

The total DFU load for the building is 38 DFUs $[(4 \times 7) + (2 \times 3) + (2 \times 1) + 2 = 38$ DFUs]. Once again, we are looking at a situation in which a reduction in the size of the building drain would be appropriate since a 3-inch building drain at $\frac{1}{4}$-inch per foot slope can carry 42 DFUs, as shown in Table 9.7. We will stick with the 4-inch building drain since we can never reduce the diameter of a drainage pipe in the direction of flow.

Keep in mind that if you are doing a complete or a partial rough-in of drainage piping for fixtures that will not be installed at this time, but might be in the future, those DFU values must be included in your sizing calculations just as if the fixtures were being installed now.

Sumps and Ejectors

The material used for sump pumps and ejector pumps, discharge piping, and other fittings shall be rated for maximum system operating pressure and temperature, and where (when?) buried in the earth they shall be suitable for that purpose.

Whenever possible, drains must drain by gravity to the building sewer. When a drain is too low to drain by gravity, it can drain into a sump where the effluent is collected to await being lifted up to the gravity building drain or building sewer by a sewage ejector pump. Sump pumps are now permitted to connect to the drainage system at the soil stack, waste stack, or horizontal branch drain. The sump must be tightly sealed and a vent must terminate either through the roof of the building or into a stack vent in the building that terminates in the open air. The pump must operate automatically. A check valve followed by a full-opening valve are required on the discharge pipe of the ejector pump. The tightly sealed sump must allow access to the pump through a removable cover. The sump itself must be a minimum of 18 inches in diameter and 24 inches deep, but must be sized according to the load. Effluent cannot rise to within more than 2 inches of the top of the sump before the automatic controls start the pump. (Refer to IPC 712.1–712.4.2; UPC 710.1–710.710.13)

A macerating toilet system must be installed according to the manufacturers' instructions.

Healthcare Drainage Piping

Healthcare facilities require specialized fixtures, and unique provisions are needed to keep high-infection-risk plumbing safe to the patients, the staff, and the public. (Refer to IPC 713.1–713.11.4)

Bedpan washers must connect to the drainage system in a similar way to a water closet and they must be vented with a **local vent**. Except when it is connected to only one fixture, the base of a local vent must have a trap and a drain connection to the waste stack, which is the same size as the local vent. The trap of a local vent must be primed at each use of the bedpan washer.

Sterilizers, steamers, and condensers must drain through an indirect waste system.

Vacuum systems must be installed so that medical personnel can easily monitor their operation. Vacuum systems must operate at all times. Therefore, backup electricity must be available, and a second pump must be installed so vacuum service is never disrupted.

Vacuum collectors must be connected directly to the drainage system. Indirect wastes are not allowed for these due to the risk of airborne pathogens.

Backwater Valves

Occasionally, a municipal sewer system can become plugged and all the sewage from your neighbors' buildings can back up into your building drain and up through your fixtures, overflowing gallons and gallons of wastewater in your building. In the determination of backwater valve protection from sewage backflow, the use of the finish flood elevation where the fixtures are installed rather than the flood-level rim of the fixture provides a new point of reference.

Chapter Review

Note: Answers may vary based on whether the IPC or UPC is applicable.

1. When is it allowable for sewage to be disposed of without treatment?
   a. in an individual system
   b. never
   c. when disposal is in a river within 1 mile of an ocean
   d. always

2. Which of these are permitted in a public sewer without treatment?
   a. #2 fuel oil
   b. sulfuric acid
   c. detergent
   d. none of the above

3. Steam exhaust can be connected to a sanitary drainage system
   a. when it is below 15 psi.
   b. when it is at a pH of 7.
   c. when it is below 200°F.
   d. never.

4. In a food service establishment, exposed drainage piping can be installed
   a. above a food preparation surface.
   b. above an eating surface.
   c. above a food storage area.
   d. under the floor of the kitchen.

5. Vitrified clay tile pipe can be used for a building drain.
   a. true
   b. false

6. Bituminous fiber pipe can only be used for a building sewer.
   a. true
   b. false

7. Lead pipe can be used for chemical waste.
   a. true
   b. false

8. Solvent welded PVC pipe can be used for a building sewer.
   a. true
   b. false

9. Concrete pipe is acceptable for an underground building drain.
   a. true
   b. false

10. ABS solvent welding is a one-step process.
    a. true
    b. false

11. Purple primer is no longer used on PVC solvent weld joints.
    a. true
    b. false

12. Approved methods for joining concrete pipe are
    a. elastomeric O-ring.
    b. 50-year silicone caulk.
    c. oakum and Portland cement.
    d. all of the above.

13. Approved methods for joining cast-iron pipe include
    a. caulked (oakum and lead).
    b. mechanical elastomeric coupling.
    c. compression gasket.
    d. all of the above.

SANITARY DRAINAGE

14. Glass pipe can be joined by heating and melting (welding) the glass together.
   a. true
   b. false

15. Accepted cement for PVC pipe includes
   a. PVC cement.
   b. PVC–ABS cement.
   c. all-purpose cement.
   d. all of the above.

16. Schedule 40 PVC pipe can be threaded to make connections to other pipe materials.
   a. true
   b. false

17. A tee wye can be used to convey waste in a horizontal to horizontal pattern.
   a. true
   b. false

18. A wye can be used to convey waste in
   a. a horizontal to horizontal pattern.
   b. a vertical to horizontal pattern.
   c. a horizontal to vertical pattern.
   d. all of the above patterns.

19. Which fittings can be used directly below a water closet connection?
   a. side-inlet quarter bend
   b. high heel inlet 90-degree bend
   c. low heel inlet 90-degree bend
   d. 4 × 3 in. coupling

20. A cleanout must be installed at every change of direction in a building drain.
   a. true
   b. false

21. In building sewers that are 8 inches or larger, a manhole is required only at changes of direction.
   a. true
   b. false

22. Cleanouts must be
   a. no more than 100 ft. apart.
   b. at changes of direction greater than 45 degrees.
   c. within 10 ft. of the junction of the building drain and the building sewer.
   d. all of the above.

23. Manholes must be installed on sewer pipes greater in diameter than
   a. 6 in.
   b. 8 in.
   c. 10 in.
   d. 12 in.

24. A bathroom group has a DFU value of
   a. 4.
   b. 6.
   c. 5.
   d. 7.

25. Most flat-bottomed fixtures have a DFU value of
   a. 2.
   b. 4.
   c. 3.
   d. 5.

26. A horizontal branch carrying 34 DFUs must have a pipe diameter of
   a. 2 in.
   b. 4 in.
   c. 3 in.
   d. 5 in.

SANITARY DRAINAGE

27. One bathroom group, one laundry tray, one kitchen sink, and a half-bath consisting of a water closet and a lavatory have a drainage load of
   a. 10 DFUs.
   b. 12 DFUs.
   c. 11 DFUs.
   d. 13 DFUs.

28. Waste that can enter a building subdrain can be the discharge from
   a. a water closet.
   b. a lavatory.
   c. a urinal.
   d. all of the above.

29. A sump pit must
   a. be at least 24 in. wide.
   b. have a plastic cover.
   c. have a 4-in. vent.
   d. be at least 24 in. deep.

30. A sewage ejector must
   a. automatically start when sump level rises.
   b. sit in a plastic sump.
   c. be cleaned once every 6 months.
   d. none of the above.

31. A local vent
   a. connects to the building vent stack.
   b. must be at least 4 in. in diameter.
   c. connects to the fixture on the building side of the trap.
   d. is the main sewer vent for a building.

32. A backwater valve is installed if the elevation of the nearest manhole is above the elevation of the flood-level rim of any fixture.
   a. true
   b. false

33. A backwater valve must be full-opening.
   a. true
   b. false

34. A 2-in. horizontal branch can carry _____ DFUs.
   a. 6
   b. 3
   c. 12
   d. 160

35. Every soil stack requires a cleanout at its base.
   a. true
   b. false

CHAPTER 10 ▶ INDIRECT AND SPECIAL WASTES

CHAPTER SUMMARY

This chapter examines the theory, purpose, and proper installation of indirect wastes and special wastes. Often, the simplest solutions to a problem are the most elegant and ideal. The indirect waste is a perfect example of a faultless solution to a problem. How do we ensure that wastewater and sewage from a drainage system do not contaminate food and related items that are being cleaned and prepared in a fixture that has to drain into that potential source of contamination? The indirect waste is the most dependable preventer of contamination devised. As long as the laws of gravity and physics remain reliable, we can always depend on the indirect waste to protect our lettuce and spatulas from sewage contamination in restaurants, among other places.

INDIRECT AND SPECIAL WASTES

Learning Objectives

- State the reasons for using indirect and special wastes
- Explain the practical installation of indirect and special wastes
- Identify places where indirect wastes are not needed
- Classify special wastes and explain their proper handling

Key Term

air gap

Objective of Indirect Waste

The indirect waste protects food that is being cleaned or otherwise prepared in a plumbing fixture. Generally, this means a sink (Figure 10.1). In a food-handling establishment, this would mean all sinks in which food preparation *may* occur. A sink, or multiple compartment sink, in a smaller food preparation establishment that is primarily used for washing dishes and pots, but could be used for food preparation, will probably be required to drain through an indirect waste in order to ensure the safety of the food prepared at the site. Larger restaurants with specialized food preparation areas and dishwashing and pot-scrubbing areas will now be required to drain the dishwashing sinks through an indirect waste. In healthcare locations, sterilizers, steamers, and rinsers must be drained through indirect wastes. Any fixture that is not required to be drained through an indirect drain *must* be directly connected to the building drainage system. (Refer to IPC 801.1–802.4; UPC 801.0–814.3)

JOB CONNECTION

Carlos, the owner of Ace Plumbing, has taken Evan, one of his installers, to his favorite diner, Dogs N Donuts, for a little visit. Carlos isn't buying lunch though. Evan is there to install the new three-compartment sink for washing pots and pans, and the new dishwashing machine. Carlos is more of an executive than a plumber, so he doesn't know what the code is for these fixture drains. Evan is on his own. See if you can decide how he should go about installing these drains. Then, go to your favorite diner and see if the owner will let you check out the drains at that establishment.

INDIRECT AND SPECIAL WASTES

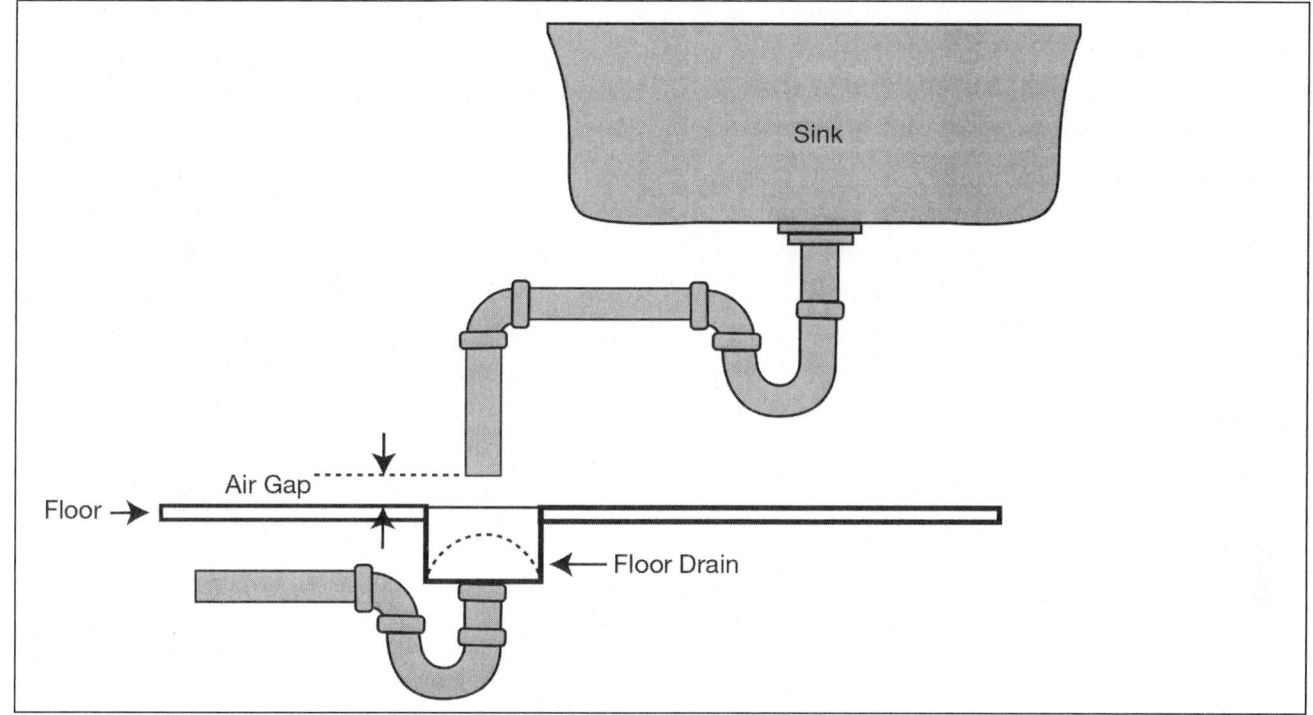

Figure 10.1 Typical Indirect Waste Configuration

Often, the sink involved is a two-compartment or three-compartment commercial stainless steel sink. The drains from the two or three compartments can be connected to each other and discharge through an **air gap** into a receptor, or they can individually drain through an air gap into a receptor or receptors. Receptors cannot be located in bathrooms or in any inaccessible place. The distance of the air gap must be at least 1 inch or twice the diameter of the waste pipe, whichever is larger. An air gap is now the only acceptable discharge means for a sink used for the washing, rinsing, or sanitizing of utensils, dishes, pots, pans, or serviceware in the preparation, serving, or eating of food.

Hints and Tips

One of the most common complaints of restaurant owners involves the overflow that occurs when an indirect waste drains into a receptor that cannot drain fast enough to accommodate the discharge of the indirect waste pipe. Sometimes owners take things into their own hands and make a direct connection that prevents the overflow but renders the backflow prevention capability of the indirect waste useless. Be sure to properly size an indirect waste to handle the full flow of the fixture drain.

INDIRECT AND SPECIAL WASTES

When the developed length of an indirect waste pipe is more than 30 inches long in developed length measured horizontally or 54 inches in total developed length, a trap is required on the indirect waste. This trap requirement reduces bacteria and odor in indirect wastes. No vent is required for this trap because the pipe is open at both ends and there is no direct connection to the building drainage system, and therefore no imbalance in air pressures can be created within the drainage system. Waste receptors cannot be located in plenums, crawl spaces, attics, or interstitial spaces above ceilings or below floors.

All clear water wastes must drain indirectly. These types of wastes include the many kinds of condensate that drain from refrigeration equipment, condensing heating equipment, water treatment equipment, and other such equipment. Wastewater from swimming pool maintenance equipment also falls under this same category and requires an indirect waste. Dishwashing machines also must drain indirectly. However, they can be drained through an air break or an air gap (Figure 10.2).

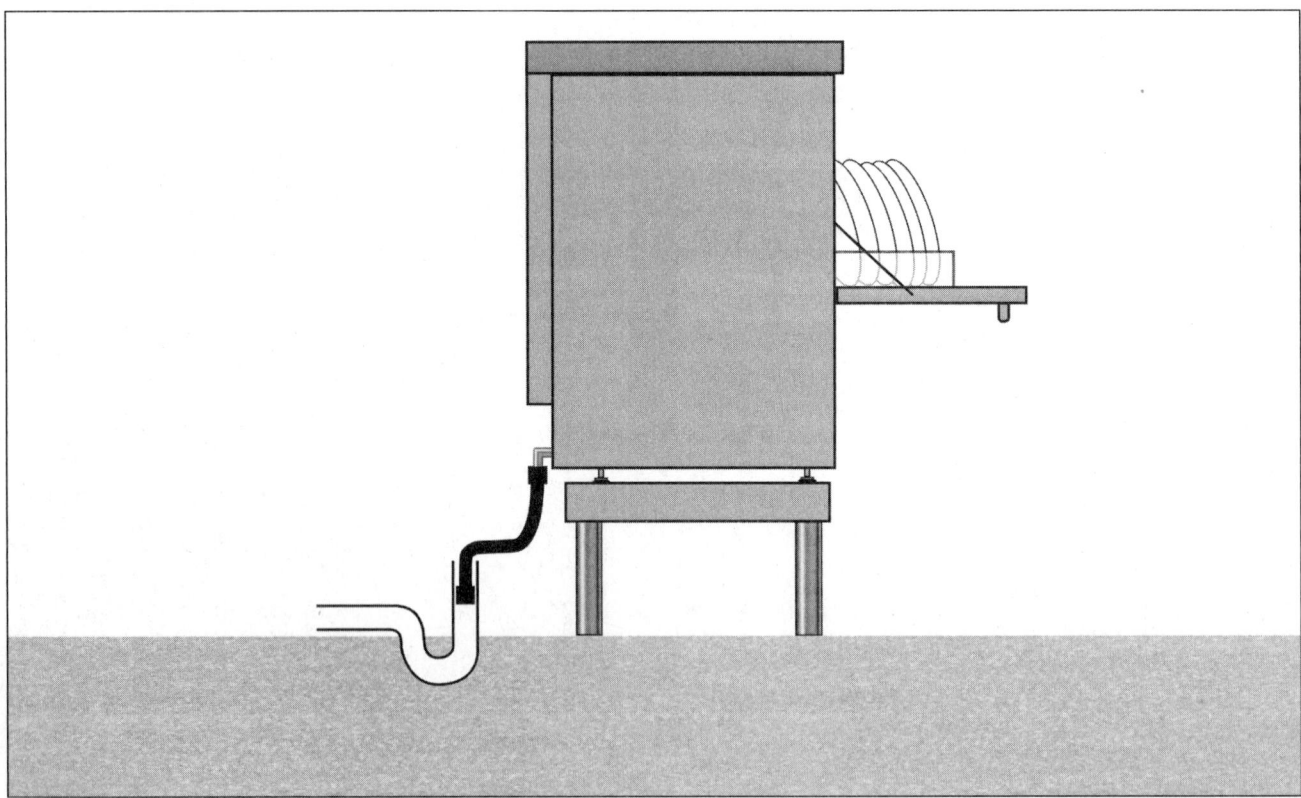

Figure 10.2 Air Break Indirect Waste for a Commercial Dishwashing Machine

INDIRECT AND SPECIAL WASTES

> **ALERT**
>
> Whenever there is a question about whether an air break or an air gap should be used, use the air gap due to the greater level of backflow prevention that an air gap affords. Remember, an air gap will always be approved in an application where an air break is accepted.

Specific Waste (Special Wastes)

Steam, water above 140°F, corrosive liquids, radioactive liquids, and toxic liquids are considered special wastes. Wastewater above 140°F is not permitted in a building drainage system. Hot wastewater must be cooled to below 140°F, usually in a retention tank, before it is allowed to flow into the building drainage system. A code official could potentially permit the use of special heat-resistant piping for water above 140°F. (Refer to IPC 803.1–803.3; UPC 810.0–813.0)

All corrosive liquids must be neutralized before flowing into a drainage system. Chemical and toxic wastes cannot enter a drainage system without being rendered harmless. Any waste that cannot be neutralized satisfactorily cannot be introduced into a building drainage system.

> **DID YOU KNOW?**
>
> An indirect waste pipe is still considered to be a part of the sanitary drainage system and must meet all the other applicable provisions of the plumbing code that is adopted. Even though the indirect waste is disconnected from the sanitary drainage system, all the code provisions must be applied to the installation.

Specific Fixtures

This section outlines some of the basic requirements for specific fixtures. Some fixtures have special designs that require unique accommodations. (Refer to IPC 802.1, 802.1.6, 802.1.7802.1.4; UPC 801.2.3, 807.4, 807.1, 813.0)

Sinks

- Any sink in which food is prepared must drain through an indirect waste.
- Any sink that is used for other purposes, but would likely be used for food preparation as well, must be drained through an indirect waste.
- Any sink that is used exclusively for washing dishes, pots, and pans and will not be used for food preparation must be directly connected to the sanitary drainage system.
- Floor sinks are receptors designed and intended to receive drainage from indirect wastes (Figures 10.3 and 10.4). Their design allows for many drains to go to one receptor, as long as the receptor and direct drainage piping connected to the floor sink is sized adequately.

Figure 10.3 Floor Sink (*Photo courtesy of Pep-Plastic*)

INDIRECT AND SPECIAL WASTES

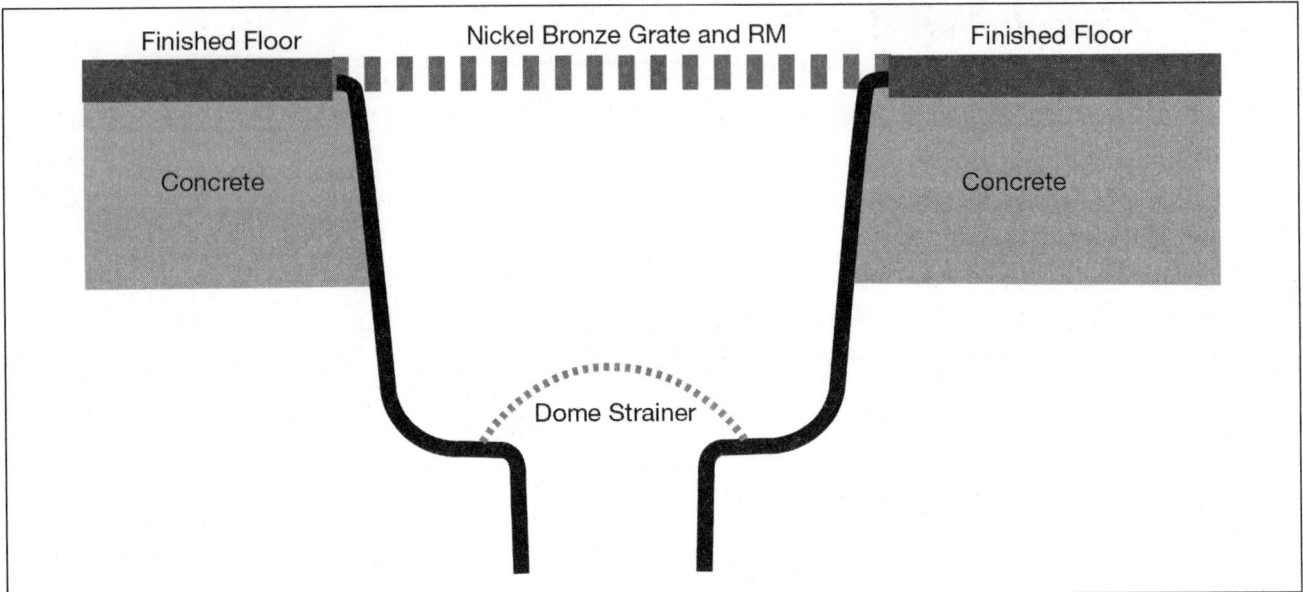

Figure 10.4 Floor Sink Diagram

Dishwashing Machines
Domestic
- In the IPC, domestic dishwashing machines can drain through a deck-mounted air gap or the drain hose must be looped up as high as possible under the countertop, following the manufacturer's instructions.
- In the UPC, all domestic dishwashing machines must be drained through a deck-mounted air gap.

Commercial Dishwashing Machines
- Commercial dishwashing machines must be indirectly drained, but they can be drained with an air break rather than an air gap.

Swimming Pools
- The discharge from pool filter backwash must be drained through an indirect waste if the filter drain is connected to the building drain.
- If a pool drain is connected to the building drain, it must pass through an indirect waste.

INDIRECT AND SPECIAL WASTES

Chapter Review

Note: Answers may vary based on whether the IPC or UPC is applicable.

1. How can a steam vent be drained to a sanitary drainage system?
   a. Condense the steam, and then connect the condensate drain to the drainage system.
   b. Connect the steam vent to the downstream side of a trap in the drainage system.
   c. Condense the steam, cool the condensate to less than 140°F, then drain the condensate indirectly into the drainage system.
   d. It cannot be done.

2. The drain from a swimming pool filter backwash can be drained
   a. outdoors.
   b. through an air break into a receptor that is trapped and is directly connected to the building drain.
   c. through an air gap into a receptor that is trapped and is directly connected to the building drain.
   d. all of the above.

3. If the discharge of nonpotable water from a boiler drains through an indirect waste, an air gap must be used.
   a. true
   b. false

4. A domestic dishwashing machine must discharge through a deck-mounted air gap.
   a. true
   b. false

5. Commercial dishwashing machines may be directly connected to the sanitary drainage system
   a. always.
   b. when connecting to the inlet of a food waste grinder.
   c. never.
   d. when the dishwashing machine cycles no more than twice per hour.

6. Indirect waste pipes that exceed _____ feet in developed length must be trapped.
   a. 3
   b. 4
   c. 5
   d. 6

7. A vent on an indirect waste is required when the developed length of the indirect waste pipe is _____ feet or longer.
   a. 10
   b. 15
   c. 0
   d. Not applicable; the vent is never necessary.

8. Condensate drains from refrigeration equipment must be indirectly drained, but may be drained with an air break.
   a. true
   b. false

9. The connection created when the discharge hose of an automatic washing machine is inserted into a standpipe is called
   a. an air break.
   b. semi-direct.
   c. disconnected.
   d. an air gap.

INDIRECT AND SPECIAL WASTES

10. Corrosive waste can be discharged into a plumbing system if
   a. it flows at a rate less than 1 GPH.
   b. the receptor is stainless steel.
   c. it is neutralized.
   d. it is cooled to below 140°F.

11. If an indirect waste is long enough to require a vent, the vent
   a. must tie into a vent from the waste and vent system in the building.
   b. must be 2 in. in diameter or larger.
   c. must vent independently through the roof.
   d. all the above.

12. Chemical drainage and vent systems must be separate from the building drain and vent system.
   a. true
   b. false

13. Indirect waste pipe materials can be
   a. ABS and copper.
   b. PVC and cast iron.
   c. cast iron and copper.
   d. all of the above.

14. Standpipes
   a. must be individually trapped.
   b. may have a common trap.
   c. are not considered receptors.
   d. can extend no more than 18 in. above the trap.

15. When an air break is specified, an air gap is acceptable.
   a. true
   b. false

16. The drain hose for a domestic dishwasher
   a. must connect to the inlet side of a trap on a tailpiece.
   b. must connect to the outlet side of a trap on a tailpiece.
   c. must be at least 1 in. in diameter.
   d. must connect to the downstream side of a trap.

17. In an indirect waste, the distance between the outlet of the indirect waste pipe and the flood-level rim of the receptor must be
   a. at least 6 in.
   b. twice the diameter of the indirect waste.
   c. twice the diameter of the receptor drainpipe.
   d. 1 in.

18. A floor drain in a walk-in freezer can drain into the sanitary drainage system through an air break only if
   a. the floor drain is trapped.
   b. the floor drain is protected with a backwater valve.
   c. the floor drainpipe is 2 in. diameter or larger.
   d. the floor drain uses a running trap.

19. A three-compartment sink used only for dishwashing must be
   a. indirectly drained.
   b. trapped.
   c. connected directly to the sanitary drainage system.
   d. drained to a properly sized receptor through an air break.

20. If draining to a sanitary drain, pool deck drains must pass through an air gap.
   a. true
   b. false

INDIRECT AND SPECIAL WASTES

21. An indirect drain from a refrigeration coil can be no longer than
  a. 4 ft.
  b. no limit.
  c. 15 ft.
  d. none of the above.

22. Any discharge from a potable water system must be indirectly drained with an air break.
  a. true
  b. false

23. All sterilization equipment
  a. must be indirectly drained through an air gap.
  b. must be directly connected to a sanitary drainage system.
  c. must have a trapped indirect waste.
  d. must have an indirect waste with a minimum diameter of 2 in.

24. An air-conditioning condensate drain can connect to a lavatory tailpiece.
  a. true
  b. false

25. A drinking fountain may be directly or indirectly connected to the sanitary drainage system.
  a. true
  b. false

26. Which fixture must be drained indirectly?
  a. public lavatory
  b. water closet
  c. food prep sink
  d. none of the above

27. The purpose of an air gap is
  a. to protect the potable water supply.
  b. to prevent contamination of food in a food prep sink.
  c. to prevent backflow into the indirect waste.
  d. all of the above.

28. All of the drains in a combination fixture that is used for food preparation must be individually trapped.
  a. true
  b. false

29. An air break provides sufficient backflow protection for an automatic clothes washer.
  a. true
  b. false

30. The materials used for standard drainage systems are high enough quality for an indirect waste.
  a. true
  b. false

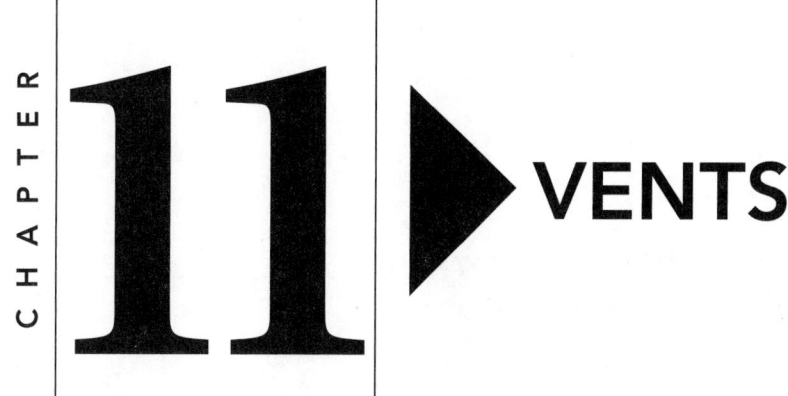
CHAPTER 11 ▶ VENTS

CHAPTER SUMMARY

Vents are an important and necessary part of a drainage system. Not only do they allow water to flow down a drain without it acting like a gallon jug turned upside down, but they also protect the water seals in the traps and help sewer gas vent out of a building.

VENTS

Learning Objectives

- Understand the primary function of plumbing vents
- Explain how vents accomplish their "mission"
- Specify vent locations
- Estimate vent sizing
- Identify weather considerations

Key Term

invert

JOB CONNECTION

Julio and Evan have been given a free hand in designing the drainage system for the new Audie's Autos car dealership since it is a time and material job. There is a women's bathroom consisting of one water closet and one lavatory, and two men's rooms, each containing one water closet, one urinal, and one lavatory. All of the bathrooms are adjoining. There is one wash sink out in the shop area and one floor drain in the indoor showroom. They know they need at least one vent through the roof. Then what? Think about minimum roof penetrations and economy of piping as you read this chapter.

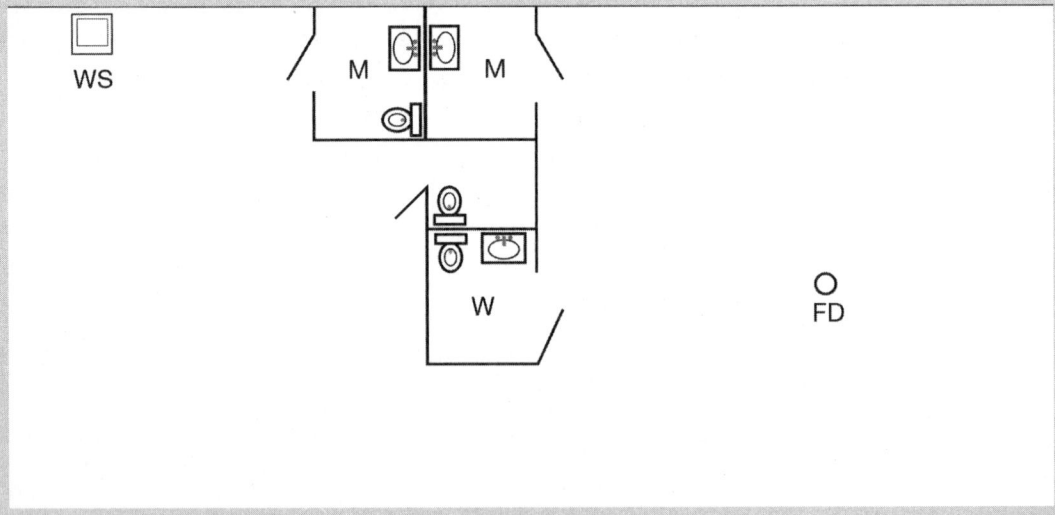

VENTS

Objective of Vents

To keep it as simple as possible, vents are placed in a drainage system to prevent the water seal from being broken in any of the traps in the system. The water seal in the trap keeps sewer gas from entering a building through the fixture drains. A typical trap has a seal depth of at least 2 inches.

Venting Theory

The goal of the vent system is to keep any air pressure in the system, whether it is positive or negative, from pushing the water out of the trap or sucking it into the drain (Figure 11.1). As water and waste move through the horizontal and vertical sections of a drainage system, air is pushed ahead of the slug of water like a string of box cars being pushed around by a yard switcher locomotive operated by an engineer whose shift ends in 8 minutes, as well as being dragged along behind it. The vent system must not allow that pressure/vacuum from exceeding more than 1 inch WC exerted in either direction on any trap connected to any fixture branch of the system. That way, with the 2-inch trap seal depth, no trap seal will be broken.

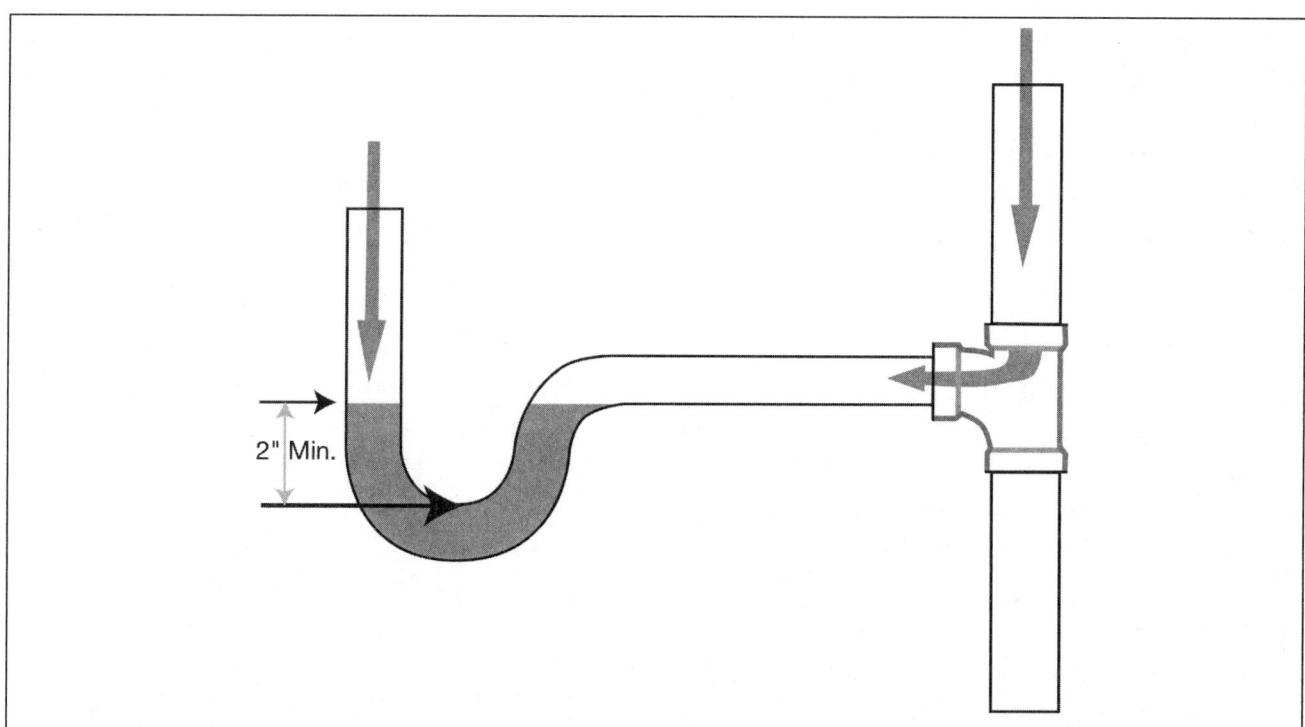

Figure 11.1 Maintaining Normal Atmospheric Pressure on Both Sides of the Trap

Chemical Waste System Vents

Chemical waste systems must be kept separate from sanitary waste systems. This separation applies to their respective venting systems as well. Any vent on a chemical waste system must be completely independent of the sanitary drainage system and must terminate above the roof independently or to an air admittance valve. Even though we are talking about pipes that carry only vapors, the possible condensation of corrosive vapors draining down the sanitary system is too great a risk to take. (Refer to IPC 901.3)

Misuse of Vents

A vent cannot be used for any other purpose than venting. A common practice is the direct connection of a condensate drain from an air-conditioning evaporator located in an attic to the vent as it passes through the attic. Often, this allows sewer gas into the building through the condensate drain, which may have an inadequate or nonexistent trap. Drains found connected in this manner should be redone properly as an indirect waste. (Refer to IPC 901.4)

Materials

Review Chapter 9 on sanitary drainage piping. The materials used for vents are applied the same way as for drainage piping. If sheet copper or sheet lead is used for a vent flashing, there are some basic specifications that must be followed. Sheet copper must weigh more than 8 ounces per square foot. For a long-lasting flashing, the copper cannot come into contact with any other type of metal or severe corrosion will occur. When sheet lead is used for a flashing, it must weigh at least 3 pounds per square foot for a custom flashing built on-site, or $2\frac{1}{2}$ pounds per square foot on a prefabricated flashing. Care must be taken when sheet lead is brought to the top of the vent termination and folded over it. The part of the lead sheet inside the vent pipe must be carefully formed to the pipe so that the lead does not tend to close the vent termination and make the vent size smaller by doing so. (Refer to IPC 902.1–902.3; UPC 903.0–903.4)

Installation

Even though the vent portion of a drainage system only carries air and sewer gas, the same degree of workmanship is necessary to maintain the correct pressures in the vent system. Remember that sewer gas is more than just a bad smell. Dangerous diseases are easily spread in sewer gas. Correct construction and diligent craftsmanship ensure a sound vent system. A combination waste and vent system sized per the code is unlimited in its horizontal length.

Vent Terminations

Every building drain must have at least one vent that extends through the roof of the building to free air. In most buildings, this means one vent because most buildings only have one building drain. This is not the case, however, in some large buildings. The minimum required size of the one vent is one-half the diameter of the building drain. This may seem odd to those raised on the old adage that "one vent must connect full size from the building drain to the vent termination." Modern venting strategies are based on actual venting demand for the fixtures served rather than an arbitrary rule of thumb. Be aware that many jurisdictions still require the "full size" vent and others may require at least a 3-inch vent. Remember that unless those vent terminations are at least 3 feet above any windows and doors, they must be 10 feet away measured horizontally. (Refer to IPC 903.1; UPC 906.0)

Sometimes it is better to terminate a vent through a sidewall rather than out through the roof. If the sidewall is selected as the vent location, it

VENTS

cannot be located under soffit vents, nor can it be within 10 feet of the property line. A sidewall vent must be protected to prevent rodents or birds from entering.

If your location is in an area where the $97\frac{1}{2}\%$ value for outdoor winter design temperature is below 0°F, vent pipes running on the exterior of a building must be insulated, and every vent extension through a roof must have a minimum diameter of 3 inches beginning at least 12 inches below the roof. This is to prevent the vent from being closed by a buildup of frost. You can imagine warm, moist air rising up a shower vent. When the water vapor reaches the vent termination, it freezes inside the pipe and continues to build. It's like making a candle by repeatedly dipping it in melted wax and pulling it out, only in reverse. When the outdoor design temperature is above 0°F, the warm air rising up the vent has an opportunity to melt away any frost buildup.

Whenever a building has five or more branch intervals, that is, stories with fixtures on each at least 8 feet apart vertically, a vent stack is required. All vent stacks and stack vents must end outdoors in the open air or connect to an air admittance valve (only under IPC governance).

All vents must go up vertically to at least 6 inches above the flood-level rim of the highest trap or trapped fixture that the vent serves before it can travel horizontally. This is to keep deposits from building up and remaining in the horizontal portion of a vent, in the event of a drain blockage and subsequent backup into the vent.

Vent Connections

When a vent stack connects to the building drain, it must connect downstream of the stack it serves. It must also be within 10 pipe diameters of the stack it serves. Whenever a dry vent connects to a horizontal drain, the connection must be above the centerline of the horizontal drainpipe. (Refer to IPC 903.4, 905.1–905.6; UPC 905.1–905.6) Horizontal braces are now permitted to connect at any point in a stack above or below a horizontal offset. Also, they can connect to the base of stacks at a point located not less than 10 times the diameter of the drainage stack downstream from the stack

Sizing

When vent stacks and stack vents are connected at the top to a header, every section of the header must be sized according to the venting load of that section. The developed length used to size the vents must be the longest vent length from the bottom of the furthest stack to the vent end in the open air (Table 11.1). Connecting vents together in this manner is common and efficient, while reducing the number of holes cut through the roof for vent extensions. (Refer to IPC 903.1.2; UPC 904.0)

TABLE 11.1 SIZE AND DEVELOPED LENGTH OF STACK VENTS AND VENT STACKS (IPC TABLE 916.1)

DIAMETER OF SOIL OR WASTE STACK (INCHES)	TOTAL FIXTURE UNITS BEING VENTED (DFU)	MAXIMUM DEVELOPED LENGTH OF VENT (FEET)[a] DIAMETER OF VENT (INCHES)										
		$1\frac{1}{4}$	$1\frac{1}{2}$	2	$2\frac{1}{2}$	3	4	5	6	8	10	12
$1\frac{1}{4}$	2	30	—	—	—	—	—	—	—	—	—	—
$1\frac{1}{2}$	8	50	150									
$1\frac{1}{2}$	10	30	100									

(Continued)

TABLE 11.1 (Continued)

DIAMETER OF SOIL OR WASTE STACK (INCHES)	TOTAL FIXTURE UNITS BEING VENTED (DFU)	MAXIMUM DEVELOPED LENGTH OF VENT (FEET)[a] DIAMETER OF VENT (INCHES)										
		$1\frac{1}{4}$	$1\frac{1}{2}$	2	$2\frac{1}{2}$	3	4	5	6	8	10	12
2	12	30	75	200	—							
2	20	26	50	150	—	—	—	—	—	—	—	—
$2\frac{1}{2}$	42	—	30	100	300							
3	10		42	150	360	1040						
3	21	—	32	110	270	810	—	—	—	—	—	—
3	53		27	94	230	680						
3	102		25	86	210	620	—					
4	43	—	—	35	85	250	980	—	—	—	—	—
4	140		—	27	65	200	750					
4	320			25	55	170	640	—				
4	540	—	—	—	50	150	580	—	—	—	—	—
5	190			—	28	82	320	990				
5	490	—			21	63	250	760				
5	940		—	—	18	53	210	670	—	—	—	—
5	1,400				16	49	190	590				
6	500					33	130	400	1000			
6	1,100	—	—	—	—	26	100	310	780	—	—	—
6	2,000					22	84	260	660			
6	2,900					20	77	240	600	—		
8	1,800	—	—	—	—	—	31	95	240	940	—	—
8	3,400					—	24	73	190	720		
8	5,600						20	62	160	610	—	
8	7,600	—	—	—	—		18	56	140	560	—	—
10	4,000						—	31	78	310	960	
10	7,200							24	60	240	740	
10	11,000	—	—	—	—	—	—	20	51	200	630	—
10	15,000							18	46	180	570	
12	7,300								31	120	380	940
12	13,000	—	—	—	—	—	—		24	94	300	720
12	20,000								20	79	250	610
12	26,000								18	72	230	500
15	15,000	—	—	—	—	—	—	—		40	130	310
15	25,000									31	96	240
15	38,000									26	81	200
15	50,000	—	—	—	—	—	—	—	—	24	74	180

For SI: 1 inch = 25.4 mm. 1 foot = 304.8 mm.

[a]The developed length shall be measured from the vent connection to the open air.

VENTS

Vent Patterns

A prime concern for the plumber (and the plumbing inspector) is the distance, or developed length, the trap weir serving a fixture is from the fixture vent serving it. These distances should be memorized (Table 11.2). Please remember that there is no restriction on the distance from a fixture that depends on siphonage to operate, such as a water closet or a urinal, to a vent. (IPC 906.1–911.5; UPC 907.0–908.0)

TABLE 11.2 MAXIMUM DISTANCE OF FIXTURE TRAP FROM VENT (IPC TABLE 906.1)

SIZE OF TRAP (INCHES)	SLOPE (INCH PER FOOT)	DISTANCE FROM TRAP (FEET)
$1\frac{1}{4}$	$\frac{1}{4}$	5
$1\frac{1}{2}$	$\frac{1}{4}$	6
2	$\frac{1}{4}$	8
3	$\frac{1}{8}$	12
4	$\frac{1}{8}$	16

Crown vents are prohibited. A crown vent is a vent that is located within 2 pipe diameters of the trap weir. You may never see a crown vent. Years ago, some traps were manufactured with the crown vent as a part of the trap. It was found that debris from flowing waste would force its way up into the crown vent with no flow down the vent to wash it out. By moving the vent at least 2 pipe diameters away from the weir, the waste and water has a chance to settle into a normal flow in the **invert** of the pipe past the chaotic turbulence of the trap area.

The common vent is simply a vent that serves more than one trap on the same floor level.

Wet venting is a permitted way of adding length to a vent. Even though drainage may flow in a vented pipe, the portion with the flow is still considered a vent to the downstream fixtures in the bathroom group. The wet vent does not begin until the vent is connected to the horizontal drain. In other words, if a water closet is upstream of the wet vent, no other fixture branch can be connected to the horizontal drain upstream of the wet vent. The water closet would create pressure and vacuum in excess of 1 inch WC on the trap seal in the fixture branch that it is flushing past. The trap seal could either be blown out or drawn out of the trap.

> **Hints and Tips**
> When roughing in a vent for a lavatory in a new home, always install the waste stack and stack vent off to the side of the center of the lavatory. The homeowner may decide to install a recessed medicine cabinet and your vent cannot go up through the middle of it. If you make it a habit to always rough in the vertical portion of the drain and vent off to the side, you will never have to be concerned about whether a recessed medicine cabinet will be installed.

DID YOU KNOW?

The complex loop vent used in an island sink can be completely replaced by simply installing an air admittance valve. Install the trap arm into the branch of a vertically mounted tee wye. The upper part of the tee wye run is extended upward a minimum of 4 inches and then is terminated with an air admittance valve. The downward oriented part of the run of the tee wye simply drops vertically into a horizontal drain. Many feet of pipe, a number of fittings, and precious time are saved. Unfortunately, this cannot be done in a jurisdiction that uses the Uniform Plumbing Code because air admittance valves are not permitted under that code.

TABLE 11.3 WASTE STACK VENT SIZE (IPC TABLE 910.4)

STACK SIZE (INCHES)	MAXIMUM NUMBER OF DRAINAGE FIXTURE UNITS (DFU)	
	TOTAL DISCHARGE INTO ONE BRANCH INTERVAL	TOTAL DISCHARGE FOR STACK
$1\frac{1}{2}$	1	2
2	2	4
$2\frac{1}{2}$	No limit	8
3	No limit	24
4	No limit	50
5	No limit	75
6	No limit	100

The waste stack vent can serve a handy purpose of venting fixtures, other than water closets and urinals, that are located on different floors. If each fixture is connected to the waste stack by itself, and there are no offsets in the waste stack, multiple floors can be vented this way. Sizing is an important consideration because it is imperative that pressures in the stack be kept at or below the critical 1 inch WC. When you study Table 11.3 you will see that the waste stack seems to be oversized, and it is, but the large sizing is necessary since there can be multiple floors of fixtures discharging into the waste stack.

Circuit venting is really a wet vent that serves up to eight fixtures on a horizontal branch. You can install up to eight fixtures in a row on the same horizontal drainage pipe, and locate the dry vent between the two fixtures at the upstream end, and you have a circuit vent. The circuit vent must be dry. No waste can be discharged into it.

Combination Drain and Vent System

Sometimes a fixture is located where it's just not possible to vent it within the proper distance. This is where the combination drain and vent system can come in handy. A common application is for a floor drain stuck out in the middle of a large warehouse or industrial plant where there are no nearby walls in which to hide a vent. An oversize drain becomes a combination drain and vent. With oversizing, there is always room for air to move above the drainage flow in the invert of the pipe. A drainage stack shall serve as a single stack vent system where sized and installed in accordance with Sections 917.2 through 917.9. The drainage stack brace piping shall be vented for the drainage system. The drainage stack shall have a stack vent. (Refer to IPC 912.1–912.3; UPC 910.1–910.3)

This piping arrangement can also be used for sinks, lavatories, and drinking fountains. The trap arm is the normal size for the fixture. The drain can then turn directly vertical without a traditional vent going up. However, the elbow that makes the turn from the horizontal trap arm to the vertical drop must increase the size of the pipe so that only the oversized pipe is dropping vertically to a horizontal fixture drain. The downstream horizontal portion of the combination drain and vent must continue at the oversize to allow for free air movement. If you recall, no drainpipe is allowed to be reduced going downstream anyway. The sizing of the combination drain and vent is done according to Table 11.4.

TABLE 11.4 SIZE OF COMBINATION DRAIN AND VENT PIPE (IPC TABLE 912.3)

DIAMETER PIPE (INCHES)	MAXIMUM NUMBER OF DRAINAGE FIXTURE UNITS (DFU)	
	CONNECTING TO A HORIZONTAL BRANCH OR STACK	CONNECTING TO A BUILDING DRAIN OR BUILDING SUBDRAIN
2	3	4
2½	6	26
3	12	31
4	20	50
5	160	250
6	360	575

A type of drain venting that can be substituted with either the combination drain and vent system or with an air admittance valve is the island fixture vent (Figure 11.2). This venting style was the accepted norm before either the combination system or the air admittance valve was accepted.

The piping is complex, yet it makes perfect sense. This special situation in which it is impossible to install a vertical rising vent is overcome with some piping ingenuity. We know that a vent is supposed to rise at least 6 inches above the flood-level rim of the fixtures before going horizontal, but that is impossible here. The reason for the 6-inch requirement is to keep waste from a backup staying in the vent pipe. In the island fixture vent, there is essentially no horizontal component at the top of the loop. The vent ties into a "real" vent before dropping back into the drain to rid itself of any moisture or debris. Cleanouts are provided on the vertical portions of the island fixture vent system to ensure a clear vent. This is truly an elegant solution to a tricky problem.

VENTS

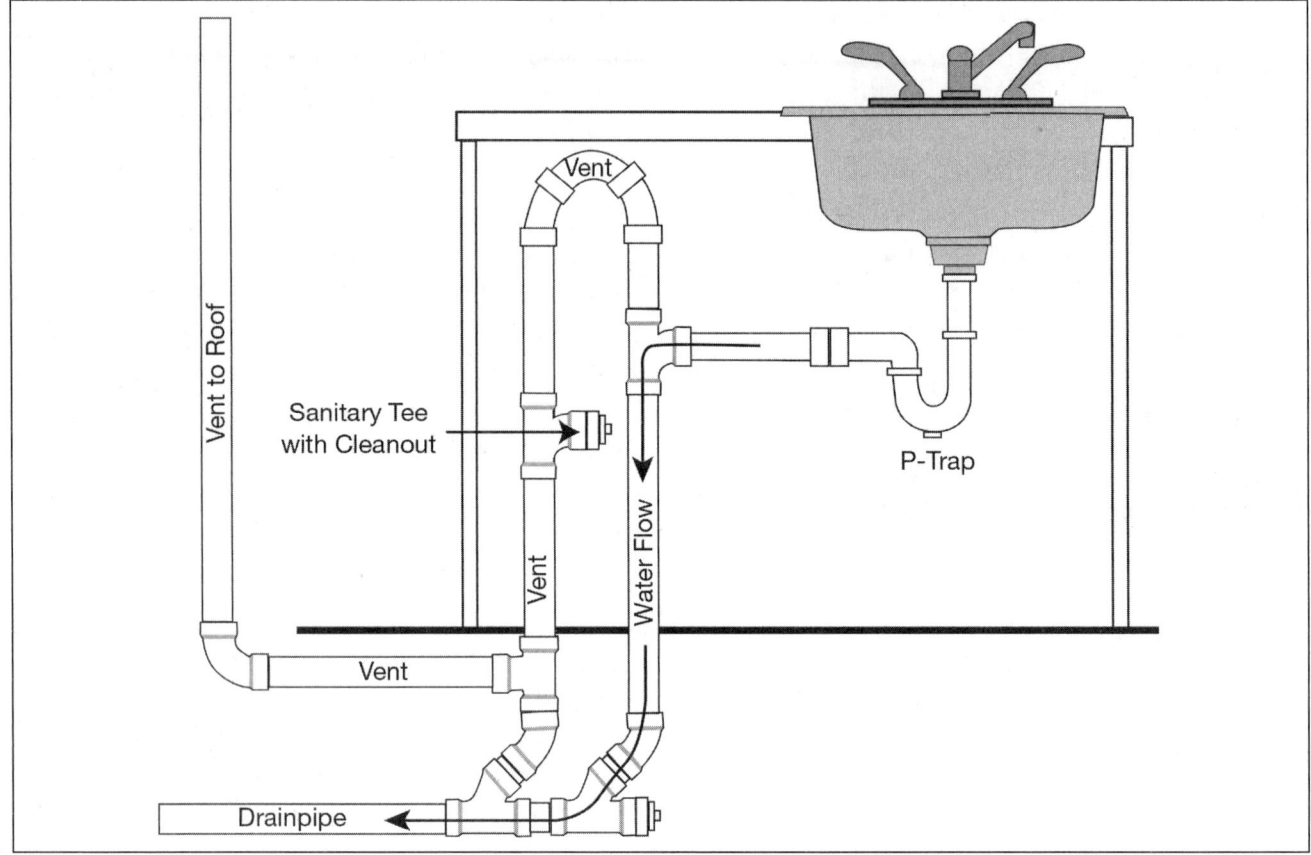

Figure 11.2 Island Fixture Venting Method

Multistory Buildings

In a building that has more than ten branch intervals (ten stories with plumbing fixtures), the codes require that a relief vent runs from the soil stack to the vent stack to relieve pressures created in the drainage system (Figure 11.3). The first floor that gets the relief vent is the top floor, and then the floors are counted from the top down by ten for the installation of the relief vent. (Refer to IPC 914.1–915.3; UPC 907)

The design of a relief vent is a wye installed in the soil stack with the relief vent pipe going upward from the wye a minimum of 3 feet above the floor to an "upside down" wye in the vent stack. The 3-feet interval keeps waste out of the relief vent. Offsets in soil stacks must have a vent on the vertical pipes above and below the horizontal portion of the soil stack that connects to the vent stack. When a branch vent has a developed length of more than 40 feet, increase the size by 1 pipe diameter.

VENTS

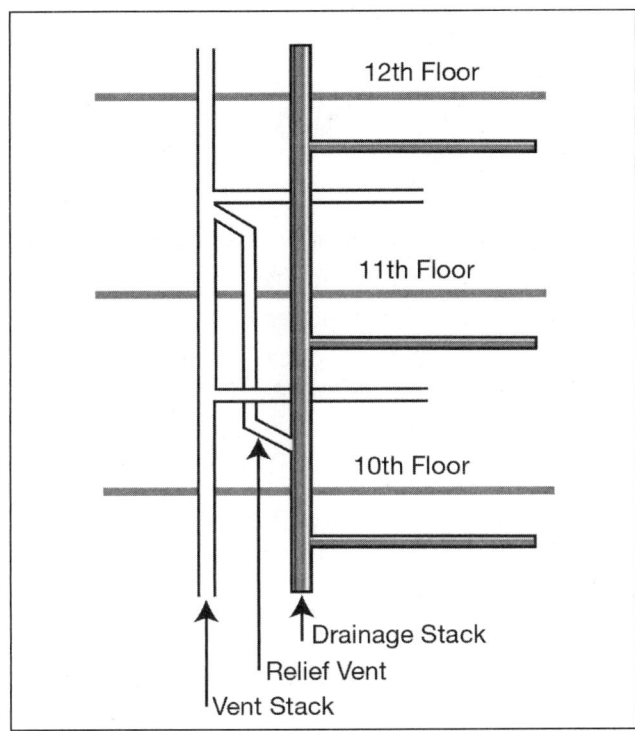

Figure 11.3 Relief Vent Model

Air Admittance Valves

Air admittance valves are only allowed in the International Plumbing Code. The Uniform Plumbing Code does not allow them. (Refer to IPC 917.1–917.8)

Air admittance valves (AAVs) are used in place of stack vents that go through the roof (Figure 11.4). The function of the air admittance valve is to allow air to be drawn into a drain and vent system to relieve vacuum created by the drainage of the served fixture or by other fixtures in the system. Air admittance valves are opened by air pressure and closed by gravity, two very dependable actuators. There are *no springs* in an air admittance valve. The mechanical vents available that allow air into a system but close with a spring are definitely not allowed. Dependability is a prime goal of any plumbing system and the light springs in the mechanical vents will ultimately fail. Gravity and air pressure will not.

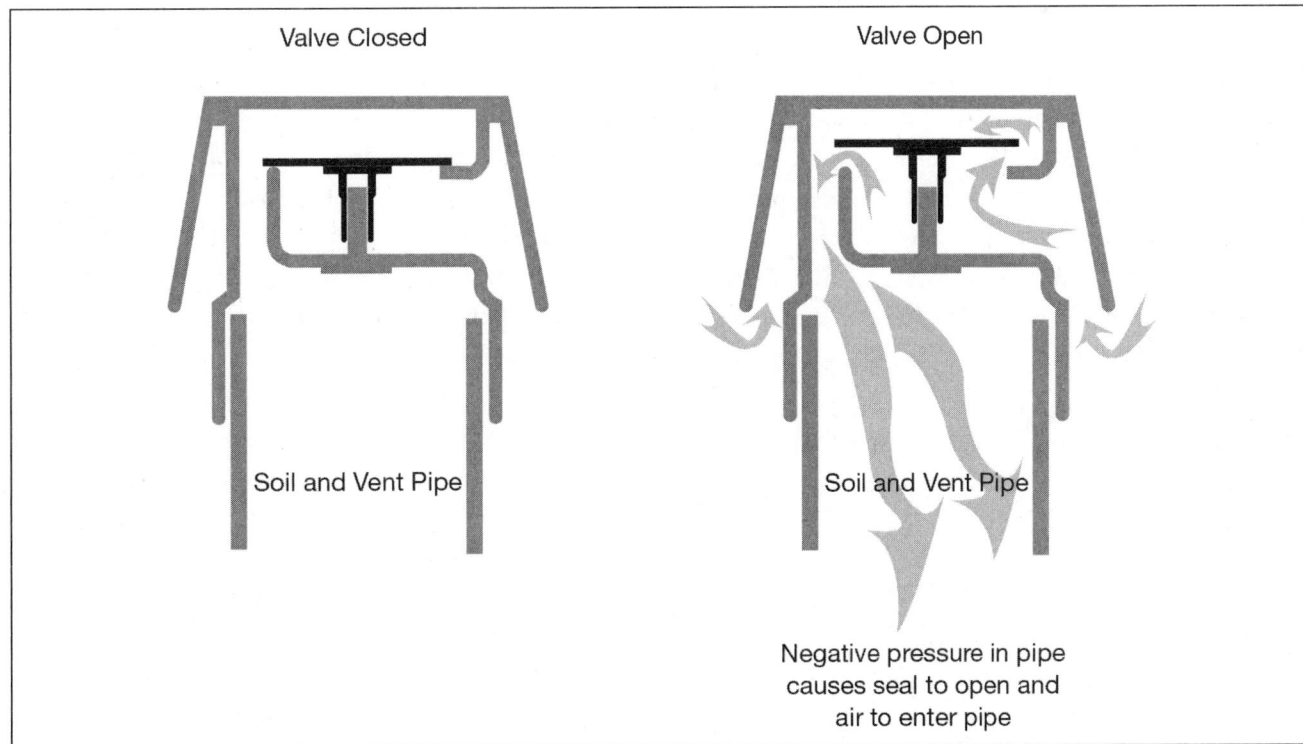

Figure 11.4 Air Admittance Valve

When an air admittance valve is used, it must be sized according to the DFU load of the drain served. Air admittance valves come in different sizes for differing loads. When AAVs are used in a system, there still must be at least one vent in the system that goes out through the roof. This is to allow pressure in the system to be released as well as the vacuum that is relieved by the AAVs. AAVs are generally installed as "single-fixture" installations, but they can serve more than one fixture as long as it is on the same floor. Stack-type air admittance valves cannot be used on stacks that are higher than six branch intervals.

> **ALERT**
>
> An air admittance valve cannot be installed as a vent in any chemical waste system that has not been made harmless to the drainage system. Volatile gasses could degenerate the integrity of the seal of an air admittance valve.

Every air admittance valve must be installed a minimum of 4 inches above the horizontal portion of the drain being served. That measurement starts at the top of the horizontal pipe and extends upward to the bottom of the AAV. Every AAV must be accessible for service and it must have access to free air. That means it must either be installed out in the open or behind a grille.

Chapter Review

Note: Answers may vary based on whether the IPC or UPC is applicable.

1. If a kitchen sink must be drained through a combination drain and vent system, what will be the pipe size of the vertical portion of the drain?

    _____
    _____
    _____
    _____

2. How high is the highest portion of an island fixture vent?

    _____
    _____
    _____
    _____

3. A combination drain and vent system is used
    a. primarily for water closets.
    b. when there are more than six branch intervals.
    c. when it is too distant to a vent.
    d. for back-to-back fixtures.

4. The primary purpose of vent piping is
    a. to let water vapor escape the drainage system.
    b. to maintain trap seals.
    c. to relieve vacuum in the city sewer.
    d. to allow smaller sizing of the drains.

VENTS

5. The greatest pressure differential allowed against a trap is
   a. 2 in. WC.
   b. $\frac{1}{2}$ in. WC.
   c. 3 in. WC.
   d. 1 in. WC

6. Self-siphonage is a problem with
   a. S-traps.
   b. relief vents.
   c. P-traps.
   d. combination drain and vent systems.

7. Sheet copper for vent pipe flashing must weigh at least
   a. 8 oz. per sq. ft.
   b. 1 lb. per sq. ft.
   c. 8 oz. per sq. yd.
   d. 1 lb. per sq. yd.

8. The materials used for vents must be selected according to the same chart as that for drainage piping.
   a. true
   b. false

9. If there is greater than 2 in. WC vacuum in a properly sized and installed vent system, what is a likely cause?
   a. building drain is too small
   b. frost closure
   c. vent is too large
   d. offset in vent

10. Horizontal vent pipes must be sloped
    a. upward from the trap being vented with no portion able to collect water.
    b. in any direction as long as no portion can collect water.
    c. downward from the trap being served with no portion able to collect water.
    d. all of the above.

11. A crown vent is
    a. a vent within 2 pipe diameters of the trap weir.
    b. a horizontal vent 6 in. above the flood-level rim of the fixture.
    c. the highest vent through the roof.
    d. permitted on traps smaller than 2 in.

12. S-traps
    a. are permitted in temporary housing.
    b. must be sized larger than P-traps.
    c. can be self-siphoning.
    d. are always brass.

13. If a roof has other uses besides weather protection, the vent termination must be _____ feet above the roof.
    a. 2
    b. 4
    c. 7
    d. 10

14. What is the smallest size vent permitted for any fixture?

    _____

    _____

VENTS

15. What is the smallest size vent permitted for a 4-in. stack?

_____

_____

_____

_____

16. How many DFUs can be drained through a 3-in. combination drain and vent pipe that connects to a stack?

_____

_____

17. How far can the weir of a water closet trap be from a vent?

_____

_____

_____

_____

18. What is the maximum venting system design target for pressures created in the vent system?
   a. 1 psi
   b. 2 psi
   c. 1 in. WC
   d. 2 in. WC

19. A trap weir that has a $1\frac{1}{4}$-in. diameter can be _____ feet from a vent.

20. A trap weir that has a $1\frac{1}{2}$-in. diameter can be _____ feet from a vent.

21. A trap weir that has a 2-in. diameter can be _____ feet from a vent.

22. A trap weir that has a 3-in. diameter can be _____ feet from a vent.

23. A trap weir that has a 4-in. diameter can be _____ feet from a vent.

24. A common application of a combination drain and vent system is
   a. in an office bathroom.
   b. a floor drain in a large warehouse.
   c. for commercial washing machines.
   d. none of the above.

25. In a multistory building with a vent stack, a relief vent must be installed every _____ branch intervals.

26. Any type of fixture can be installed using a combination drain and vent.
   a. true
   b. false

27. A circuit vent must be connected to the waste line between the fixture furthest upstream and the next fixture downstream.
   a. true
   b. false

28. A waste stack must remain the same diameter throughout its length.
   a. true
   b. false

29. There is no limit on the number of fixture units that can be discharged into one branch interval of a 3-in. waste stack as long as the load for the stack is not exceeded.
   a. true
   b. false

VENTS

30. A waste stack must have a stack vent.
   a. true
   b. false

31. A urinal can be connected to a combination drain and vent pipe.
   a. true
   b. false

32. Air admittance valves are approved for use on kitchen sinks.
   a. true
   b. false

33. Under the IPC, an air admittance valve can be installed below the flood-level rim of a sink.
   a. true
   b. false

34. A vent stack can terminate with an air admittance valve.
   a. true
   b. false

35. It is not permitted to drain an attic condensate pipe into a vent pipe.
   a. true
   b. false

CHAPTER 12: TRAPS, INTERCEPTORS, AND SEPARATORS

CHAPTER SUMMARY

All fixtures need to be trapped, but there are many fixtures that require special traps or have special provisions concerning their traps. Restaurants, manufacturing facilities, vehicle service buildings, medical offices, and dental offices have fixtures with uses that are not found in typical residential plumbing. This chapter explores the varieties of traps outside of the home, as well as their uses and applicable code requirements.

TRAPS, INTERCEPTORS, AND SEPARATORS

Learning Objectives

- Identify trap design details
- State the purpose and design requirements of interceptors
- Discuss separator theory and application
- Explain the need for and application of backwater valves

Key Term

trap seal

Trap Requirements

By the time you get to this point in your work experience and your preparation for the licensing test, you know that every fixture needs a trap to keep out sewer gas. Still, there are some fine points and some exceptions you need to know about. (Refer to IPC 1002.1–1002.10; UPC 1001.0–1008.0)

The vertical distance between a fixture outlet to the trap weir cannot be greater than 24 inches. In other words, waste cannot fall straight down farther than 2 feet before it lands in the trap. This protects the trap from the possibility of self-siphoning. The accelerated velocity of drainage falling from a drop higher than 24 inches can sometimes cause siphon through a trap. The trap cannot be more than 30 inches horizontally away from the fixture outlet. This keeps the length of the inlet side of the trap to a minimum because bacteria will grow in this piping and can create odors.

The standpipe from an automatic clothes washer would appear to be an exception to the 24-inch vertical distance rule; however, the standpipe is a receptor for an indirect waste. The washing machine is not directly connected to the standpipe.

Combination fixtures, including three-compartment sinks, can be trapped with a single trap. A fixture trap is not needed if waste from the fixture drops vertically no more than 30 inches and horizontally no more than 60 inches and then flows into a grease interceptor. The grease interceptor functions as the trap for the fixture, whether it is a single-compartment fixture or a multiple-compartment fixture of no more than three compartments.

JOB CONNECTION

The steel grease interceptor at Burger Bonanza has completely rusted out. Julio, from Ace Plumbing Company, has been dispatched to size up the situation. The owner reported problems with the interceptor long before it rusted out. He complained that it had to be cleaned out too often. Julio suspects that it is undersized. What information must Julio gather so that the replacement grease interceptor can be properly sized?

DID YOU KNOW?

The first known use of a trap was in 1775. Alexander Cummings created an S-trap for use under a water closet. It used a sliding valve to initiate a flush. The function of that trap more than 200 years ago was no different than that of modern traps—keeping sewer gas out of the building.

Trap Configurations

(Refer to IPC 102.3–102.10; UPC 1003.0–1006.0)

- Any trap that depends on a moving part, whether it is a floating ball, a flapper, or any other moving device, is illegal.
- Bell traps are illegal traps found most commonly in floor drains. The drainpipe extends up through the connection point at the bottom of the floor drain body, which creates a moat within the floor drain body. Then, a bell-shaped casting is hung from the strainer by a bolt. When the cover is dropped into its place on the body, the edge of the bell gets its feet wet in the moat. This creates a trap, which prevents sewer gas from entering the building. In a perfect world, this would be fine, except most of the world does not have an address called "Utopia." The **trap seal** depth of the bell trap is generally only about $\frac{1}{2}$ inch, far short of the 2-inch depth requirement. The large exposed water surface area of the bell trap promotes evaporation of the trap seal, further exacerbating the small depth of trap seal. The bell trap also fills with debris in the bottom of the trap causing a failure that defeats the whole purpose of having a floor drain (Figure 12.1 and Figure 12.2).

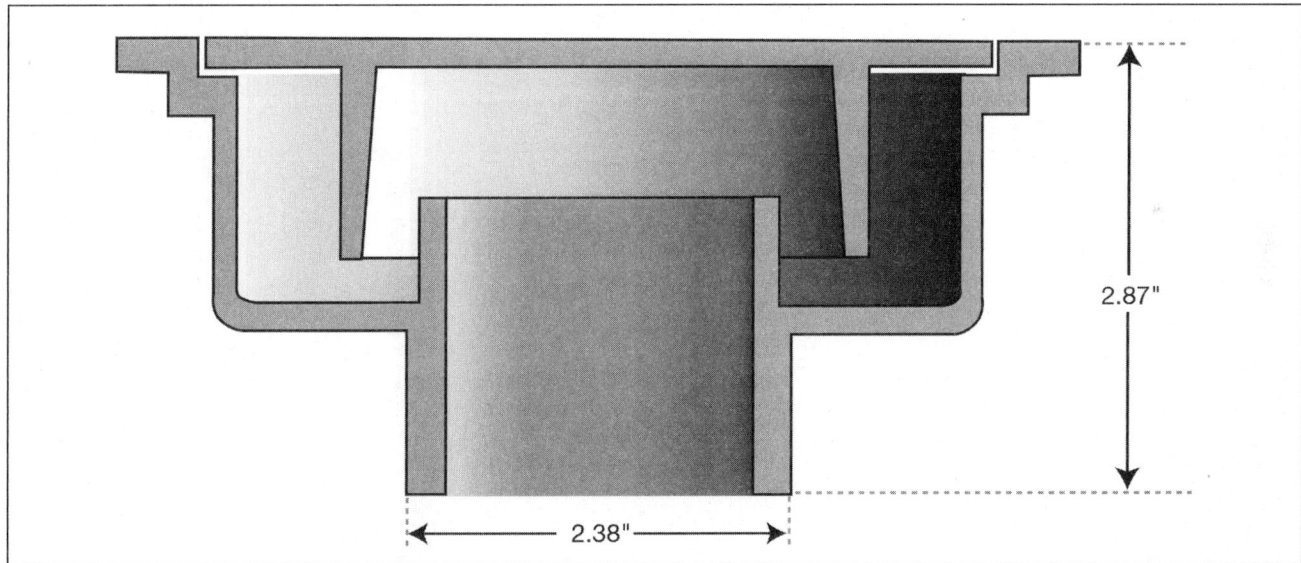

Figure 12.1 Bell Trap Cutaway

TRAPS, INTERCEPTORS, AND SEPARATORS

Figure 12.2 Bell Trap

- Crown-vented traps are illegal because waste tends to sling up into the crown vent, then accumulates, hardens, and plugs the vent (Figure 12.3).
- Any trap that is not an integral part of a fixture and has a partition to create a seal is illegal, unless it is built of an approved material that is highly resistant to corrosion.
- All S-traps are illegal (Figure 12.4). This trap, which was the mainstay of earlier plumbing systems, has a strong tendency to self-siphon, breaking the trap seal.

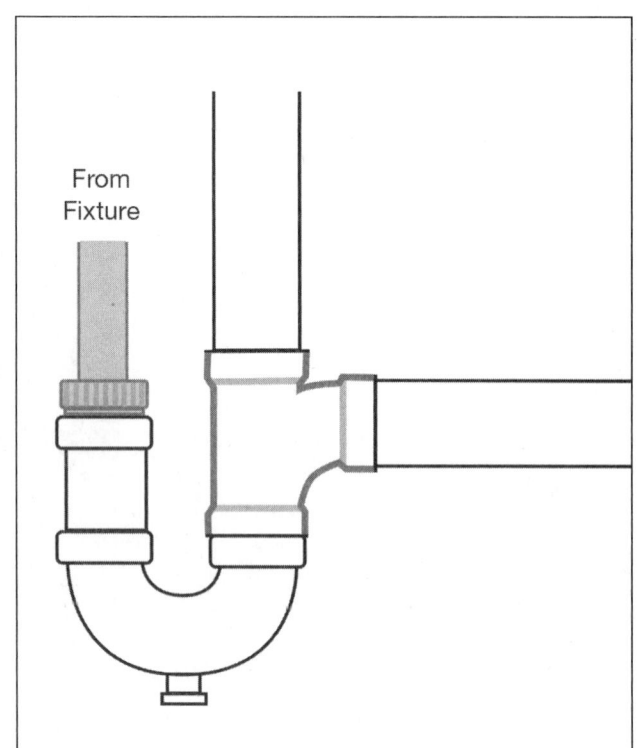

Figure 12.3 Crown-Vented Trap

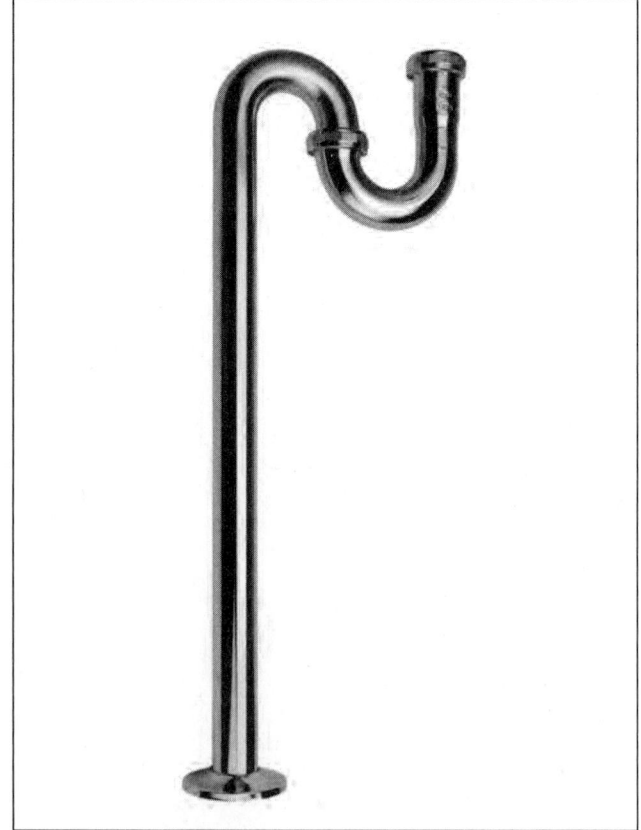

Figure 12.4 S-Trap

TRAPS, INTERCEPTORS, AND SEPARATORS

- Drum traps, except for the special ones used as interceptors of solid waste such as hair, precious metals, or plaster, are not allowed (Figure 12.5). Sometimes a drum trap configuration is used as an acid neutralizer. Illegal drum traps are found mostly on old bathtub installations.
- The seal of all traps must be a minimum depth of 2 inches with a common maximum depth of 4 inches. In certain installations deeper trap seals are allowed.
- If a trap seal is likely to be lost due to evaporation, such as in a little-used floor drain, a trap primer must be installed.

ALERT

It's not uncommon on many large commercial projects to find that the architect has specified a building trap when one is not necessary. Be sure to check with the architect and the Authority Having Jurisdiction before going ahead with installing a building trap. The inspector may look at an unnecessary building trap as a second trap, which would place every fixture in violation of the provision prohibiting any fixture to be double-trapped.

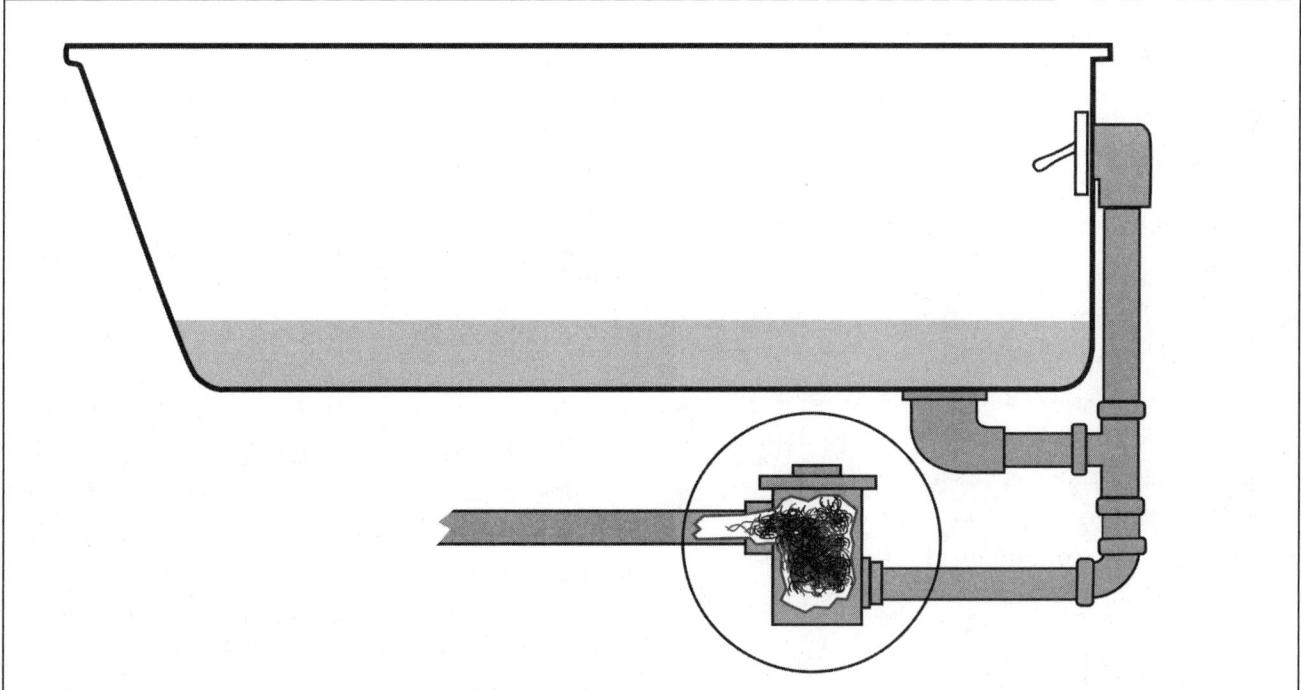

Figure 12.5 Drum Trap

TRAPS, INTERCEPTORS, AND SEPARATORS

- Building traps, otherwise known as house traps, are prohibited unless the local jurisdiction requires it because of conditions in the municipal sewer system. If local conditions require a building trap, the trap must have a cleanout and must have a vent on the inlet side of the trap. The vent must terminate above grade on the outside of the building.
- All traps must be installed level. A trap that is not level cannot maintain the trap seal depth for which the trap was designed.
- Traps must be protected from freezing.
- On slab construction, the recess created to make room for the bathtub or shower trap must be waterproof, rat-proof, and termite-proof.
- If a trap constructed of brittle material is installed underground, it must be encased in cement that extends 6 inches below and around the trap. This type of installation could occur if a vitrified clay trap is installed for chemical waste.
- Traps and pipes in any mental health center cannot be exposed. This provision is designed to protect patients and their caregivers from possible injury by preventing plumbing parts from being used as weapons.

Hints and Tips
When testing the integrity of the seals on a slip joint trap, run hot water through it. The heat tends to soften the seals and will encourage leaking, if it is going to happen at all. Also, before leaving the fixture, fill the bowl at least halfway and then let the water go out. Sometimes the head pressure of a partially filled bowl will root out a leak before it becomes a callback.

Interceptors and Separators

When there are materials entering the plumbing system that can be harmful to the system, an interceptor must be installed. Interceptors and separators can be located downstream of the building drain. You know the kind of "junk" I'm talking about—grease, oil, sand, chemicals, and solids. (Refer to IPC 1003.1–1003.10; UPC 1014.1–1017.2) Plumbing appurtenances must be installed in the sanitary drainage system to remove this "junk," and it is called a grease interceptor.

A grease interceptor is now called a hydromechanical grease interceptor by IPC.

Design
Separators must be installed following the instructions of the manufacturer. You must ensure that any type of discharge that does not need to pass through the interceptor does not get into it. That would only serve to overload the interceptor and possibly allow waste that is supposed to stay in the interceptor to pass through it.

Grease interceptors are the most common type in use and justify some special attention (Figure 12.6). Standards for the sizing and installation of grease interceptors are not consistent among manufacturers. Therefore, follow the manufacturer's instructions for the particular grease interceptor you are installing.

It is always best to keep the grease interceptor as close to the source as possible; this will ensure that the grease is warm enough to allow for the best separation in the interceptor. All the fixtures in a commercial kitchen that handle food grease must flow to the grease interceptor. If a food waste grinder drains to a grease interceptor, a solids interceptor must be installed before the grease interceptor to keep it clear of food waste solids. Grease interceptors are not required for private homes.

TRAPS, INTERCEPTORS, AND SEPARATORS

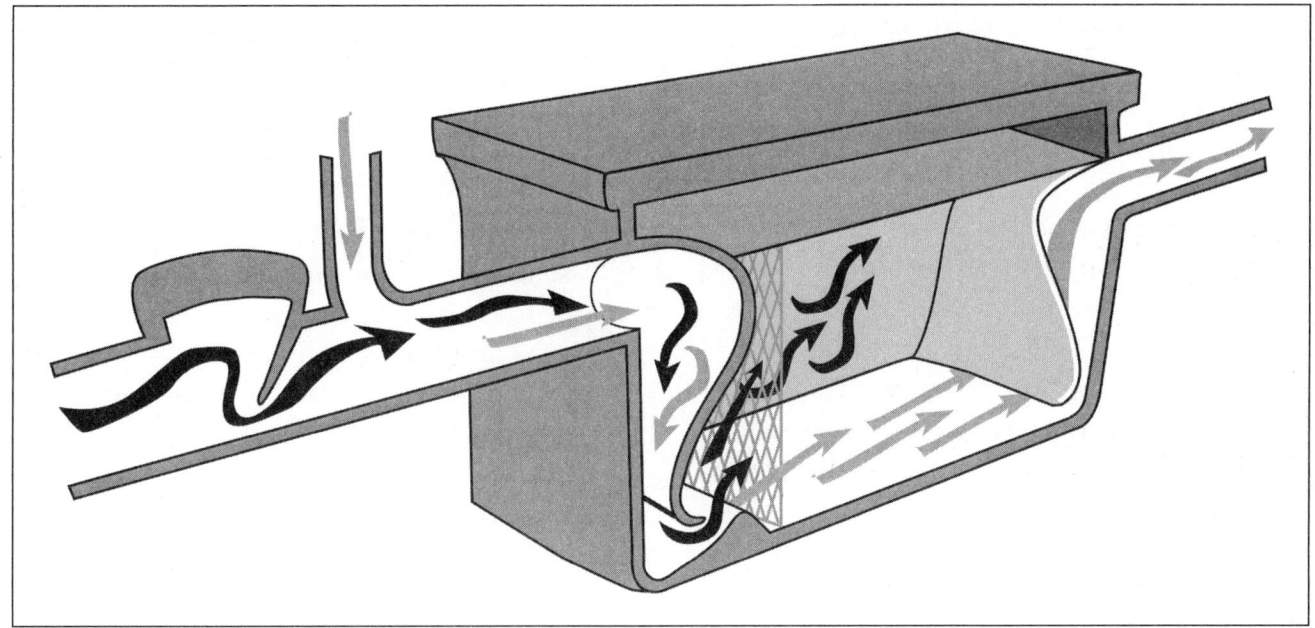

Figure 12.6 Typical Grease Interceptor

Sizing

Most manufacturers offer sizing assistance when you purchase a new grease interceptor. Manufacturers understand how their specific products will perform in the field. Never undersize a grease interceptor to bid lower than your competition. A selection like that will haunt you in maintenance until you install the correct size. As a general guide, Table 12.1 provides information from the IPC's Table 1003.3.4.1 for sizing grease interceptors. If due to renovation an existing hydromechanical grease interceptor has become too small to carry the new load, additional interceptors can be added upstream of the existing interceptors.

TABLE 12.1 CAPACITY OF GREASE INTERCEPTORS (IPC TABLE 1003.3.4.1)	
TOTAL FLOW-THROUGH RATING (GPM)	GREASE RETENTION CAPACITY (POUNDS)
4	8
6	12
7	14
9	18
10	20
12	24
14	28
15	30
18	36
20	40
25	50
35	70
50	100
75	150
100	200

TRAPS, INTERCEPTORS, AND SEPARATORS

Other Interceptor Requirements

- Oil separators must function by slowing the flow of drainage enough to allow oils to separate from other liquids so they can be skimmed at the top into a receiver. An oil separator's depth cannot be any less than 2 feet below the bottom (invert) of the discharge pipe, with no less than an 18-inch water seal. An oil separator is sized according to the square feet of the service area of a garage area. A minimum of 6 cubic feet of capacity for the first 100 square feet is required, plus another cubic foot of capacity for every extra 100 square feet of building area.
- Sand interceptors must be accessible for cleaning and servicing.
- Any commercial laundry must have a separator that catches all solids larger than $\frac{1}{2}$ inch in diameter. The filter basket must be readily accessible for cleaning.
- Bottling plants need a separator to catch any broken glass that may find its way into the drainage system.
- Slaughterhouses need separators to prevent a variety of products from entering the drainage systems that could cause stoppages in sewage systems.
- Separators must be vented as their design demands. Follow the manufacturer's instructions. When an interceptor cover is tightly sealed, venting must allow for changes in liquid levels in the separator without causing pressure changes.
- Interceptors and separators must remain accessible for servicing.

Materials

The design of interceptors and separators must not cause reductions in flow, other than the flow restrictor that may be provided with the interceptor. Any ledges or shoulders within the construction of the interceptor or separator can obstruct flow. The materials used must follow the guidelines for drainage piping. (Refer to IPC 1004)

Chapter Review

Note: Answers may vary based on whether the IPC or UPC is applicable.

1. The purpose of a trap is
   a. to keep solids out of the drainage system.
   b. to keep sewer gas out of the building.
   c. to maintain less than 1-in. WC air pressure in the system.
   d. to catch valuables.

2. Define integral trap.

   _____
   _____
   _____
   _____

3. A grease interceptor can serve as a fixture trap.
   a. true
   b. false

4. Drains for a combination fixture can be served by one trap as long as the trap is within _____ inches of the drain outlet.
   a. 24
   b. 12
   c. 32
   d. 30

5. The vertical distance from a fixture outlet to the trap weir shall be no more than
   a. 24 in.
   b. 12 in.
   c. 32 in.
   d. 30 in.

TRAPS, INTERCEPTORS, AND SEPARATORS

6. A single trap can serve _____ compartments on a combination fixture.
    a. up to 4
    b. up to 2
    c. up to 3
    d. any number of

7. A P-trap is the only trap design allowed.
    a. true
    b. false

8. The minimum depth of a trap seal is _____ inch(es).
    a. 4
    b. 2
    c. 3
    d. 1

9. Slips joints are only allowed on the trap inlet and within the trap seal.
    a. true
    b. false

10. Crown venting means a vent connected within _____ of the trap weir.
    a. 2 in.
    b. 1 pipe diameter
    c. 2 pipe diameters
    d. 1 in.

11. Drum traps are allowed for
    a. bathtubs.
    b. floor sinks.
    c. solid interceptor applications.
    d. all of the above.

12. A trap seal primer must be installed
    a. when a trap is prone to evaporation.
    b. on all floor drains.
    c. never.
    d. in commercial applications.

13. A trap can be no larger than
    a. the tailpiece of the fixture.
    b. the building drain.
    c. the tailpiece nut.
    d. the drainpipe it empties into.

14. Building traps are permitted
    a. always.
    b. when specified by the architect.
    c. never.
    d. when specified by the local jurisdiction.

15. A building trap must have
    a. a diameter of 3 in.
    b. cast-iron construction.
    c. a relief vent.
    d. a concrete support.

16. Traps require leveling when installed.
    a. true
    b. false

17. A vitrified clay trap must be set in _____ when installed underground.
    a. concrete
    b. sand
    c. rubble
    d. clay

18. Grease interceptors are installed only if a business owner wants to protect his or her drainage system.
    a. true
    b. false

19. The simplest way to size a grease interceptor is to
    a. use the manufacturer's sizing service.
    b. use the building owner's idea.
    c. install the same size that was there.
    d. use the 2 to 1 rule.

20. Fixtures that must flow to a grease interceptor include
   a. prerinse sinks.
   b. soup kettles.
   c. pot sinks.
   d. all of the above.

21. Food waste grinders can connect to a grease interceptor if
   a. the outlet is no larger than 2 in.
   b. no kitchen waste goes in it.
   c. it first drains to a solids interceptor.
   d. ABS pipe is used.

22. Individual dwellings require grease interceptors with a capacity of less than 10 pounds.
   a. true
   b. false

23. A grease interceptor must have the flow control from the manufacturer installed on the outlet side within 6 in. of the outlet.
   a. true
   b. false

24. Oil separators are required at
   a. repair garages.
   b. factories with oily waste.
   c. car washes.
   d. all of the above.

25. The main difference between an interceptor and a separator is that the interceptor retains waste while the separator sends waste out to a storage tank.
   a. true
   b. false

26. A grease interceptor in an area using the IPC, with a capacity of 50 pounds, can handle a flow-through capacity of
   a. 100 GPM.
   b. 25 GPM.
   c. 5 GPM.
   d. 10 GPM.

27. What number of drainage fixture units can a grease interceptor in an area using the UPC, with a flow capacity of 50 GPM handle?
   a. 20
   b. 5
   c. 10
   d. 13

28. An accepted trap design for a sink used in the making of plaster casts and denture grindings would be a(n)
   a. drum trap.
   b. $\frac{3}{4}$ rap.
   c. S-trap.
   d. bag trap.

29. A building trap should be installed in all public buildings.
   a. true
   b. false

30. The soil pipe draining a water closet must have a trap sized a minimum of 3 inches in diameter.
   a. true
   b. false

CHAPTER 13 ▶ STORM DRAINAGE

CHAPTER SUMMARY

Before humans began settling in communities, storm drainage was a straightforward issue. People scrambled to high ground, stayed away from the gullies, and waited for the land to drain of its own accord. Normal precipitation and erosion formed a natural drainage system that was self-managing, with no concern or accommodation for the new humans. As civilizations developed and cities were built, the early peoples of the Earth realized that storm runoff had to be directed and managed if they hoped to keep their buildings and civilization from crumbling. Clearing the land, building, paving, and concrete were all human inventions, and the problems they often created required humans to come up with solutions.

This chapter will help you to identify the storm drain system exclusive of the sanitary drainage system, as well as the drains and fixtures that are permitted or not permitted to drain into the storm drain system. Removal of storm runoff, which belongs in the storm drain systems, from the sanitary drainage systems has been an ongoing concern in many communities. It is critical that municipalities not have their sewage treatment plants be overburdened by excess storm water. The sewage treatment plants would have to be unnecessarily oversized or untreated sewage would overflow into the environment.

STORM DRAINAGE

Learning Objectives

- Recognize where storm drains are required
- Identify the materials allowed for storm drains
- Explain when secondary roof drains are required and state their design parameters
- Name the requirements for subsoil drains, sumps, and sump pumps

Key Terms

conductor
flashing
primary roof drain
scupper
secondary roof drain
subsoil drain
sump
sump pump

General Requirements

All of the flat, man-made areas we have created must be drained of storm water in an efficient and reliable manner. Almost all localities require that all roofs, paved areas, yards, courts, and courtyards must drain to an approved location. (Refer to IPC 1101.1–1104.3; UPC 1101.1–1101.11.1)

No storm water can drain into a sanitary sewer system that is intended only for sewage. This includes the majority of the sewers in American cities today. This also involves disconnecting all of those illegally connected sump pumps and roof drains that are presently connected to the sanitary sewer.

Some areas still have combination sewers that are designed for sewerage as well as storm water runoff. These combination sewers can become overtaxed, more often than not, when it rains heavily. Raw sewage ends up flowing to waterways, and into lakes and oceans, without being properly treated.

Conductors, which are the pipes inside a building that move storm water from roof drains, must be tested to the same standards as a sanitary drainage system. During heavy rain or combination rain and snowmelt, the conductors and building storm sewer are subjected to a more strenuous load than any sanitary system would be. Imagine the damage that would occur if a conductor failed inside a building.

JOB CONNECTION

For years now, every time it rains hard, the Springfield Shopping Mall has had water flowing out of the **secondary roof drains**. The maintenance department is at a loss as to how to correct the problem. The new owners of the mall want a quote from Ace Plumbing to correct this problem, which has the potential to cause a catastrophic roof cave-in. The mall owners expect the repairs to be costly, but they want the roof fixed correctly once and for all. The reputation of Ace Plumbing Company will be on the line with this very visible job. What are the challenges that Carlos and Sophia face in getting this job done right the first time?

ALERT

If a building is in an area prone to heavy thunderstorms, careful attention needs to be paid to adequate sizing of the system. An average per hour rainfall amount will do little good when all of that average hourly rain falls in a 5-minute period. The system must be sized to handle the highest rainfall rate for a short period of time.

STORM DRAINAGE

Like sanitary sewers, no reductions in pipe diameter are allowed in a storm system. All the fittings used must be approved drainage fittings applicable to the piping material being installed.

Both the UPC and the IPC require that every roof be designed to withstand the weight of the maximum water depth that could ever pool on it. This requirement will also be found in all local building codes. This must be calculated assuming that all of the primary roof drains are plugged, and the roof has to rely on the secondary drains with the depth elevated enough to have water flow out through the secondary drains. So what does roof design have to do with plumbers? This rule is obviously for builders, but as a plumber, you must be aware of it and alert to the potential for **primary roof drains** to become plugged, and what needs to be done to ensure the drainage system works.

The primary system is the roof drains, which drain through the conductors to storm sewer system. The secondary system, which must be completely independent of the primary system, simply drains through **scuppers** or over the roof edge to the ground below. The secondary drainage must flow where it is obvious to building personnel so they are made aware that the primary system is not working and needs attention.

The storm drain system needs cleanouts placed within it in a similar manner as the sanitary sewer. Subsurface drains, which drain water away from building footings and foundations, are an exception to this rule.

If a storm sewer is susceptible to backflow, a backwater valve must be installed. This is especially important for combination sewers.

The materials and installation methods used for conductors and underground storm drainpipe must meet the same requirements as those for sanitary drainage. The materials used in the storm sewer, which is the portion of the main storm drain that is outside the building, must be of the types and grades found in Table 13.1.

The materials used for subsoil drainage systems must be those found in Table 13.2. **Subsoil drains** must be made of either perforated pipe or horizontally split pipe, or installed with open joints.

TABLE 13.1 ALLOWABLE MATERIALS FOR BUILDING STORM SEWER PIPE (IPC TABLE 1102.4)

MATERIALS	STANDARD
Acrylonitrile butadiene styrene (ABS) plastic pipe	ASTM D 2661; ASTM D 2751; ASTM F 628; CSA B181.1; CSA B182.1
Asbestos-cement pipe	ASTM C 428
Cast-iron pipe	ASTM A 74; ASTM A 888; CISPI 301
Concrete pipe	ASTM C 14; ASTM C 76; CSA A257.1M; CSA A257.2M
Copper or copper-alloy tubing (Type K,L,M or DWV)	ASTM B 75; ASTM B 88; ASTM B 251; ASTM B 306
Polyvinyl chloride (PVC) plastic pipe (Type DWV, SDR26, SDR35, SDR 41, PS50 or PS100)	ASTM D 2665; ASTM D 3034; ASTM F 891; CSA B184.4; CSA B181.2; CSA B182.2
Vitrified clay pipe	ASTM C 4; ASTM C 700
Stainless steel drainage systems, Type 316L	ASME A112.3.1

STORM DRAINAGE

TABLE 13.2 ALLOWABLE MATERIALS FOR SUBSOIL DRAINPIPE (IPC TABLE 1102.5)

MATERIALS	STANDARD
Asbestos-cement pipe	ASTM C 508
Cast-iron pipe	ASTM A 74; ASTM A 88; CISPI 301
Polyethylene (PE) pipe	ASTM F 405; CSA B182.1; CSA B182.6; CSA B182.8
Polyvinyl chloride (PVC) plastic pipe (type sewer pipe, PS25, PS50, or PS100)	ASTM D 2729; ASTM F 891; CSA B182.2; CSA B182.4
Stainless steel drainage systems, Type 316L	ASME A112.3.1
Vitrified clay pipe	ASTM C 4; ASTM C 700

If a storm drain is installed in a combined sewer with a sanitary system, a trap must be installed in the storm system just before it connects with the sanitary system. This one trap is simply intended to keep sewer gas out of the storm drain system. Trap sizing for a storm system is simple: Make the trap the same size as the storm drain. And you always install it on the building side of the sanitary system.

Sanitary systems and storm drain systems must remain completely independent except for cases where there is a combination sewer. Even when a combination sewer is involved, the piping systems must still be completely independent of each other, except for the point at which the storm drain connects to the sanitary system. This connection must be made using a wye fitting, and it must be at least 10 feet downstream of any stack. This connection procedure enables the sanitary system to function normally, even during heavy rainfall.

Floor drains cannot be connected to a storm drainage system. If a below-grade window well requires a drain, it will not be considered a floor drain so that it can be connected to the storm drain system. A parking garage that has no chance of being used for automobile repairs can have the drains in the floor connected to the storm drain system. Floor drains in multilevel parking garages are no longer required to have individual traps, provided the drains are connected to a main trap before discharging to a combined sewer.

Roof Drains

Roof drains shall be installed in accordance with the manufacturer's instructions. The inside opening for the roof drain shall not be obstructed by the roofing membrane material. Table 11–1 (Dome Area Comparison Table) must be used for sizing purposes of the strainers. Roof drains need strainers that are at least 4 inches tall. The diameter of the strainer must be at least $1\frac{1}{2}$ times the diameter of the roof drainpipe. Think of all the stuff that lands on a roof—leaves, paper scraps, plastic grocery bags, to name a few. All of these items must be kept far enough away from the drain so that rain can drain down the pipe. If the roof is used for human activities or parking, the strainer will be flat, but it must now be increased to at least two times the diameter of the drain. If there is a collection of debris on the drain, we must assume the folks who use the roof will notice it and clear it. (Refer to IPC 1105.1–1110.4; UPC 1105.3, 1106.4, 1108.0)

There is a new technology, called symphonic roof drainage systems, which originated in Europe and is now being used with much success throughout the world. It is made up of roof drains with air baffles that comply with ASME 112.6.9. All types of roof drains must have a leakproof **flashing** to prevent leakage into the building.

Sizing for the storm-water conductors and leaders, both horizontal and vertical, must be done

STORM DRAINAGE

according to rainfall charts available from the National Weather Service. These charts are also found in your plumbing codebook, which is handy since you can most likely take it with you into your test.

> **Hints and Tips**
> Expansion joints are an important part of a roof drain system. The height of a roof changes constantly due to temperature changes and the settling of the building. The base of a conductor is generally firmly mounted to a storm drain that is buried in the earth. An ideal place for an expansion joint is immediately beneath a roof where the roof drain body connects to the leader. This configuration allows the roof assembly to move without putting physical stress on any part of the storm water drainage system.

> **DID YOU KNOW?**
> The first building code concerning roof construction was written by King Hammurabi of Babylon about 1780 B.C. The law concerning building construction was as follows: "If a builder builds a house for someone, and does not construct it properly, and the house which he built falls in and kills its owner, then that builder shall be put to death." The possibility of a roof collapse is a real threat. If a flat roof with an area of 5,000 square feet is covered with 2 inches of water, that means it is holding 833.33 cubic feet of water on it. That amount of water weighs approximately 26 tons!

The slope for the building storm drain and storm sewer must be between $\frac{1}{8}$-inch per foot and $\frac{1}{2}$-inch per foot.

Keep in mind that whenever a vertical wall is in a position that forces the rain that strikes it to land on the roof, one-half of the surface area of that vertical wall must be added to the area of the roof being drained.

Now, let's get back to those secondary roof drains mentioned earlier. If the design of a roof creates a surface that will pool with water when the roof drains are plugged, then secondary roof drains are necessary. Commonly, scuppers are installed as part of the construction of the upper area of the building wall. The area of the openings is three times the area of the primary roof drains. Any secondary roof drainage system must direct the rainwater to an area above grade that will be noticed so the primary roof drains can be cleaned or repaired.

Subsoil

As stated earlier, subsoil drains must be open-jointed, horizontally split, or perforated pipe. This design allows for an almost continuous area of drainage for the subsoil area. The minimum pipe size for a subsoil drain is 4 inches in diameter. The subsoil drain can drain to an approved grade area or to a **sump** where it can be pumped out to an appropriate drainage area. The sump pit for subsoil drains does not require the gastight cover and vent that a sanitary sump does since there is no sewer gas to contend with. (Refer to IPC 1111.1; UPC 1101.5.1–1101.5.6)

STORM DRAINAGE

Building Subdrains

If a building subdrain is below the level of the storm sewer or other approved area of disposal, it must drain to a sump where it can be pumped out to an approved disposal location. The **sump pump** must be sized to handle the anticipated inflow and it must be capable of overcoming the head needed to lift the water to a place of disposal at the anticipated flow rate. (Refer to IPC 1112.1–1113.1.4)

The sump pit must be at least 18 inches in diameter and 24 inches deep. It must be made of an approved material, and its floor must be able to be a permanent support for the pump and the piping attached to it. This rule prevents anyone from simply digging a hole and dropping a pump into it. Neither soil nor gravel can be used as a permanent base for the pump.

The discharge pipe from the sump pump must be at least as large as the pump outlet, and the pipe must also incorporate a full flow valve and a check valve close to the pump.

Chapter Review

Note: Answers may vary based on whether the IPC or UPC is applicable.

1. The drain in a garage floor can be connected to
   a. the sanitary drainage system.
   b. either the sanitary or the storm system.
   c. the storm drainage system.
   d. only a combination system.

2. Storm runoff can drain to
   a. the storm sewer.
   b. streets.
   c. lawns.
   d. all of the above.

3. Storm water cannot be drained into sewers intended for
   a. sewage only.
   b. gray-water drainage.
   c. soil waste.
   d. all of the above.

4. Conductors are required to be tested in the same manner as
   a. sanitary drainage
   b. backflow preventers.
   c. distribution piping.
   d. all of the above.

5. Conductors cannot be reduced in size at any point.
   a. true
   b. false

6. What type of fittings must be used in a storm drainage system?
   a. drainage
   b. cast iron
   c. schedule 40
   d. malleable

STORM DRAINAGE

7. Roofs must be designed to carry the depth of water load up to
   a. the edge of the parapet.
   b. the edge of the roof drain.
   c. the height allowed by the design of the secondary roof drain.
   d. 12 in.

8. Cleanouts are required in a storm drain system
   a. every 100 ft.
   b. never.
   c. just like in the sanitary system.
   d. if the conductor is over 4 in.

9. Materials for a storm drainage system must comply with the same rules as those for a sanitary drainage system.
   a. true
   b. false

10. When does a storm drainage system need a trap?

    _____

    _____

    _____

    _____

11. Storm water traps must be sized
    a. the same as the pipe its connected to.
    b. according to Table 1103.5.
    c. one size larger than the pipe.
    d. like a building trap.

12. Gray water may be drained into conductors when connecting to a combination sewer.
    a. true
    b. false

13. Roof drains need strainers that are
    a. 4 in. high.
    b. 2 in. high.
    c. 3 in. high.
    d. 1 in. high.

14. A roof used as a deck shall have a flat drain _____ times the area of the conductor.
    a. 2
    b. 3
    c. $1\frac{1}{2}$
    d. $2\frac{1}{2}$

15. The maximum slope for a horizontal storm drain is
    a. $\frac{1}{8}$-in. per ft.
    b. $\frac{1}{2}$-in. per ft.
    c. $\frac{1}{4}$-in. per ft.
    d. 1-in. per ft.

16. If a vertical wall diverts rainwater onto a roof, _____ percent of its area must be added to the roof area for sizing the roof drains.
    a. 30
    b. 50
    c. 40
    d. 75

17. Secondary roof drain conductors can connect with the primary storm drainage system
    a. 12 in. below the roof.
    b. at a vertical section of piping downstream of the horizontal portion below the roof.
    c. never.
    d. as long as they are sized the same.

18. Secondary storm drains are sized the same as primary storm drains.
    a. true
    b. false

STORM DRAINAGE

19. Subsoil drains must be
   a. open jointed.
   b. perforated.
   c. horizontally split.
   d. any of the above.

20. A subsoil drain sump does not require a gastight seal.
   a. true
   b. false

21. A sump pump discharge pipe must
   a. discharge on a concrete splash block.
   b. be at least the same diameter as the pump outlet.
   c. be made of PVC.
   d. not have a threaded end.

22. All sump pump discharge pipes require
   a. a check valve.
   b. a 2-in. pipe.
   c. a gate valve.
   d. all of the above.

23. A sump must have a diameter of _____ inches.
   a. 18
   b. 12
   c. 24
   d. 15

24. A sump must have a depth of _____ inches.
   a. 12
   b. 18
   c. 24
   d. 15

25. Roof drain sizing is based on
   a. the 100-year, 1-hour rainfall rate.
   b. tables found in the plumbing code.
   c. National Weather Service tables.
   d. any of the above.

26. A drain located in a window well can be connected to a storm drain.
   a. true
   b. false

27. A roof drain can be made of plastic.
   a. true
   b. false

28. A subsoil drain must have a minimum diameter of _____ inches.
   a. 2
   b. 3
   c. 4
   d. 6

29. A backwater valve MUST be installed in a subsoil drain that could be subject to backflow.
   a. true
   b. false

30. Only roofs that have a design that causes ponding during rain require a secondary roof drainage system.
   a. true
   b. false

CHAPTER 14 ▶ SPECIAL PIPING AND STORAGE SYSTEMS

CHAPTER SUMMARY

This short chapter covers the immensely important task of monitoring the safe installation and subsequent long-term use of certain gas and vacuum systems. To prevent a disaster, there are very specific rules that must be followed for the safe installation and use of these systems.

This chapter focuses on nonflammable medical gas systems and nonmedical oxygen systems. These include medical oxygen, inhalation anesthetic, vacuum piping, and nonmedical oxygen. The maintenance and operation of these systems is found under the regulations of the International Fire Code (IFC) and the National Fire Protection Association (NFPA).

SPECIAL PIPING AND STORAGE SYSTEMS

Learning Objectives

- Understand safe installation of medical gas and nonmedical oxygen systems
- Develop awareness of associated codes
- Recognize proper marking of piping systems

Key Terms

ASME
oxidation
station outlet

Nonflammable Medical Gasses

Common nonflammable medical gasses include oxygen, nitrous oxide, compressed air, carbon dioxide, helium, and nitrogen. (Refer to IPC 1201.1; UPC 1301.0–1302.0, 1309.0–1328.4)

A medical gas system extends from the source, whether it is bulk storage, a compressor, a vacuum pump, or a manifold, to the **station outlet** where the gas is available to the area in which it is installed (Figure 14.1). All nonflammable medical gasses must be conducted through type L or K copper tubing that has been cleaned and capped. Only tubing that carries the imprinted designation OXY, OXY/MED, ACR/OXY, or ACR/MED can be used in a medical gas system. All fittings and valves used in a medical gas system must be cleaned and sealed in airtight containers that are oil-free and have been purged with nitrogen.

JOB CONNECTION

Ace Plumbing has built a reputation for excellence, thanks to the dedicated employees Carlos has hired and kept over the years. Carlos's demanding yet benevolent leadership has created a company that his workers are proud to be a part of. Carlos's team has given him the confidence to bid and win the contract for the new Jackson Wing of County General Hospital. The head supervisor, Emil Meconi, will have his hands full organizing the project and keeping it on schedule. In addition to all of the sanitary waste piping and water supply piping work, Ace Plumbing must extend all of the medical gas piping into the new wing from the existing systems in the main hospital.

Only Dharini, Ace's new employee from Bhopal, India, has experience with medical gas systems. Carlos made her the supervisor for all of the medical gas piping because she had been the assistant maintenance manager at the AIIMS medical college and hospital there. Bhopal is where the Union Carbide plant leaked deadly methyl isocyanate gas in 1984 killing thousands of people, including Dharini's grandfather. Carlos expects Dharini to be especially diligent with the installation of the medical gas systems.

SPECIAL PIPING AND STORAGE SYSTEMS

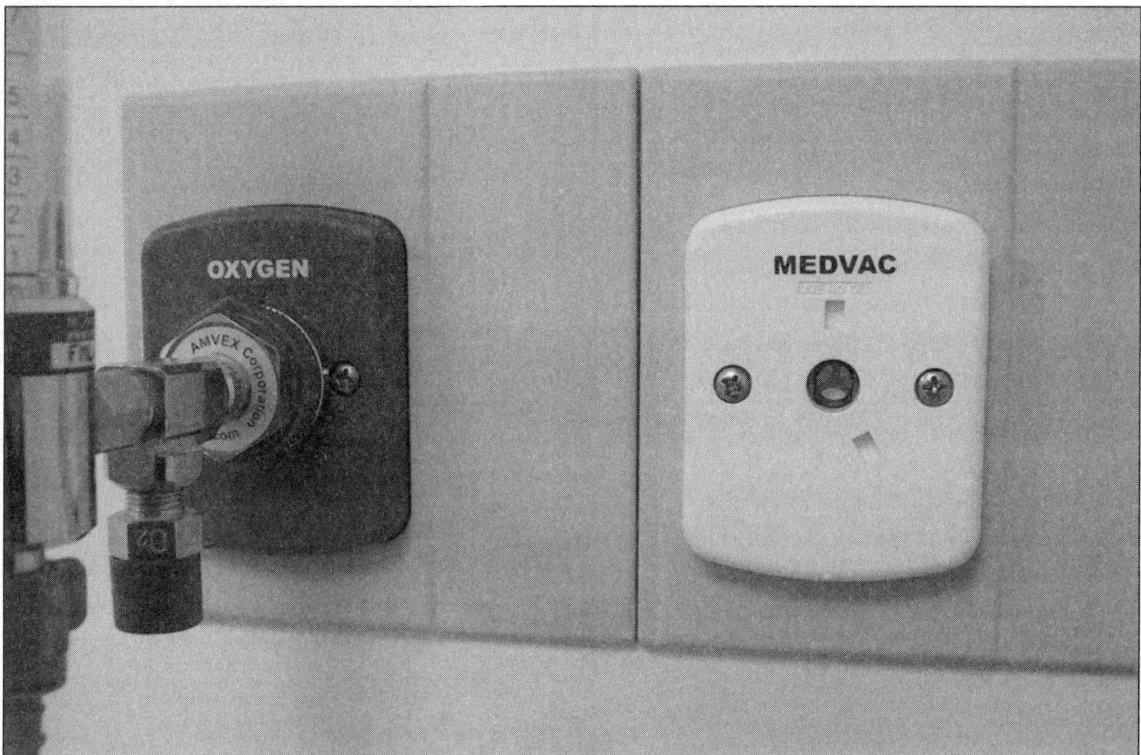

Figure 14.1 A Station Outlet in a Medical Gas System

The UPC requires all valves to close and open with no more than a quarter turn. Connections must be brazed or soldered while nitrogen is flowing through the tubing to prevent any **oxidation** scale from forming inside the fittings or valves. The piping system must be tested to $1\frac{1}{2}$ times the working pressure of the system. In addition to an inspection by the Authority Having Jurisdiction, a medical gas system must be certified by a third-party agency that carries a certification from the Medical Gas Verifiers Professional Qualifications Standard.

The UPC specifies the need for piping plans to be drawn up and in the hands of the installer so that the piping is installed correctly to exact specifications the first time. In addition, a copy must be on-site at all times to allow for review by the Authority Having Jurisdiction and must remain in the possession of the building owner after installation.

ALERT

Take nothing for granted when working on medical gas systems. Ensuring the perfect operation of these systems is critically important. The lives of the patients and all of the people in the building are at stake. All of the gasses that flow through a medical gas system must be pure with no chance of contamination from any source, and there is an elevated chance that contamination could occur during sloppy installation procedures. Also, you must keep in mind that even though oxygen is considered nonflammable, it supports combustion very well, meaning that anything that can burn will burn very violently in an oxygen-rich environment.

SPECIAL PIPING AND STORAGE SYSTEMS

> **Hints and Tips**
> Always follow the manufacturer's installation instructions for all equipment that is connected to any medical gas system. Many pieces of equipment will be foreign to installers and will have certain installation idiosyncrasies that must be followed for the equipment to operate as designed.

All medical gas piping needs to be clearly marked to prevent cross-connections with other systems and to ensure that there is no interruption of service from these critical systems. Table 14.1 shows the markings and colors used for these piping systems.

TABLE 14.1 COLORS FOR PIPES AND LETTERING USED ON MEDICAL GAS SYSTEMS IN THE UNITED STATES

MARKING COLORS

GAS	BACKGROUND	LETTERS
Medical air	Yellow	Black
Nitrogen	Black	White
Nitrous oxide	Blue	White
Oxygen	Green	White
Vacuum	White	Black
Helium	Brown	White
Carbon Dioxide	Gray	Black or white

Medical Oxygen

Oxygen is one of the most common nonflammable medical gasses. That's right, nonflammable. Oxygen *supports* combustion, but it does not burn. Burning, or combustion, is a rapid form of oxidation. Oxidation is the chemical process that occurs when a substance combines with oxygen. Therefore, since oxygen does not combine with oxygen (with the exception of ozone formation), it does not combust. So, it is not flammable. (Refer to IPC 1203.1)

> **DID YOU KNOW?**
> Our normal atmosphere contains 21% oxygen. An oxygen-enriched atmosphere occurs when the oxygen content of the air increases to as little as 24%. Oxygen-enriched atmospheres most often occur from leaks in an oxygen system. Cracked hoses, loose fittings, or damaged equipment are likely culprits. In an oxygen-enriched atmosphere, clothing, hair, and all other combustible materials can very easily catch fire and it becomes very difficult, if not impossible, to extinguish the flames. Never use an open oxygen hose to dust yourself or anything else off.
>
> Some materials, such as grease, oil, and certain metal alloys can burst into flames even without a source of ignition in an oxygen-enriched atmosphere. Never use oil or grease as a lubricant in an oxygen system and only use thread tape or pipe dope that is explicitly approved for oxygen use. Even many soaps and detergents contain grease and should not be used near an oxygen system. In short, only use materials and compounds that are specifically approved for oxygen service.

Obviously, however, oxygen must be handled very carefully because it is a key player in combustion. That status justifies special handling. Figure 14.2 shows a medical gas distribution center.

When installing and inspecting medical gas systems, properly defining systems covered under medical gas is critical. Piping systems covered in medical gas have their own definitions and language, which need to be defined for both inspector and installer to verify installations meet current standards.

SPECIAL PIPING AND STORAGE SYSTEMS

Figure 14.2 A Medical Gas Distribution Center

Figure 14.3 Nonmedical Oxygen Storage

SPECIAL PIPING AND STORAGE SYSTEMS

Nonmedical Oxygen

This section applies to facilities that have more than 20,000 cubic feet of oxygen stored on premises, but are not producers of oxygen (Figure 14.3). Bulk oxygen storage tanks must be **ASME** Boiler and Pressure Vessel rated. They also must be corrosion resistant. An alternative design for the tanks is to build them to U.S. DOT 4L container regulations. In either case, the tank(s) must be individually equipped with a pressure relief valve. (Refer to IPC 1202.1–1203.1)

All of the pipes and tubing must conform to standards for oxygen service. The joints must be of the brazed, flanged, threaded, socket, slip, or compression design.

When oxygen is used in conjunction with acetylene or another fuel gas for welding and cutting purposes, the equipment must have protective devices that prevent flow of the oxygen into the fuel gas and prevent flashback of the flame into the fuel gas. As with other piping systems, the piping must be tested to $1\frac{1}{2}$ times the working pressure of the finished system as it will operate.

Vacuum System Piping

Vacuum systems are not held to as high a standard as the medical gasses that are supplied to patients and medical equipment. However, they still must be constructed of corrosion-resistant materials, which includes copper water tube type K, L, or M; ACR tubing; copper medical gas tubing; galvanized steel pipe; or stainless steel tubing. All copper connections must still be brazed. Galvanized steel joints must be flanged or threaded. (Refer to IPC 1202.1; UPC 1326.1–1326.5)

Medical Air Compressors

Medical air compressors shall be sufficient to serve the peak calculated demand with the largest single compressor out of service. In no case shall there be less than two compressors.

The vacuum piping must be clearly marked "Medical-Surgical Vacuum," or MED/SURG VAC, no more than 20 feet apart. Medical vacuum systems must have two vacuum pumps equipped with isolation valves in order to ensure the continuous operation of the system to a reasonable degree. Testing and inspecting shall be carried out and reference made to NFPA99.

SPECIAL PIPING AND STORAGE SYSTEMS

Chapter Review

Note: Answers may vary based on whether the IPC or UPC is applicable.

1. Specify two of the markings that must appear on piping to show that it is allowed to be used in a medical gas system.

   _____

   _____

   _____

2. The only material usable for medical gas systems is _____.

3. _____ gas must be flowing through a medical gas system when brazing joints.

4. Medical gas piping systems must be pressure tested to _____ times the working pressure of the system.

5. What is the purpose of the two pumps on a medical vacuum system?

   _____

   _____

   _____

6. What else is needed after the installer pressure tests a medical gas system and the Authority Having Jurisdiction inspects the system before it can be put into operation?

   _____

   _____

   _____

7. Fittings for installation of medical gas systems must be cleaned and stored in airtight containers that have been purged with _____.

8. What special installation procedures must be used in rooms occupied by psychiatric patients?

   _____

   _____

   _____

9. In addition to approved copper tubing, what kind of piping can be used in a medical vacuum system?

   _____

   _____

   _____

10. On medical gas piping with a diameter of less than $2\frac{1}{2}$ in., all valves must be of the _____ type.

11. Copper-to-copper joints must be
   a. brazed.
   b. joined with flare fittings.
   c. silver soldered.
   d. tested with water.

12. In a medical vacuum system, the exhaust from the vacuum pumps must terminate
   a. outdoors, turned up.
   b. indoors, turned up.
   c. outdoors, turned down.
   d. indoors, turned down.

13. A medical vacuum system can be converted for use as a medical gas supply system.
   a. true
   b. false

14. The design of a medical gas system is the responsibility of the installer as long as an adequate supply of gas or adequate vacuum is available at each station outlet.
   a. true
   b. false

15. The best place to use as a chase for medical gas systems is an elevator shaft.
   a. true
   b. false

CHAPTER 15 ▶ GRAY-WATER RECYCLING

CHAPTER SUMMARY

Gray-water recycling may be the next big thing for our water-hungry nation. The technology has been developing in fits and starts, but these systems haven't achieved widespread popularity as of yet. Gray water from bathing, laundry, and dish washing is estimated to comprise 50% to 80% of residential wastewater. Inspectors must ensure that onsite treated non-potable systems are only being used in installations that are specifically allowed by the authority having jurisdiction.

Installers and inspectors must have a clear understanding of the fixtures allowed in a gray water system. Due to the growing popularity of water conservation, there is an increased demand in the plumbing industry to install systems that promote sustainability. The potential water savings are obviously substantial. Think about it: with a full-fledged gray-water recycling system, you could flush all the water closets and urinals and still supply the water you need for landscaping without putting any demand on the potable water system—it's like "free" water. The main drawback is the added expense of installing such a system and the extra room it takes up. However, it is quite possible that gray-water recycling systems will become more prevalent with the continued "greening" of America.

GRAY-WATER RECYCLING

Learning Objectives

- Identify applications for gray-water recycling
- State the requirements of a safe and effective gray-water recycling system
- Apply system sizing guidelines
- Distinguish the differences and similarities of water closet and urinal flushing systems as compared to landscape irrigation systems

Key Term

black water

What Is Gray-Water Recycling?

The two basic types of drainage water that go into the sanitary drainage system of a building are often referred to as gray water and black water. Gray water is the drainage from tubs, showers, lavatories, and laundry facilities. **Black water** is the discharge from water closets, bidets, and urinals. Black water contains human bodily waste. Gray water does not have the disease-carrying capability of black water, even though it is certainly not potable. The discharge from kitchen sinks and other food preparation fixtures cannot be included in the gray-water category because of the organic (food) waste that it carries. Food waste decomposes and would foul a gray-water recycling system. (Refer to IPC Commentary Appendix C: Introduction)

Gray-water recycling systems are simply an effective means of saving precious water resources by reusing the water that runs down the drains from showers, baths, lavatories, automatic clothes washers, and laundry trays. Even though this is not potable water, it is not contaminated to an excessive or dangerous degree. After filtering to prevent clogging of any parts of the gray-water distribution system, gray water flushes fixtures and waters landscaping just as well as water drawn from the potable water system. Figure 15.1 shows how a typical gray-water recycling system is set up. This type of system shall be designed by a person registered or licensed to perform plumbing design work. Components, piping, and fittings used in an alternate water source system such as this one shall be listed.

JOB CONNECTION

Julio and Pat are roughing in a new house that will have a gray-water recycling system. This is the first time Ace Plumbing Company is installing one of these systems, but it is a field in which Carlos, Ace's owner, sees growth potential. Green construction techniques will continue to develop. Which fixtures should Pat and Julio drain to the gray-water reservoir? Sophia, the Ace Plumbing Company estimator, has sized the gray-water reservoir to hold 100 gallons for this four-bedroom colonial, which will have two full bathrooms. How can the installers double check to be sure the system is sized correctly? After all, this is the first time for Sophia, too.

GRAY-WATER RECYCLING

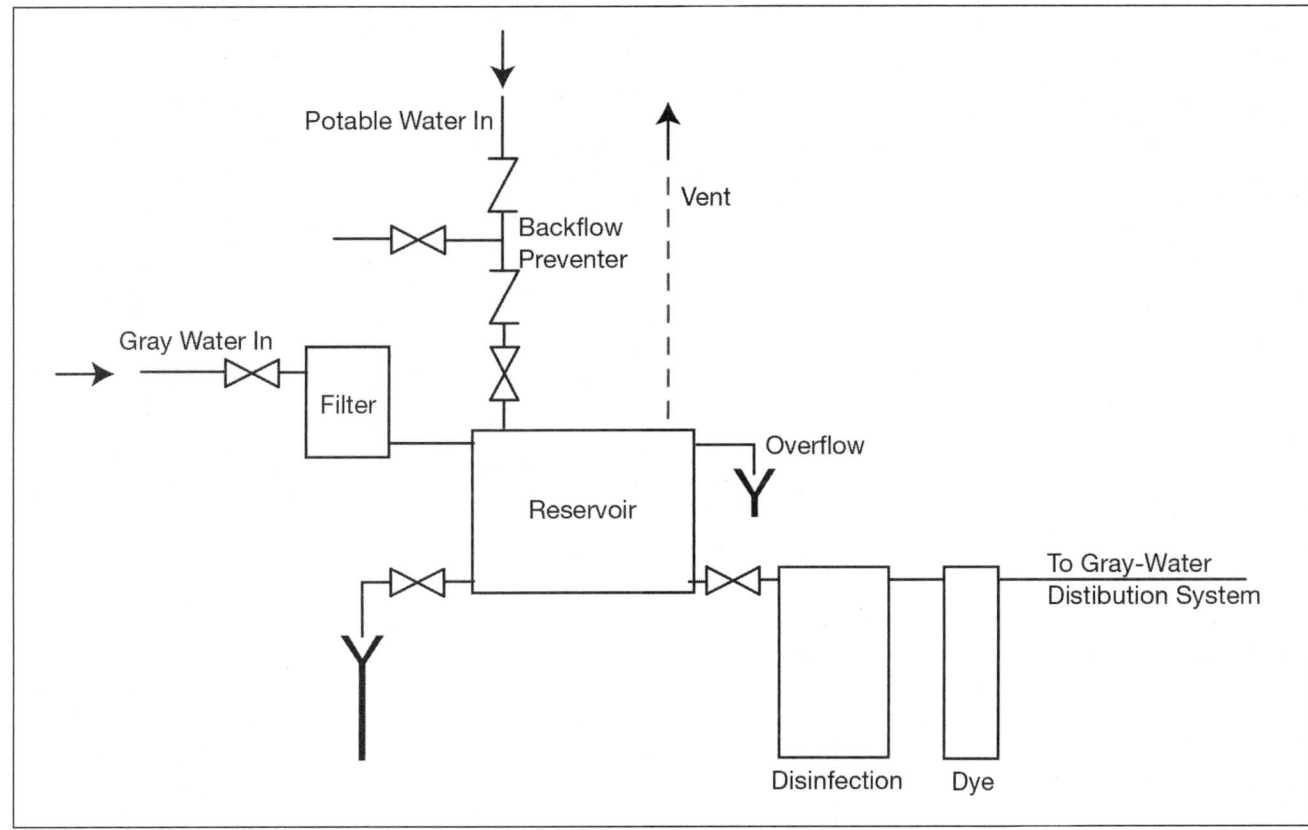

Figure 15.1 Typical Gray-Water Recycling System

DID YOU KNOW?

The U.S. Environmental Protection Agency (EPA) estimates that 15% of the fresh, potable water used in the United States is used for irrigation. Gray water can easily be used for irrigation purposes and save some of that potable water for human and animal consumption.

Connections

This section of the code (Ref: UPC 1602.3) makes it very clear that under no circumstances shall a gray water system be directly connected to a potable water supply. In the IPC, Section 301.3 requires all fixtures to be connected to the sanitary system of the building. This rule precludes the installation of a gray-water recycling system because some of a building's fixtures must be connected to the gray-water drainage system and completely avoid the sanitary drainage system. If the jurisdiction you are testing in supports gray-water recycling, the local code must have amended this rule to allow for gray-water recycling. As with any installations, it is important to first check your local code. (Refer to IPC 301.3, C101.1–C101.14)

Alternate water sources for non-portable applications require inspections at regular intervals. A maintenance log is required to have a permit and shall be maintained by the property owner and be available for inspecting. In general, all the same rules apply to gray-water piping systems as standard piping systems. Permitting rules are similar. The materials used must meet the same quality and design criteria as potable water distribution systems and sanitary drainage systems. The testing of systems and inspections by the Authority Having Jurisdiction are all the same.

All gray-water distribution piping must be labeled as such using black letters on a yellow background. Whenever non-potable waters systems are installed, including outside of the building, the piping must be identified. Also, the water drained from appropriate fixtures must be gathered and stored in a reservoir so a supply is available for pumping to the fixtures that will use the gray water.

Gray water carries debris with it that must be filtered out before the water can be pumped to any fixtures. This can be accomplished through a variety of filtration methods. Since we can expect that the dirt load will be fairly heavy, the system must have a back-washing feature, and the discharge from the backwash filter must drain directly to the sanitary drainage system. In order for a backwashing procedure to be effective and safe, the gray-water drain must have a full-opening valve in it just before the gray-water filter. The valve must be closed before backwashing or maintenance can be carried out. In addition, all fixtures must be tagged as "out of service" to prevent an overflow while the valve is closed.

Water Closets

(Refer to IPC C102.1–C102.6)

Storage

Sizing a gray-water reservoir for a water closet and urinal flushing system requires some thought and advanced planning. You want to ensure that there will be enough water to meet the flushing needs of the building, without storing it for so long that organic debris in the water begins to decay and cause odors. The target for a maximum retention time is 72 hours. (Refer to IPC C102.1)

The IPC requires the gray-water reservoir to hold twice the daily requirement of water sufficient to flush the fixtures, with a minimum size of 50 gallons.

Therefore, we must aim for a reservoir size that falls between a 2-day supply and a 3-day (maximum) supply. To calculate the requirements for a residence, you count the bedrooms and estimate that there will be two people in one bedroom and one person in each of the other bedrooms. Multiply the number of people by 7 to get an approximation of the total number of flushes per day. Multiply the "flushes per day" number by the gallons per flush of the fixture(s). Multiply that number by 2. That will be the reservoir capacity required, as long as it is at least 50 gallons.

Let's look at an example using a single-family home that has four bedrooms and 1.6 GPF water closets.

One bedroom: 2 people
Other 3 bedrooms: +3 people
5 people
5 × 7 = 35 flushes
35 × 1.6 GPF = 56 gallons
56 × 2 = **112-gallon reservoir minimum**
56 × 3 = **168-gallon reservoir maximum**

The installation, construction, alteration, or repair of rainwater catchment systems, intended to supply uses such as water closets, urinals, trap primers for floor drains and floor sinks, irrigation, industrial processes, water features, cooking tower makeup and other uses, shall be approved by the authority having jurisdiction.

Disinfection

The water that is stored for flushing water closets and urinals must be disinfected after the filter. Disinfection reduces the possibility of a buildup of bacteria in the reservoir and prevents odors. The disinfection process is not intended to make the gray water potable. (Refer to IPC C102.2)

Dye

Gray water must be dyed blue or green before it is pumped to the water closet(s) and/or urinal(s) it supplies. This identification will make it clear to any

GRAY-WATER RECYCLING

user of the fixture that the water is not intended for consumption. The dye must be food-grade to ensure that no chemical enters the system that could damage any part of the system. (Refer to IPC C102.4) Gray water systems shall be marked in accordance with this section with the words, "Caution: Nonpotable gray water, do not drink" in yellow letters.

Makeup Water

The reservoir must have makeup water introduced into the reservoir as necessary from the potable water supply. Preventing gray water from migrating into the potable water supply system is critical. Therefore, the best way to prevent this is with a reduced pressure principle backflow preventer on the potable water supply. (Refer to IPC C102.3)

> ### Hints and Tips
> Before you begin the installation of a gray-water recycling system, discuss the job with the local Authority Having Jurisdiction. Even though the owner may have a permit for the installation, some localities limit the allowable uses for the gray water. Some jurisdictions limit gray-water use to underground irrigation. Others may permit surface irrigation, but only by the flood method, with no spraying allowed. There are very small quantities of human waste in bath water, which causes concern if the water is untreated.

Landscape Irrigation

The size of a reservoir for a subsurface irrigation system must be calculated estimating retention of the gray water for no more than one day. Sizing is done in the same general way as for a water closet and urinal flushing system, except the capacity stays at the single-day volume. Since disinfection of irrigation water is not required, bacteria and odors would grow quickly in water retained longer than a day. If there is excess gray water, it can be wasted to the sanitary drainage system or the rate of irrigation can be increased. If there is not enough gray water produced, watering can be reduced. (Refer to IPC C103.1–C103.10)

This type of system would be used where the irrigation is controlled by a timer. If the crop or landscaping is estimated to need 50 gallons per day, then a 50-gallon tank would be the size required. Remember, gray water for irrigation does not require treatment so the tank must be emptied daily, either to the irrigation system or, when rainfall or changing irrigation needs decrease, to the sanitary drainage system.

> ### ALERT
> Gray water, especially from laundry sources, tends to be alkaline (high pH) in nature because of the soaps and detergents. Many plants and crops thrive in an alkaline environment, but others grow better in acidic soil. Before installing a gray-water system, find out the preferred pH level for the crop(s) that will be irrigated so that the pH can be adjusted through treatment, if necessary.

The gray-water piping and reservoir must be labeled as nonpotable water to prevent anyone from piping a cross-connection by mistaking the gray water for a potable water supply. There is no requirement for dye to be added to a subsurface irrigation system since the water is not visible or available for the public to use mistakenly as potable water.

On the outlet of the reservoir, a full-opening valve and a check valve must be installed to prevent backflow from entering irrigation pipes. No makeup water is required on irrigation systems. If makeup water is connected, it must be done in a way that prevents backflow. As with the gray-water recycling system from the water closet, the reduced pressure principle backflow preventer is the best choice.

GRAY-WATER RECYCLING

Chapter Review

Note: Answers may vary based on whether the IPC or UPC is applicable.

1. Recycled water for a subsurface irrigation system must be
   a. colored blue or green.
   b. clear.
   c. colored red or orange.
   d. none of the above.

2. Recycled water for a water closet and urinal flushing system must be
   a. colored blue or green.
   b. clear.
   c. colored red or orange.
   d. none of the above.

3. Gray water is drained from
   a. urinals.
   b. kitchen sinks.
   c. showers.
   d. water closets.

4. Filtering is required on the outlet side of the reservoir before the gray water is pumped to the distribution system.
   a. true
   b. false

5. Makeup water is added to a water closet and urinal flushing gray-water recycling system
   a. never.
   b. through a reduced pressure principle backflow preventer.
   c. at the disinfection point.
   d. with the dye.

6. A recycled gray-water subsurface landscape irrigation system reservoir is sized to retain gray water for
   a. 72 hours.
   b. 48 hours.
   c. 24 hours.
   d. no specified time.

7. Drainage pipes from fixtures that deliver gray water to the gray-water recycling system
   a. must meet the regulations for sanitary drainage pipes.
   b. can be connected with slip joints.
   c. must be one size larger than usual.
   d. must be no-hub cast iron.

8. For sizing, it is estimated that each person discharges _____ gallons of gray water per day.
   a. 33
   b. 60
   c. 10
   d. 40

9. The color code for nonpotable water is
   a. blue letters on red.
   b. red letters on yellow.
   c. yellow letters on black.
   d. black letters on yellow.

10. Gray water must be disinfected
    a. when used for fixture flushing.
    b. never.
    c. for all applications.
    d. when connected to a sanitary drain.

11. The minimum reservoir size for a fixture-flushing gray-water system is _____ gallons.
    a. 300
    b. 50
    c. 100
    d. no minimum

GRAY-WATER RECYCLING

12. A reservoir for a fixture-flushing gray-water system is designed to retain gray water for a maximum of _____ hours.
    a. 36
    b. 36
    c. 24
    d. 72

13. The drain at the bottom of a gray-water reservoir must connect to the sanitary drainage system
    a. indirectly.
    b. directly.
    c. never.
    d. above ground.

14. Gray water may be disinfected with
    a. chlorine.
    b. iodine.
    c. ozone.
    d. all of the above.

15. If the inflow pipe for a gray-water reservoir is 3 in. in diameter, the overflow pipe must be at least _____ in diameter.
    a. 3 in.
    b. 2 in.
    c. $1\frac{1}{2}$ in.
    d. 4 in.

16. The overflow pipe for a gray-water reservoir must connect to the sanitary drainage system
    a. indirectly.
    b. directly.
    c. never.
    d. above ground.

17. The main drain from all of the gray-water fixtures must have a valve installed in it.
    a. true
    b. false

18. The gray-water reservoir must be
    a. steel.
    b. open.
    c. gastight.
    d. none of the above.

19. What is the maximum reservoir size allowed for a fixture-flushing gray-water system for a family of four?
    a. 50 gallons
    b. 200 gallons
    c. 480 gallons
    d. 1,000 gallons

20. Makeup water for a subsurface gray-water recycling system
    a. is unnecessary.
    b. must be dyed blue or green.
    c. must fill through an air gap.
    d. must be the full volume needed for the system regardless of the amount of gray water that will be produced.

CHAPTER 16 ▶ VACUUM DRAINAGE SYSTEMS

CHAPTER SUMMARY

Vacuum drainage systems are beginning to increase in application in areas where water conservation is a high priority. Little attention has been paid to them from a code-writing standpoint because of their specialized nature. We are sure to see more detailed code provisions for vacuum drainage systems as they become more common and perhaps even required for use in buildings in some areas.

VACUUM DRAINAGE SYSTEMS

Learning Objectives

- Gain familiarity with vacuum drainage system design
- Identify vacuum drainage system applications
- Explain the importance of manufacturer's instructions when installing vacuum drainage systems

Key Term

vacuum drainage system

What Is a Vacuum Drainage System?

Vacuum drainage systems have been in use in airplanes, trains, and other public transportation for many years. The main advantage of vacuum drainage systems in such applications is their compact design and the fact that they are self-contained. Concern about water conservation and legislation are beginning to fuel the move toward adapting some of these systems as permanent fixtures. (Refer to IPC G101.1–G101.4)

In Europe, vacuum drainage systems have been in limited use in buildings for about 40 years. The premium on space has been the main reason for their popularity. Occasionally, special building designs that require no penetrations through a slab floor have necessitated the use of a vacuum drainage system.

Most vacuum drainage systems are capable of lifting waste approximately 15 feet, so waste can be transported in overhead pipes. Drainage pipes can be a smaller diameter in a vacuum drainage system because they have the force of atmospheric pressure ramming the waste through the drainage piping.

This is not a plumbing configuration that most of us are used to seeing. Consequently, following the manufacturer's instructions to the letter is the clarion call for vacuum drainage system installations. The IPC expressly states that you do so. Even the piping materials used must be those recommended by the system manufacturer as long as they appear in Table 702.1 of the IPC, which lists the materials acceptable for above-ground drainage piping.

JOB CONNECTION

Carlos, the owner of Ace Plumbing Company, has another new adventure for his crew. He has just won the contract to service, repair, and/or replace the vacuum drainage system components for all of the Springfield city buses. You get the honor of doing the first one because Carlos has a lot of faith in your work. You must find out all you can about these systems before the job starts to maintain your reputation. You are sure to learn more as you tear into your first system.

DID YOU KNOW?

The first vacuum drainage systems were used in Prague, Czechoslovakia, and in the Dutch city of Amsterdam as early as 1882. These early systems were powered by steam engines.

Only about 1 to $1\frac{1}{2}$ quarts of water are used to flush a water closet in a vacuum drainage system, and that is used mostly for the final rinse and refill (Figure 16.1). At first look, a vacuum water closet does not appear to be much different from a standard gravity water closet. On closer inspection, one will find that there is no tank and only a small push button or a foot pedal for flushing. In the bottom of the bowl you may be able to see the ball of a large ball valve. Some vacuum water closets have the discharge out the back and the ball valve, or in some cases the slice valve, is not visible.

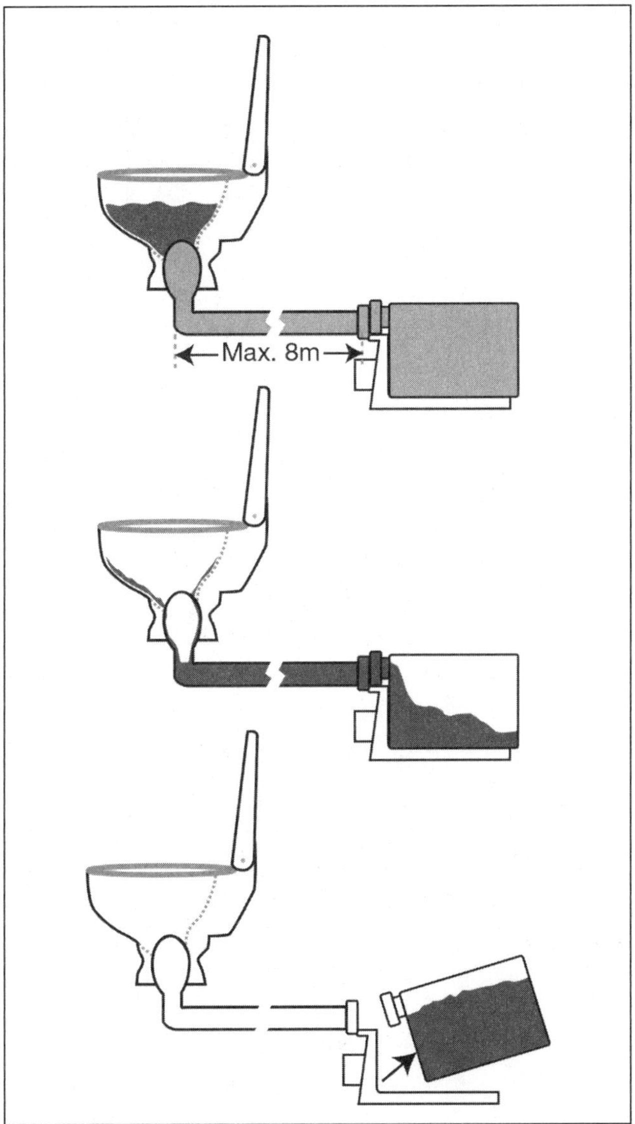

atmospheric pressure to force the waste and water through the discharge pipe. The bowl refills and the cycle is ready to start again. The water closet is the only special fixture in a vacuum drainage system (Figure 16.2).

Figure 16.2 Pedal-Operated Vacuum Drainage System Water Closet (Photo courtesy of Conrad Anderson Company)

Figure 16.1 Vacuum Drainage System Water Closet Function Cycle Cutaway

ALERT

The correct installation of a vacuum drainage system depends on the installer carefully following the manufacturer's instructions. The plans and specifications must be approved before the installation begins and must be followed diligently.

When the pneumatic button or the foot pedal is pushed, a water valve opens to initiate a rinse of the bowl, and the large discharge valve opens to allow the

Other fixtures drain to a small receiving tank that automatically opens a discharge valve when a predetermined tank level is reached. The water supply fixture unit is 1 for a vacuum-type water closet. There is generally no vent on a vacuum system because there is a positive seal between the fixture and the wastewater. When standard gravity draining fixtures

VACUUM DRAINAGE SYSTEMS

are used, traps and cleanouts are used according to the code.

When a vacuum drainage system is installed, it must be tested by subjecting it to a vacuum of 19 inches of mercury. In operation, the vacuum system will have from 16 to 20 inches of mercury placed on it by the vacuum pump. The fixtures are flushed by air pressure forcing the waste down the drain toward the holding tank at a rate of 15 to 18 feet per second.

Vacuum drainage systems are built for large developments as well as for individual systems. Coastal and lakefront developments, where gravity systems cannot function due to lack of slope or are prohibited because of environmental concerns, are prime areas where vacuum systems are considered (Figure 16.3).

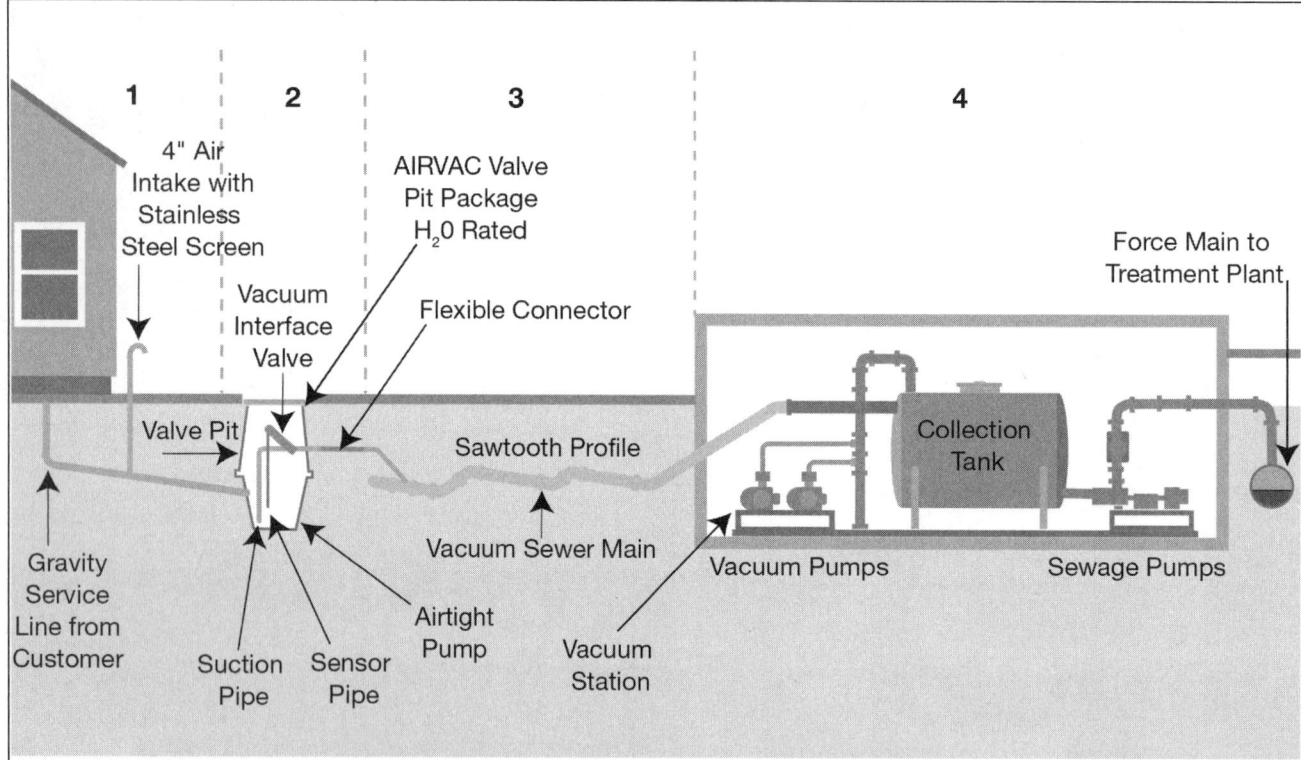

1. A traditional gravity line carries wastewater from the customer to an AIRVAC valve pit package.

2. When 10 gallons of wastewater collect in the sum, the AIRVAC valve opens and differential pressure propels the contents into the vacuum main.

3. Wastewater travels at 15 to 18 FPS in the vacuum main, which is laid in a sawtooth fashion to insure adequate vacuum levels at the end of each line. Wastewater enters the collection tank. When the tank fills to a predetermined level, sewage pumps transfer the contents to the treatment plant via a force main.

4. Vacuum pumps cycle on and off as needed to maintain a constant level of vacuum on the entire collection system.

Figure 16.3 Vacuum Drainage System for a Housing Development

VACUUM DRAINAGE SYSTEMS

Hints and Tips

More information is available online, primarily from the manufacturers of vacuum drainage system equipment. Some of the manufacturers are AcornVac, Airvac, European Vacuum Drainage Systems LTD, and Jetvac, Inc.

Chapter Review

Note: Answers may vary based on whether the IPC or UPC is applicable.

1. The water supply fixture unit value of a vacuum drainage system water closet is
   a. 1.
   b. 2.
   c. 3.
   d. 4.

2. Testing of vacuum drainage systems must be done to
   a. 20 in. mercury.
   b. 19 psi.
   c. 19 in. WC.
   d. 19 in. mercury.

3. A vacuum water closet is flushed by
   a. opening a valve.
   b. a flushometer valve.
   c. a 2-qt. bucket.
   d. gravity.

4. The materials accepted for vacuum drainage piping are
   a. stainless steel and cast iron.
   b. copper, type K and L.
   c. found in IPC Table 702.1.
   d. brass and cast iron.

5. The vacuum maintained in the system by the vacuum pump is
   a. between 16 in. and 20 in. mercury.
   b. actually pressure in the system of 9.3 psi.
   c. intermittent.
   d. 10 in. mercury.

6. The piping in a vacuum drainage system is
   a. smaller.
   b. larger.
   c. flexible.
   d. oval shaped.

7. The special fixture(s) needed for use in a vacuum drainage system is/are
   a. lavatory.
   b. shower.
   c. water closet.
   d. sink.

8. In the United States, the primary reason for increased interest in vacuum drainage systems is due to
   a. self-containment.
   b. concern about conserving water.
   c. smaller piping.
   d. improved vacuum pump designs.

9. Any ordinary fixtures used in a vacuum drainage system must have traps.
   a. true
   b. false

10. All plans and specifications for vacuum drainage systems must
   a. be drawn out by the installer.
   b. be approved by the Authority Having Jurisdiction before installation.
   c. be submitted by the manufacturer.
   d. be in pdf form.

POSTTEST 1

You are now familiar with the kinds of question and answer formats you will see on a plumber's licensing exam. Take this posttest to identify any areas that you may need to review in more depth before the test day. When you are finished, check the answers on page 271 carefully to assess your results. Each answer in the Answers and Explanations section provides a cross-reference to the plumbing code addressed in the question. Use the correlation guide in Appendix A to connect these codes to specific chapters in this book as well as in the IPC or UPC codebooks. Remember to do the following:

- Work carefully.
- Use estimation to eliminate answer choices or to check your work.
- Answer every question.
- Check to make sure your answers are logical.

To simulate the test conditions, use the time constraints of an official plumber's licensing exam. Remember, on the official exam, an unanswered question is counted as incorrect, so it's always better to make a good guess.

POSTTEST 1

1. It is appropriate to use DWV copper tubing for
   a. irrigation piping.
   b. building drains.
   c. all aboveground plumbing applications.
   d. all belowground plumbing applications.

2. The abbreviation _____ designates the type of threads cut in pipes that forms a tapered thread commonly used in plumbing.
   a. DFU
   b. PSI
   c. PVC
   d. NPT

3. For a test pressure of 50 psi, a gauge with increments of _____ psi must be used.
   a. 1
   b. 2
   c. 5
   d. 10

4. _____ cannot be put into a building drainage system.
   a. 40°F water
   b. 120°F water
   c. Naphtha
   d. Human waste

5. Which of the following is considered a plumbing fixture?
   a. dishwasher
   b. clothes dryer
   c. floor drain
   d. none of the above

6. Concrete is considered corrosive to pipes.
   a. true
   b. false

7. If you drill a hole with a radius of $2\frac{1}{4}$ in., what is the area of that hole?
   a. 7.065 sq. in.
   b. 15.896 sq. in.
   c. 19.625 sq. in.
   d. 21.195 sq. in.

8. A pitch of $\frac{1}{4}$-in. per foot must be maintained on a drain pipe that is being hung under floor joists. On the upstream end of the pipe, it measures $14\frac{1}{2}$-in. from the underside of the floor above to the centerline of the pipe. What will the measurement be from the underside of the floor to the centerline of the pipe 20 ft. downstream?
   a. $3\frac{5}{8}$ in.
   b. $14\frac{7}{8}$ in.
   c. $18\frac{7}{16}$ in.
   d. $19\frac{1}{2}$ in.

9. A plumbing distribution system has 275 ft. of 1-in. copper pipe. What size bucket, in gallons, would you need to drain the whole system?
   a. 0.935 gallons
   b. 11.214 gallons
   c. 22.429 gallons
   d. 44.857 gallons

10. All condensate from the steam piped to a sterilizer must be kept out of a sterilizer.
    a. true
    b. false

11. How many lavatories are required in the men's room in a movie theater seating 500 people?
    a. 1
    b. 2
    c. 3
    d. 4

POSTTEST 1

12. A continuous waste cannot be made of PVC.
   a. true
   b. false

13. How high must a vent rise above the flood-level rim of a trapped fixture?
   a. 2 in.
   b. 4 in.
   c. 6 in.
   d. 5 in.

14. Hot water piping can be bundled together with cold water piping.
   a. true
   b. false

15. When superheated water flashes to steam, approximately how much does it expand in volume?
   a. 60 times
   b. 300 times
   c. 850 times
   d. 1,600 times

16. In retail stores, service sinks are not required if the occupant load does not exceed how many occupants?
   a. 20
   b. 22
   c. 18
   d. 15

17. The relief valve outlet pipe must end within how many inches of the floor?
   a. 6
   b. 8
   c. 10
   d. 14

18. The drain from a swimming pool filter backwash must always be drained through an air gap into a receptor that is trapped and that is directly connected to a building drain.
   a. true
   b. false

19. When is it safe to directly connect a commercial dishwashing machine to the sanitary drainage system?
   a. always
   b. never
   c. only when connecting to the inlet of a food waste grinder
   d. only when the dishwashing machine cycles at least twice per hour

20. Water from bathtubs, showers, lavatories, clothes washers, and laundry tubs is called _____ water.
   a. black
   b. potable
   c. gray
   d. none of the above

21. The drain hose for a domestic dishwasher must
   a. connect to a wye tailpiece on the inlet side of a trap.
   b. be 1 in. in diameter.
   c. connect to a wye tailpiece on the outlet side of a trap.
   d. form a trap below the bottom of the dishwasher.

22. A _____ must be drained indirectly.
   a. food prep sink
   b. public lavatory
   c. water closet
   d. all of these are correct

POSTTEST 1

23. The use of gray water as an alternative water source is becoming highly desirable due to the fact that
    a. it is cleaner.
    b. it saves significant amounts of water.
    c. it can be stored for long periods of time.
    d. it does not contain alkalis.

24. The maximum venting system design target for pressures created in the vent system is _____ in. WC.
    a. 1
    b. $1\frac{1}{2}$
    c. 2
    d. 4

25. Toilet facilities shall not open directly into a room used for the preparation of food for service to the public.
    a. true
    b. false

26. Where are slip joints allowed?
    a. on the trap inlet
    b. on the trap outlet
    c. within the trap seal
    d. all of these

27. A vent should be connected at least what distance from the trap weir?
    a. 2 in.
    b. 3 in.
    c. 2 pipe diameters
    d. 3 pipe diameters

28. In water closet compartments containing a wall-hung water closet, the minimum depth of the compartment should be 56 inches.
    a. true
    b. false

29. When are you allowed to use a building trap?
    a. always
    b. never
    c. only when the trap will be located at least 20 ft. above the local water table
    d. only when specified by the local jurisdiction

30. An oil separator would be required for which of these businesses?
    a. library
    b. car wash
    c. computer repair shop
    d. all of these

31. What size must a storm water trap be?
    a. the same size as the building trap
    b. one size larger than the building trap
    c. the same size as the pipe it is connected to
    d. one size larger than the pipe it is connected to

32. The supply lines and fittings for every plumbing fixture shall be installed
    a. with cast-iron material only.
    b. with PVC material only.
    c. so as to prevent backflow.
    d. so that they are exposed for repairs.

33. When sizing a roof drain, you should refer to which of the following?
    a. National Weather Service tables
    b. tables found in the plumbing code
    c. the 100-year, 1-hour rainfall rate
    d. any of these

34. Which gas must be flowing through a medical gas system when brazing joints?
    a. oxygen
    b. nitrogen
    c. carbon dioxide
    d. nitrous oxide

POSTTEST 1

35. Which two materials can be used in a medical vacuum system?
  a. stainless steel or PVC
  b. PVC or lead
  c. copper or stainless steel
  d. copper or lead

36. Where and how must the exhaust from vacuum pumps in a medical vacuum system terminate?
  a. indoors, turned up
  b. indoors, turned down
  c. outdoors, turned up
  d. outdoors, turned down

37. An installed fixture is required to be connected to a water and drainage system if
  a. the structure is designed for industrial use.
  b. the structure is designed for nonindustrial use.
  c. the building is intended for occupancy.
  d. the fixture is likely to be used frequently.

38. A water closet, urinal, lavatory, or bidet shall not be set closer than how many inches from its center to any side wall, partition, vanity, or other obstruction?
  a. 20 in.
  b. 10 in.
  c. 25 in.
  d. 15 in.

39. What is the normal demand rate for a bathtub faucet?
  a. 2 GPM
  b. 4 GPM
  c. 8 GPM
  d. 12 GPM

40. If the water pressure from the street is inadequate to maintain fixture flow pressure, a booster pump must be installed.
  a. true
  b. false

41. If a water tank has a capacity of up to 1,500 gallons, what size must its drain pipe be?
  a. $\frac{1}{2}$ in.
  b. $1\frac{1}{2}$ in.
  c. 2 in.
  d. $2\frac{1}{T}$ in.

42. Hospitals must have at least _____ water service pipe(s).
  a. 1
  b. 2
  c. 3
  d. 5

43. Is a waste outlet required for emergency shower and eyewash stations?
  a. Yes
  b. No

44. It is safe to install exposed drainage piping _____ in a restaurant.
  a. under the kitchen floor
  b. overhead in a food storage area
  c. above a heated food preparation surface
  d. above a nonheated food preparation surface

45. Solvent welded PVC should never be used for a building sewer.
  a. true
  b. false

46. Which of the following methods must you use to join sections of glass pipe?
   a. welding
   b. caulking
   c. compression gasket
   d. elastomeric coupling

47. A low heel inlet 90-degree bend can be used directly below a water closet connection.
   a. true
   b. false

48. What is the DFU value of a bathroom group with a 1.6 GPF water closet?
   a. 3
   b. 5
   c. 6
   d. 9

49. A building subdrain can receive waste from
   a. a lavatory.
   b. a urinal.
   c. a water closet.
   d. any of these

50. What is the minimum size of a shower stall?
   a. 32 in. × 32 in.
   b. 30 in. × 30 in.
   c. 28 in. × 28 in.
   d. 34 in. × 34 in.

51. A 3-in. diameter pipe carrying potable water must be identified with a color field that is _____ long.
   a. 6 in.
   b. 8 in.
   c. 12 in.
   d. 24 in.

52. What is the water supply fixture unit value of a vacuum drainage system water closet?
   a. 1
   b. $1\frac{1}{2}$
   c. 3
   d. 5

53. In a vacuum drainage system, the vacuum pump maintains an operating range of _____ of mercury vacuum.
   a. 6 to 10 in.
   b. 10 to 15 in.
   c. 16 to 20 in.
   d. 21 to 25 in.

54. Unless the local code requires a deeper burial, an outdoor buried water supply must be buried at least _____ below grade.
   a. 6 in.
   b. 8 in.
   c. 12 in.
   d. 14 in.

55. Which of the following can be placed in an elevator shaft?
   a. a floor drain
   b. pipes carrying potable water
   c. pipes carrying nonpotable water
   d. all of these

56. To prevent rodents from entering the plumbing system, no opening on a drain inlet may be larger than _____ in the least dimension.
   a. $\frac{1}{8}$ in.
   b. $\frac{1}{4}$ in.
   c. $\frac{1}{2}$ in.
   d. $\frac{3}{4}$ in.

57. Copper tubing with a diameter of $1\frac{1}{2}$ in. hung horizontally must be supported at least every
   a. 2 ft.
   b. 5 ft.
   c. 6 ft.
   d. 10 ft.

58. To connect plastic pipe to another material, you must use
   a. an adapter.
   b. a pressure-lock connection.
   c. an increaser.
   d. a reducer.

59. For drain pipe sizes greater than 4 in., restraints must be provided at all changes in direction.
   a. true
   b. false

60. When conducting a water test of the entire drainage and vent system,
   a. all openings in the system must be completely open.
   b. all openings in the system must be completely closed.
   c. all but the lowest opening must be completely closed.
   d. all but the highest opening must be completely closed.

61. After installation, hot and cold water supply piping must be subjected to a pressure test of not less than its working pressure.
   a. true
   b. false

62. Where do you plug the sewer in a forced sewer test?
   a. at each drain in the building, regardless of size
   b. at each drain in the building with a diameter larger than 4 in.
   c. at the point of connection with the public sewer
   d. anywhere along the building's main sewer pipe

63. Seats on water closets in public restrooms may be open front or closed front, depending on the preference of the building owner.
   a. true
   b. false

64. How many water closets for students are required for a school with 250 children?
   a. 3–4
   b. 5–6
   c. 10–11
   d. 25–26

65. The circulation piping on a whirlpool bath must
   a. be made of brass, copper, or PVC.
   b. remain heated when not in use.
   c. drain completely after each use.
   d. retain at least 50% of its water to prime the pump during the next use.

66. Fixtures used for therapy must be able to withstand water temperatures as high as 120°F.
   a. true
   b. false

POSTTEST 1

67. A closet flange must be attached to
   a. the water closet base.
   b. the structure beneath the water closet.
   c. both the water closet base and the structure below.
   d. either the water closet base or the structure below.

68. In a covered mall building, toilet facilities for store employees must be available
   a. on the same floor where they work.
   b. within one floor of where they work.
   c. within two floors of where they work.
   d. within 500 ft. of where they work.

69. What size is stipulated by code for the access openings for the pumps on a whirlpool?
   a. 10 in.
   b. 12 in.
   c. 14 in.
   d. 8 in.

70. Once installed, a whirlpool bath pump need not be accessible.
   a. true
   b. false

71. Specialty fixtures can be made of
   a. stainless steel.
   b. soapstone.
   c. chemical stoneware.
   d. any of these.

72. When multiple water closets are installed in the same room, they must be located no closer than _____ apart, center-to-center.
   a. 15 in.
   b. 24 in.
   c. 30 in.
   d. 42 in.

73. All piping that contains steam traps, valves, relief valves, check valves, vacuum breakers, or other similar devices should be installed with
   a. identification labels.
   b. accessibility.
   c. union connections.
   d. copper piping.

74. What should the discharge water temperature from a bidet fitting be?
   a. 90°F
   b. 110°F
   c. 100°F
   d. 120°F

75. What height must the discharge line from the relief valve on the water heater be from the floor or waste receptor?
   a. 8 in.
   b. 6 in.
   c. 10 in.
   d. none of the above

76. Floor drains must have a minimum diameter of
   a. 1 in.
   b. 2 in.
   c. 4 in.
   d. 6 in.

77. When a hot water storage tank is installed in a location where water leakage from the tank will cause damage, the tank must be
   a. insulated.
   b. piped with galvanized pipe.
   c. supported securely.
   d. installed in a galvanized pan.

78. A clinical sink must include a flushing rim.
   a. true
   b. false

79. What should you see when you open the lid of a flushometer tank?
   a. water
   b. a set of interior tanks
   c. an actuator level
   d. all of these

80. In addition to the thermostat on the water heater, a _____ valve is required to control the temperature of water in a public restroom.
   a. flushometer
   b. mixing
   c. relief
   d. fill

81. When facing a faucet, the control for hot water will always be on the left, and the control for cold water will always be on the right.
   a. true
   b. false

82. The open area of a strainer must be _____ the area of the tailpiece.
   a. larger than
   b. smaller than
   c. equal to
   d. larger than or smaller than

83. How much water is delivered to a water closet by a flushometer valve?
   a. 1.6 GPF
   b. 2.8 GPF
   c. 3.0 GPF
   d. 3.5 GPF

84. What is the maximum flow rate allowed for a showerhead?
   a. 1.6 GPM
   b. 2.5 GPM
   c. 3.5 GPM
   d. 4.0 GPM

85. What is the minimum distance allowed between a water service and a building sewer?
   a. 2 ft.
   b. 4 ft.
   c. 1 ft.
   d. 5 ft.

86. What is the minimum depth for a water heater drain pan?
   a. 1 in.
   b. $1\frac{1}{2}$ in.
   c. $2\frac{1}{2}$ in.
   d. 3 in.

87. What is the slope of horizontal drainage pipes sized 3 in. to 6 in.?
   a. $\frac{1}{4}$ in. per foot
   b. $\frac{1}{8}$ in. per foot
   c. $\frac{1}{16}$ in. per foot
   d. none of the above

88. The portion of a water heater designed for the attachment of a venting system is called the
   a. chimney.
   b. draft hood.
   c. vent connector.
   d. flue collar.

89. What is the heat transfer fluid in a direct connection solar heating system?
   a. a type of antifreeze
   b. potable water
   c. nonpotable water
   d. oil

90. A blowout water closet with a flushometer valve has the lowest flow rate for fixtures.
 a. true
 b. false

91. Which of the following requires a full-opening valve?
 a. a water heater outlet pipe
 b. a water closet supply pipe
 c. a water heater supply pipe
 d. all of these

92. Horizontal drains in buildings must have cleanouts installed every _____ feet.
 a. 50
 b. 75
 c. 100
 d. 25

93. A water pump must only be protected from freezing if it will be used throughout the winter months.
 a. true
 b. false

94. Which of the following cannot be used as an individual potable water supply unless treated?
 a. a spring
 b. a stream
 c. a lake
 d. a driven well

95. If a building uses a cistern for its water supply, how far from the cistern must its septic or treatment tank be?
 a. 25 ft.
 b. 50 ft.
 c. 75 ft.
 d. 100 ft.

96. What is the minimum flow rate for a residential sink?
 a. 2 GPM
 b. 2.5 GPM
 c. 3 GPM
 d. 4 GPM

97. In an area where the pressure of the street water main fluctuates, a building's water distribution system must be designed for the maximum pressure available.
 a. true
 b. false

98. _____ is the flow of liquids in potable water distribution piping in reverse of their intended path.
 a. Backpressure
 b. Back-siphonage
 c. Backflow
 d. Cross-connection

99. The effluent in a sump must not rise within how many inches of the invert of the gravity drain inlet into the pump?
 a. 1 in.
 b. $1\frac{1}{2}$ in.
 c. 3 in.
 d. 2 in.

100. How long should a solvent-cement joint be allowed to set before you apply working pressure?
 a. 1 hour
 b. 3 hours
 c. 12 hours
 d. 24 hours

POSTTEST 1

101. According to the IPC, a residential automatic clothes washer, one kitchen sink, and one bathroom group (consisting of a 1.6 GPF water closet, single lavatory, and bathtub/shower) have a drainage load of ____ DFUs.
  a. 3
  b. 7
  c. 9
  d. 11

102. Grinder pumps or grinder ejectors that receive discharge from water closets must have a discharge opening not less than
  a. 2 in.
  b. $1\frac{1}{4}$ in.
  c. 4 in.
  d. 1 in.

103. According to the IPC, a 2-in. trap has a drainage fixture unit value of _____.
  a. 1
  b. 2
  c. 3
  d. 4

104. When the discharge hose of an automatic washing machine is inserted into a standpipe, it creates a connection called
  a. a branch vent.
  b. a stack vent.
  c. an air gap.
  d. an air break.

105. If an indirect waste requires a vent, the vent must vent independently through the roof.
  a. true
  b. false

106. If the floor drain in a walk-in freezer _____, it can drain into the sanitary drainage system through an air break.
  a. is at least 2 in. in diameter
  b. is not protected with a backwater valve
  c. uses a running trap
  d. uses any kind of trap

107. Connections between local vents serving bedpan washers or sterilizer vents and normal sanitary plumbing systems
  a. are permitted.
  b. are prohibited.
  c. must be installed 2 ft. apart.
  d. are tied in above 10 ft.

108. The highest portion of an island fixture vent is
  a. at least 18 in. from the floor.
  b. no higher than 18 in. from the floor.
  c. at least to above the drain outlet of the fixture being served.
  d. no higher than the drain outlet of the fixture being served.

109. Which of the following self-siphons?
  a. an S-trap
  b. a P-trap
  c. a relief vent
  d. a combination drain and vent system

110. According to the UPC, a horizontal vent pipe can be sloped in any direction as long as no portion can collect water.
  a. true
  b. false

POSTTEST 1

111. A crown vent is a vent within _____ of the trap weir.
   a. 4 in.
   b. 8 in.
   c. 2 pipe diameters
   d. 4 pipe diameters

112. A trap weir that is 11 in. in diameter can be how many feet from a vent?
   a. 2 ft.
   b. 5 ft.
   c. 6 ft.
   d. 10 ft.

113. According to the IPC, in a multistory building with a vent stack, relief vents must be installed every _____ branch intervals.
   a. 5
   b. 10
   c. 15
   d. 20

114. A waste stack can vary in diameter throughout its length as the DFU load requires.
   a. true
   b. false

115. Where a vent terminates on a roof that is to be used for any purpose other than weather protection, the vent shall terminate not less than _____ above the roof.
   a. 6 ft.
   b. 5 ft.
   c. 8 ft.
   d. 7 ft.

116. Where would a wet vent most likely be located?
   a. between a residential kitchen and bathroom
   b. between a commercial kitchen and bathroom
   c. between two bathrooms on different floors
   d. between two bathrooms on the same floor

117. A vent terminal shall not be within 20 ft. of an air intake into a building unless it is located more than _____ above the top of the air intake opening.
   a. 4 ft.
   b. 5 ft.
   c. 3 ft.
   d. 6 ft.

118. The juncture of each vent pipe with the roofline must be made water-tight by
   a. furnace cement.
   b. silicone.
   c. an approved flashing.
   d. none of the above

119. A food waste grinder can connect directly to a grease interceptor.
   a. true
   b. false

120. The vertical rise of a vent shall be _____ above the flood level rim of the highest trap or trapped fixture being vented.
   a. 12 in.
   b. 6 in.
   c. 42 in.
   d. 36 in.

121. In an area using the UPC, a grease interceptor with a flow capacity of 150 GPM can handle how many DFUs?
   a. 20
   b. 35
   c. 172
   d. 216

122. Fixture traps must be self-cleaning.
   a. true
   b. false

123. Which of the following businesses would require an oil separator?
    a. a car wash with undercarriage cleaning capability
    b. a quick-service oil change company
    c. an automobile transmission repair facility
    d. all of these

124. Storm runoff must run to a storm sewer.
    a. true
    b. false

125. Storm drain system cleanouts are required
    a. only in locations where annual precipitation exceeds 50 in.
    b. every 100 ft., regardless of the direction of the pipe.
    c. every 100 ft., unless the pipe changes direction.
    d. at the same intervals as required for the sanitary system.

126. What is the maximum slope for a horizontal storm drain?
    a. $\frac{1}{2}$-in. per ft.
    b. $\frac{3}{4}$-in. per ft.
    c. 1-in. per ft.
    d. $1\frac{1}{2}$-in. per ft.

127. A subsoil drain sump requires a gastight seal.
    a. true
    b. false

128. A sump pump discharge pipe must have a diameter
    a. of at least 1 in.
    b. of at least 2 in.
    c. of no more than 6 in.
    d. at least the same diameter as the pump outlet.

129. According to the UPC, a subsoil drain must have a minimum diameter of
    a. 3 in.
    b. 4 in.
    c. 6 in.
    d. 8 in.

130. Annealed copper is also known as hard copper.
    a. true
    b. false

131. Which of the following is the main vent in a DWV system?
    a. stack vent
    b. vent stack
    c. soil vent
    d. artery vent

132. Pipes for a medical gas system can pass through a kitchen, but not an elevator shaft.
    a. true
    b. false

133. A specially designed _____ is required in a vacuum drainage system.
    a. automatic dishwasher
    b. shower head
    c. lavatory
    d. water closet

134. All plans and specifications for vacuum drainage systems must be approved by the Authority Having Jurisdiction before installation.
    a. true
    b. false

135. Firestop systems for piping must have an F rating of at least _____ minutes.
   a. 20
   b. 30
   c. 60
   d. 90

136. Define closet ring.
   a. The connecting pipe between the flushing tank or valve and the closet bowl.
   b. An internal expanding device used to make connections to the flush water part of the water closet bowl.
   c. The projection of the discharge spud or hollow base of the water closet.
   d. The closet ring is the gasket between the base of the water closet bowl and the closet flange.

137. Each plumbing fixture trap must have a liquid seal measuring at least
   a. 4 in.
   b. 2 in.
   c. 1 in.
   d. 3 in.

138. On a water heater, a full-opening valve is required on
   a. the relief valve discharge pipe.
   b. the cold water inlet pipe.
   c. the hot water outlet pipe.
   d. the tank drain valve.

139. How far apart must supporting hangers be for $1\frac{1}{4}$-inch steel pipe?
   a. 6 ft.
   b. 10 ft.
   c. 21 ft.
   d. 12 ft.

140. Cleanouts must be installed in a storm drainage system.
   a. true
   b. false

141. The following fixture(s) must be connected to the building water distribution system.
   a. water closet
   b. laundry tray
   c. automatic clothes washer
   d. all of the above

142. According to the UPC, a vent stack is required when there are _____ or more branch intervals.
   a. 3
   b. 5
   c. 7
   d. 10

143. What is the purpose of a percolation test?
   a. to find water underground
   b. to check the permeability of soil
   c. to identify a source of contamination
   d. none of the above

144. Frost closure is a concern for vents when the 97.5% outdoor design temperature is _____°F or below.
   a. 32
   b. 25
   c. 10
   d. 0

145. A trap must keep its seal under a positive or negative pressure of no more than
   a. 0.5" WC.
   b. 1" WC.
   c. 2" WC.
   d. 5" WC.

146. Which fixture can be connected to a combination drain and vent system?
   a. water closet
   b. laundry tray
   c. automatic clothes washer
   d. urinal

147. The maximum length of the vertical pipe in a combination drain and vent system is
   a. 2 ft.
   b. 3 ft.
   c. 4 ft.
   d. 8 ft.

148. A water supply pipe can be installed in the same trench as a sewer main as long as it is _____ inches above the sewer main.
   a. 12
   b. 18
   c. 6
   d. 24

149. Any discharge from a potable water supply system must go to a waste through an air gap.
   a. true
   b. false

150. Strainers on roof drains must be _____ inches above the roof elevation.
   a. 2
   b. 4
   c. 6
   d. 12

POSTTEST 2

You are now familiar with the kinds of question and answer formats you will see on a plumber's licensing exam. Take this posttest to identify any areas that you may need to review in more depth before the test day. When you are finished, check the answers on page 278 carefully to assess your results. Each answer in the Answers and Explanations section provides a cross-reference to the plumbing code addressed in the question. Use the correlation guide in Appendix A to connect these codes to specific chapters in this book as well as in the IPC or UPC codebooks. Remember to do the following:

- Work carefully.
- Use estimation to eliminate answer choices or to check your work.
- Answer every question.
- Check to make sure your answers are logical.

To simulate the test conditions, use the time constraints of an official plumber's licensing exam. Remember, on the official exam, an unanswered question is counted as incorrect, so it's always better to make a good guess.

POSTTEST 2

1. In what situation would you use a DFU?
   a. when capping a pipe that went to a dead end fixture
   b. when determining drainage pipe size
   c. when pressure-testing a drain and vent system
   d. when installing a bypass around an inaccessible damaged hot water pipe

2. What is a bathroom group?
   a. the number of bathrooms in one building
   b. ten water closets in one bathroom
   c. a group of fixtures consisting of a water closet, a lavatory, a bathtub or shower, and possibly a bidet in one bathroom
   d. none of the above

3. The smallest openings in strainer plates on drain inlets must be
   a. no smaller than $\frac{1}{4}$ in.
   b. no smaller than $\frac{1}{2}$ in.
   c. no larger than $\frac{1}{4}$ in.
   d. no larger than $\frac{1}{2}$ in.

4. When installing 1-in. horizontal copper tubing, it must be supported at least every
   a. 8 in.
   b. 18 in.
   c. 2 ft.
   d. 6 ft.

5. What type of testing equipment does the inspector provide when testing a new drainage system?
   a. gauges
   b. test plugs
   c. smoke
   d. no testing equipment

6. When you add $\frac{7}{8}$ in. and $\frac{1}{2}$ in., it equals
   a. $\frac{3}{8}$ in.
   b. $1\frac{1}{8}$ in.
   c. $1\frac{3}{8}$ in.
   d. $1\frac{5}{8}$ in.

7. If the top of a water heater measures 20 in. across and 60 in. high, how much water can it hold?
   a. 18.840 gallons
   b. 25.609 gallons
   c. 81.558 gallons
   d. 326.234 gallons

8. Paolo must offset a 2 in. copper pipe a distance of 16 in. using 45-degree elbows. How long must the diagonal pipe be if the fitting allowance for a 2 in. copper 45-degree elbow is $\frac{3}{4}$ in.?
   a. 16 in.
   b. $18\frac{5}{8}$ in.
   c. $21\frac{1}{8}$ in.
   d. $22\frac{7}{8}$ in.

9. When installing a drainage pipe for an automatic washing machine, you must use a pipe that is at least
   a. 1 in.
   b. $1\frac{1}{2}$ in.
   c. 2 in.
   d. $2\frac{1}{2}$ in.

10. Overflows are always required for lavatories.
   a. true
   b. false

11. Where are trough urinals allowed?
   a. in prisons
   b. in large office buildings
   c. in facilities designed to be used for three months or less
   d. in none of these places

12. Showers can be supplied with higher water temperatures than bathtubs.
   a. true
   b. false

13. A passageway to a water heater in an attic can be no longer than _____ feet.
   a. 6
   b. 8
   c. 20
   d. 35

14. When installing a temperature and pressure relief valve on a tank-type water heater, where must it be installed?
   a. on top of the water heater
   b. the bottom of the water heater
   c. toward the top of the body of the water heater
   d. toward the bottom of the body of the water heater

15. What is a fixture combining one sink and laundry tray or a two- or three-compartment sink called?
   a. kitchen fixture
   b. combination fixture
   c. plumbing utilities
   d. drainage fixtures

16. Homeowners are allowed to decide whether seismic supports need to be installed.
   a. true
   b. false

17. The outlet pipe from a T & P relief valve must be smaller in diameter than the T & P valve outlet.
   a. true
   b. false

18. The potable water supply is at risk from the release of water from a boiler pressure relief valve.
   a. true
   b. false

19. Condensate drains from refrigeration equipment
   a. must always be directly drained.
   b. can be either directly or indirectly drained.
   c. must be indirectly drained without an air break.
   d. must be indirectly drained, but may be drained with an air break.

20. Any discharge from a potable water system must be drained with an air gap.
   a. true
   b. false

21. What is the purpose of an air gap?
   a. to prevent backflow into the indirect waste
   b. to prevent contamination of food in a food prep sink
   c. to protect the potable water supply
   d. all of these

POSTTEST 2

22. What is the smallest size vent permitted for a 4 in. soil stack?
   a. 1 in.
   b. 2 in.
   c. 3 in.
   d. $3\frac{1}{2}$ in.

23. Combination drain and vent systems can be used for food waste grinders and clinical sinks.
   a. true
   b. false

24. What is the primary purpose of a trap?
   a. to catch any valuables accidentally dropped in the drain
   b. to separate solids and liquids in the drainage system
   c. to prevent sewer gas from entering the building
   d. to remove contaminants from the draining water

25. What is the maximum number of compartments on a combination fixture that can be served by a single trap?
   a. 3
   b. 4
   c. 5
   d. 6

26. What is the edge of the receptacle from which water overflows called?
   a. drain overflow
   b. flood-level rim
   c. fixture overflow
   d. the level the water rises to in a toilet

27. A building trap must always have a relief vent.
   a. true
   b. false

28. When must a grease interceptor be installed?
   a. always
   b. never
   c. when the discharge contains grease
   d. when lubricant is used in the installation of the plumbing system

29. Which type of trap should be used in a sink in a facility where plaster casts are regularly made?
   a. drum trap
   b. S-trap
   c. bag trap
   d. $\frac{3}{4}$ trap

30. The strainer on a roof drain must be _____ high.
   a. 2 in.
   b. 4 in.
   c. 6 in.
   d. 8 in.

31. What is the largest-size solid waste allowed through a strainer plate?
   a. 1 in.
   b. $\frac{3}{4}$ in.
   c. $\frac{1}{2}$ in.
   d. $\frac{1}{4}$ in.

32. When must you include a secondary roof drainage system?
   a. always
   b. never
   c. only when the design of the roof causes ponding
   d. only when the design of the roof prevents ponding

33. The dimensions of a urinal partition shall be _____ from the floor, extending up to _____.
   a. 12 in., 16 in.
   b. 8 in., 70 in.
   c. 5 in., 65 in.
   d. 9 in., 75 in.

34. Copper to copper joints must be joined with flare fittings.
   a. true
   b. false

35. Which of these is considered a branch?
   a. a main
   b. a riser
   c. a stack
   d. none of these

36. Water must be supplied to a building at _____ of the demand rate.
   a. 80%
   b. 100%
   c. 120%
   d. 200%

37. An approved type water-pressure reducing valve must be installed on a water service to a residential building if the water pressure exceeds _____ psi.
   a. 90
   b. 100
   c. 110
   d. 80

38. Which of these is not an approved type of water hammer arrestor?
   a. piston
   b. diaphragm
   c. air chamber
   d. compression element

39. Hot water temperature maintenance is required for pipes longer than 50 ft.
   a. true
   b. false

40. Which of the following types of material should not be used in potable water distribution piping?
   a. copper
   b. galvanized
   c. ductile iron
   d. carbon steel

41. An overflow for a 200-gal. water supply tank must be _____ in diameter.
   a. 1 in.
   b. 2 in.
   c. $2\frac{1}{2}$ in.
   d. 3 in.

42. It is a priority to protect a potable water supply against which of the following?
   a. leaks
   b. freezing
   c. cross-connection
   d. undersizing

43. Vitrified clay tile pipe can be used for a building sewer.
   a. true
   b. false

44. _____ primer is used on PVC solvent weld joints.
   a. Green
   b. Yellow
   c. Purple
   d. White

45. When is it appropriate to thread schedule 40 PVC pipe to make connections?
   a. always
   b. never
   c. when the two pipes being connected are both schedule 40 PVC pipe
   d. when the two pipes being connected are made of different materials

46. In building sewers that are 8 in. or larger, manholes are required at any connection point in the system.
   a. true
   b. false

47. In buildings where potable and nonpotable water piping is installed, the pipes must be
   a. supported.
   b. separated.
   c. insulated.
   d. identified.

48. A sump pit must be at least _____ deep.
   a. 12 in.
   b. 18 in.
   c. 24 in.
   d. 36 in.

49. Water drained from a _____ is considered gray water.
   a. bathtub
   b. kitchen sink
   c. storm drain
   d. all of these

50. Gray water need not be disinfected if it is used for fixture flushing.
   a. true
   b. false

51. To ensure a vacuum-tight system, you must test a vacuum drainage system to _____ of mercury.
   a. 12 in.
   b. 14 in.
   c. 15 in.
   d. 19 in.

52. Disposing of untreated liquid industrial waste into the public sewer system can result in
   a. a blocked sewer pipe.
   b. explosion.
   c. corrosion of a sewer pipe.
   d. any of these

53. What is an air gap?
   a. the distance between a hot and cold water line
   b. the height of a vent in a DWV system
   c. the distance between a fixture outlet and the flood-level rim of a fixture
   d. the height of the weir in a trap

54. Backflow preventers must be protected against which of the following to ensure that they do not malfunction?
   a. vertical installation
   b. horizontal installation
   c. freezing
   d. oversizing

55. If an opening is made in a floor to allow a pipe to pass through, you are required to protect that opening from rodents by
   a. filling the hole with plaster.
   b. installing a metal collar.
   c. stuffing it with steel wool.
   d. installing a drain inlet.

56. Cast-iron piping running vertically from floor to floor must be supported
   a. at least every 15 ft.
   b. at least every 18 ft.
   c. midway between the floor and ceiling, regardless of distance.
   d. only at each ceiling entrance point.

57. With new construction, when a pipe must pass through a concrete or block wall, the pipe should be embedded in the wall for maximum stability.
   a. true
   b. false

58. In public restroom facilities, the interior walls and surfaces must be
   a. soft and water-resistant.
   b. smooth and absorbent.
   c. hard and smooth.
   d. water-resistant and rough.

59. Test gauges requiring a pressure of greater than 10 psi but less than or equal to 100 psi must use a testing gauge having increments of
   a. 0.10 psi or less.
   b. 0.5 psi or less.
   c. 1 psi or less.
   d. 5 psi or less.

60. What backflow device is used on an automatic fire sprinkler system?
   a. check valve
   b. RPZ
   c. air gap
   d. double check detector valve

61. Condensate from appliances can be acidic.
   a. true
   b. false

62. A condensate disposal system must not include materials made of
   a. galvanized steel.
   b. PVC1.
   c. cast iron.
   d. brass.

63. If a theater has a unisex bathroom with two water closets in addition to separate restroom facilities for men and women, the extra two water closets do not apply to the overall number of water closets required for the building.
   a. true
   b. false

64. Building sewers should be provided with cleanouts every _____ feet of horizontal straight run.
   a. 50
   b. 75
   c. 25
   d. 100

65. How many water closets are required for a dormitory with 24 male residents?
   a. 2
   b. 3
   c. 4
   d. 6

66. The base of a water closet must have a watertight seal with the floor.
   a. true
   b. false

67. The seat on a water closet must be
   a. smooth.
   b. nonabsorbent.
   c. designed to fit the closet bowl.
   d. all of these.

68. A water closet can have a concealed or hidden trap as long as the walls of the bowl are washed down completely with each flush.
   a. true
   b. false

69. Sheet lead to be used as a shower pan must weigh at least _____ per square foot.
   a. 12 oz.
   b. 2 lbs.
   c. 4 lbs.
   d. 6 lbs.

70. Employees and the public may share the same toilet facilities in restaurants and nightclubs.
   a. true
   b. false

71. Closet bolts must be made of which material?
   a. lead
   b. cast iron
   c. stainless steel
   d. brass

72. Slip joints may only be installed
   a. on trap outlets.
   b. on trap inlets.
   c. within trap seals.
   d. all of these.

73. Which of these is not an approved material for a bathtub?
   a. enameled cast iron
   b. wood with a sheet metal lining
   c. porcelain-enamel formed steel
   d. formed plastic

74. Plastic bathtubs are subjected to an ignition test.
   a. true
   b. false

75. Which of the following can be used to restrict the clear opening of the waste outlet in a lavatory?
   a. a pop-up stopper
   b. a crossbar
   c. a strainer
   d. any of these

76. Shower walls must extend to a height of
   a. 5 ft. above the shower floor.
   b. 6 ft. above the shower floor.
   c. 5 ft. above the floor of the room.
   d. 6 ft. above the floor of the room.

77. The _____ water closet bowl is designed for maximum water conservation.
   a. siphon wash
   b. blowout
   c. reverse trap
   d. washdown

78. When a water service and a building sewer are installed within 5 feet of each other, the water service has to be installed _____ in. above the building sewer.
   a. 18
   b. 12
   c. 10
   d. 15

79. The UPC requires an exposed tubular trap to be constructed of
   a. cast iron.
   b. PVC.
   c. 20-gauge drawn brass.
   d. any material deemed acceptable as drainage piping.

POSTTEST 2

80. Drainage piping serving fixtures that are installed below the level of a building sewer main must be
   a. drained into a septic system.
   b. drained into an ejector pit with a pump.
   c. pitched up to the main sewer line.
   d. none of the above.

81. The plumbing code does not regulate handheld showers beyond the standard regulations for standard shower heads.
   a. true
   b. false

82. What is the minimum diameter allowed for a drain to which a food waste grinder is connected?
   a. 1 in.
   b. $1\frac{1}{2}$ in.
   c. 2 in.
   d. $2\frac{1}{2}$ in.

83. A tank-type water heater installed in a garage must have all parts that are a source of ignition
   a. directly on the floor.
   b. at least 6 in. above the floor.
   c. at least 12 in. above the floor.
   d. at least 18 in. above the floor.

84. The plumbing code allows water heaters to have either an automatic or manual temperature control.
   a. true
   b. false

85. A valve is required on the _____ pipe of a water heater.
   a. inlet
   b. outlet
   c. inlet and outlet
   d. inlet or outlet

86. A hot water storage tank must be insulated so that it loses no more than
   a. 10 BTUs per hour.
   b. 15 BTUs per hour.
   c. 18 BTUs per hour.
   d. 20 BTUs per day.

87. All water heaters must be third-party certified.
   a. true
   b. false

88. You are required to obtain a permit from the authority having jurisdiction whenever you plan to _____ a water heater.
   a. remove
   b. replace
   c. install
   d. all of these

89. A water service pipe is safe lying at least _____ from a septic tank.
   a. 4 ft.
   b. 5 ft.
   c. 8 ft.
   d. 15 ft.

90. Is a sump pump ejector bit required to have a vent?
   a. always
   b. never
   c. sometimes with exceptions
   d. only if located under grade

91. Combination stop-and-waste valves are not approved for underground water service.
   a. true
   b. false

92. An overflow for a 75-gallon water supply tank must be _____ in diameter.
   a. 1 in.
   b. 2 in.
   c. $2\frac{1}{2}$ in.
   d. 3 in.

93. A water pump in a basement
   a. can be installed directly on the basement floor.
   b. must be mounted at least 8 in. above the basement floor.
   c. must be mounted at least 12 in. above the basement floor.
   d. must be mounted at least 18 in. above the basement floor.

94. If a building uses a cistern for its water supply, how far from the cistern must its absorption field be?
   a. 25 ft.
   b. 50 ft.
   c. 75 ft.
   d. 100 ft.

95. What do vents protect from being broken?
   a. the plumbing fixture
   b. the building trap
   c. the trap seal
   d. the building sewer

96. If the water pressure in the public main is insufficient to supply the minimum pressure in a building, a _____ must be used to bring the water system into compliance with the code.
   a. gravity tank
   b. hydro-pneumatic pressure booster system
   c. water pressure booster pump
   d. any of these

97. A(n) _____ can be used to protect an irrigation system from backflow.
   a. air gap
   b. barometric loop
   c. antisiphon-type fill valve
   d. atmospheric vacuum breaker

98. How are vent sizes determined?
   a. by the size of the building
   b. the location of the building
   c. the number of fixture units
   d. the number of floors in the building

99. Cleanouts must be installed at _____ _____ in a building drain.
   a. every change of direction
   b. every change of direction less than 45 degrees
   c. every change of direction greater than 45 degrees
   d. every change of direction greater than 90 degrees

100. A sump pit can be made of
   a. steel.
   b. concrete.
   c. plastic.
   d. any of these.

101. Glass fittings are approved for drainage pipes.
   a. true
   b. false

102. To drain a steam vent to a sanitary drainage system, you must first
   a. heat the steam.
   b. condense the steam.
   c. dissipate the steam with a fan.
   d. vacuum the steam.

103. Which of the following cannot be used as an indirect waste pipe material?
   a. PVC
   b. copper
   c. concrete
   d. cast iron

104. An air gap is an acceptable substitute for an air break.
   a. true
   b. false

105. According to the UPC, no air gap is required for a pool deck drain that drains to a sanitary drain.
   a. true
   b. false

106. What is the maximum length allowed for an indirect drain from a refrigeration coil?
   a. 8 ft.
   b. 12 ft.
   c. 24 ft.
   d. There is no limitation on the maximum length.

107. A combination drain and vent system is used when it is too distant for a standard vent.
   a. true
   b. false

108. What is the primary purpose of vent piping?
   a. to eliminate the need for trap seals
   b. to maintain trap seals
   c. to allow smaller drain sizes
   d. to allow larger drain sizes

109. What is the minimum weight for sheet copper that will be used for vent pipe flashing?
   a. 8 oz. per sq. ft.
   b. 12 oz per sq. ft.
   c. 1 lb. per sq. ft.
   d. 1 lb. per sq. yd.

110. Where a roof is being used only for weather protection, the vent shall terminate _____ feet above the roof.
   a. 1
   b. 3
   c. 4
   d. 7

111. A trap weir that is 3 in. in diameter can be _____ from a vent.
   a. 5 ft.
   b. 6 ft.
   c. 8 ft.
   d. 12 ft.

112. No more than 12 fixture units can be discharged into one branch interval of a 3-in. waste stack.
   a. true
   b. false

113. A _____ is a vent pipe that acts as both a vent and a drain.
   a. double wye
   b. sanitary tee
   c. wet vent
   d. vertical vent

114. Which type of offsets are allowed with a waste stack?
   a. vertical
   b. horizontal
   c. both vertical and horizontal
   d. neither vertical nor horizontal

115. Fixtures that vent must transition horizontally a minimum of _____ in. above the flood level rim of the fixtures.
   a. 4
   b. 6
   c. 42
   d. 8

116. Traps that depend on moving parts to maintain the seal are prohibited.
  a. true
  b. false

117. When installing a vitrified clay trap underground, it must be set in
  a. potting soil.
  b. sand.
  c. concrete.
  d. none of these.

118. One trap shall be permitted to serve a set of not more than three single-compartment sinks as long as the waste outlets are not more than _____ in. apart.
  a. 25
  b. 30
  c. 35
  d. 40

119. The main difference between an interceptor and a separator is that the interceptor sends waste out to a storage tank, while the separator retains waste.
  a. true
  b. false

120. Each fixture trap generally must have a liquid seal of
  a. at least 1 in., but not more than 4 in.
  b. at least 2 in., but not more than 4 in.
  c. at least 1 in., but not more than 6 in.
  d. at least 2 in., but not more than 6 in.

121. A sand interceptor is used to separate and retain undesirable matter from normal wastes, while allowing normal sewage or liquid waste to pass through.
  a. true
  b. false

122. Storm water cannot be drained into a sewer designed for sewage only.
  a. true
  b. false

123. All distances from the fixture outlet to the trap weir must not exceed _____ inches.
  a. 20
  b. 24
  c. 28
  d. 32

124. When installing a storm drainage system, you must use
  a. cast-iron materials.
  b. PVC materials.
  c. the same materials as required for a sanitary system.
  d. any materials that seem appropriate for the job.

125. Only storm water may drain into conductors when connecting to a combination sewer.
  a. true
  b. false

126. A trap must not be _____ the trap arm to which it is connected.
  a. larger than
  b. smaller than
  c. the same as
  d. indirectly connected to

127. All sump pump discharge pipes require a _____ valve.
  a. check
  b. globe
  c. release
  d. trap

POSTTEST 2

128. What is the minimum diameter for a sump pit?
   a. 10 in.
   b. 15 in.
   c. 18 in.
   d. 20 in.

129. Building traps must be provided with
   a. an identification sign.
   b. a fresh air intake.
   c. a pitch.
   d. none of the above.

130. Which of the following markings must appear on piping to show that it is allowed to be used in a medical gas system?
   a. UL
   b. DWV
   c. MED
   d. none of these

131. A medical vacuum system can never be converted for the delivery of any medical gas.
   a. true
   b. false

132. Gas pipes that run underground must generally be installed at least _____ below the surface.
   a. 6 in.
   b. 8 in.
   c. 12 in.
   d. 18 in.

133. What is the primary reason for the increasing interest in vacuum drainage systems in the United States?
   a. Vacuum drainage systems use better pumps.
   b. Vacuum drainage systems use larger pipes.
   c. Vacuum drainage systems require less water.
   d. Vacuum drainage systems are self-contained.

134. Where a trap seal is subject to loss by evaporation due to infrequent use, such as a floor drain, _____ must be installed.
   a. a hose bibb
   b. a trap primer
   c. a water supply line
   d. a drain to the trap

135. A shower base must have a minimum diameter of
   a. 28 in.
   b. 30 in.
   c. 24 in.
   d. 36 in.

136. A water main can be laid in the same trench with a sewer
   a. if the sewer main is on a ledge 12 inches above the water main.
   b. if the water main is on a ledge 12 inches above the sewer main.
   c. if the trench is 5 feet wide and the pipes are 5 feet apart.
   d. never.

137. What is the minimum size trap allowed for a floor drain?
   a. $1\frac{1}{4}$ in.
   b. $1\frac{1}{2}$ in.
   c. 2 in.
   d. 3 in.

138. Grease interceptors are installed in establishments with fixtures that are used for
   a. washing clothes.
   b. food preparation.
   c. furniture manufacturing.
   d. gray water recycling.

139. Pipe sizes are expressed in the codes as
   a. I.D. (inside diameter).
   b. nominal size.
   c. standard size.
   d. all of the above

140. A trap seal must be a minimum of ____ inch(es) deep.
   a. 1
   b. 2
   c. 3
   d. 4

141. An air admittance valve
   a. is permitted under the IPC.
   b. is permitted under the UPC.
   c. is never permitted.
   d. must be installed on the inlet side of a trap.

142. Sewage pump discharge must connect to the building drainage system
   a. directly, through a wye fitting on top of the drainage pipe.
   b. directly, through a wye fitting on the side of the drainage pipe.
   c. indirectly, through a wye fitting on top of the drainage pipe.
   d. indirectly, through a wye fitting on the side of the drainage pipe.

143. Where are storm drains required?
   a. roofs
   b. paved areas
   c. courtyards
   d. all of the above

144. An automatic clothes washer can drain to a drywell
   a. only in a residential application.
   b. if the drywell has a capacity of at least 300 gallons.
   c. if the drainage area is 500 square feet or more.
   d. never.

145. What is used to prevent backflow in a storm drain?
   a. a check valve
   b. a backflow preventer
   c. a backwater valve
   d. a double detector check

146. Gray water recycling systems shall receive only waste discharge from
   a. bathtubs.
   b. clothes washers.
   c. lavatories.
   d. all of the above.

147. ASME stands for
   a. Associated Sanitary Mechanical Engineers.
   b. American Society of Mechanical Engineers.
   c. Association of State Mechanical Experts.
   d. Allied Systems Manufacturing Executives.

148. If an air-conditioning condensate drain connects to the building drainage system, it must
   a. connect directly to a vent pipe.
   b. connect directly to a trap arm.
   c. be indirectly connected through a trap.
   d. have a discharge of less than one gallon per hour.

149. A 2-inch vent can be used on a drain carrying no more than
   a. 12 DFU.
   b. 20 DFU.
   c. 24 DFU.
   d. 40 DFU.

150. Gray water should be disinfected with chlorine, iodine, or ozone for use in
   a. showers.
   b. urinals.
   c. bidets.
   d. dishwashers.

ANSWERS AND EXPLANATIONS

ANSWERS AND EXPLANATIONS

Pretest

1. **c.** An access panel is a fastened cover with pipes or valves behind it (IPC 202).
2. **c.** The lowest portion of a drainage system extending within the building to 30 in. outside the building (IPC 202).
3. **b.** The depth of water that would have to be removed from a full trap before air could pass through (IPC 202).
4. **c.** A pipe carrying rainwater inside a building (IPC 202).
5. **d.** The officer or other designated authority charged with the administration and enforcement of the plumbing code (IPC 202).
6. **b.** The drain from the trap of a fixture to a connection with any other pipe (IPC 202).
7. **c.** Hazardous to human health (IPC 202).
8. **b.** The inside diameter of the smallest orifice leading to a fixture (IPC 202).
9. **b.** A device to prevent backflow (IPC 202).
10. **c.** A vacuum is defined as any pressure less than atmospheric pressure (IPC 202).
11. **b.** A tank or pit that receives sewage or liquid waste, located below the normal grade of the gravity system, and that must be emptied by mechanical means (i.e., a pump) is a sump (IPC 202).
12. **b.** A fitting or device that provides a liquid seal to prevent sewer gasses from entering the building is a trap (IPC 202).
13. **c.** Water with a temperature higher than 110°F is called hot (IPC 202).
14. **d.** A pressure-actuated valve held closed by a spring or other means and designed to relieve pressure automatically at the pressure at which such valve is set is a relief valve (IPC 202).
15. **a.** Water that is safe for drinking, cooking, and personal use as determined by the board of health is potable (IPC 202).
16. **c.** The edge of the receptacle from which water overflows is the flood-level rim (IPC 202).
17. **a.** That part of the lowest piping of a drainage system that receives the discharge from soil, waste, and other drainage pipes inside and that extends 30 in. in developed length of pipe beyond the exterior walls of the building and conveys the drainage to the building sewer is the building drain (IPC 202).
18. **b.** A removable plate, usually secured by bolts or screws, to permit access to a pipe or pipe fitting (e.g., valve) for the purposes of inspection, repair, or cleaning is an access cover (IPC 202).
19. **c.** The hottest water allowed to be discharged into a building drain is 140°F (IPC 701.7).
20. **b.** In a factory, 400 workers require 1 drinking fountain (IPC Table 403.1 number 4).
21. **b.** Every dry vent must rise vertically to a minimum of 6 in. above the flood-level rim of the highest trap or trapped fixture being vented (IPC 905.4).
22. **a.** The only approved position for a tee wye carrying wastewater is a horizontal branch with a vertical run (IPC Table 706.3).
23. **d.** for IPC; **c.** for UPC. The maximum distance allowed between manholes on a sanitary sewer is 400 feet (IPC 708.3.2); 300 feet is the maximum distance allowed between manholes on a sanitary sewer (UPC).
24. **a.** Fixtures with concealed slip-joint connections must have an access panel or utility space (IPC 405.8).
25. **b.** The building sewer is sized in the same way as the building drain, according to the DFU load with pipe no smaller than 3 in. if a water closet is part of the load. Therefore, it is possible to have a 3-in. building sewer (IPC 710.1[1]).
26. **a.** A shower has a DFU of 2 with a minimum trap size of $1\frac{1}{2}$ in. (IPC Table 709.1).

241

ANSWERS AND EXPLANATIONS

27. **c.** When a roof is used for human activities, any vent termination must be extended at least 7 ft. above the roofline to keep sewer gas above the area (IPC 904.1).
28. The correct response for all four figures is for vent purposes only; the only requirement is that all portions of the vent system drain back to the drainage system (IPC 905.2).
29. **b.** S-traps are prohibited in any installation because they tend to self-siphon (IPC 1002.3).
30. **b.** There is no requirement for a drawing of concealed valve locations be given to a building owner.
31. **a. for IPC; b. for UPC.** Within each plumbing system, a minimum of one stack vent or vent stack shall extend outdoors to the open air (IPC 917.7). There is no provision in the UPC allowing the use of air admittance valves.
32. **b.** A building drain is the lowest gravity-draining horizontal drainage pipe in the building, receiving drainage from the other drains in the building (UPC 202).
33. **a.** The building sewer extends from the end of the building drain to the connection with the private disposal system, public disposal system, or other disposal system (IPC 202).
34. **b.** Contaminated water is a health hazard. Polluted water, although considered non-potable, does not present a health hazard (IPC 202).
35. **b.** A conductor carries storm water inside a building (IPC 202).
36. **b.** A cistern is a tank usually used to store rainwater. The water is generally for nonpotable uses (IPC 202).
37. **d.** The air gap on a faucet is the vertical distance between the end of the spout and the flood-level rim of the fixture (IPC 202).
38. **d.** Water at a temperature of 110°F or warmer is considered hot water (IPC 202).
39. **b.** An indirect waste pipe is waste piping not directly connected with the drain system, discharging through an air gap into a trap or fixture (IPC 202).
40. **a.** A stack vent is the extension of a soil stack (IPC 202).
41. **b.** A plumbing appurtenance places no water supply or drainage load on a plumbing system (IPC 202).
42. **a.** Polluted water is not hazardous to health (IPC 202).
43. **a.** Any pressure less than atmospheric is a vacuum (IPC 202).
44. **a.** A pipe that is 45 degrees above horizontal is included as a vertical pipe (IPC 202).
45. The expansion and increase in temperature of water through heating.
46. A yoke vent is a pipe that connects upward from a soil or waste stack to a vent stack for the purpose of preventing pressure changes in the stacks.
47. In a high-risk backflow application such as a mortuary or a lab researching waterborne pathogens (IPC 202).
48. **a.** Mechanically formed tee fittings must be brazed in accordance with Section 605.14.1 (IPC 605.1.2).
49. **b.** PVC pipe can only be used for cold water distribution (IPC 601–612).
50. **b.** The minimum water service pipe permitted is $\frac{3}{4}$ in. (IPC 603.1).
51. **b.** $V = \frac{\pi r^2 h}{231} \times 0.75$
52. **c.** $V = \frac{\pi r^2 h}{231}$
53. **a.** Clockwise, due to Earth's rotation.
54. **b.** A 2-inch horizontal fixture branch can carry 6 DFUs (IPC 710.1[2]).
55. **b.** A 2-inch stack that is three branch intervals or less can carry 10 DFUs (IPC 710.1[2]).

ANSWERS AND EXPLANATIONS

56. **b.** A 3-inch horizontal branch interval can carry 20 DFUs (IPC 710.1[2]).
57. **The slope is 4 inches** (IPC 710.1[2]).
58. **b.** An automatic washing machine requires a 2-inch trap and drain (IPC 709.1).
59. **b.** A shower has a DFU of 2 (IPC 709.1).
60. **c.** A bathroom group has a DFU of 5 (IPC 709.1).
61. **d.** Bituminous fiber pipe is permissible for use never. Neither the IPC nor UPC has any provision allowing the use of bituminous fiber pipe (i.e., Orangeburg pipe).
62. **a.** Asbestos cement pipe is permissible for use as a water main (IPC 605.3).
63. **a.** A cleanout must be located within 10 feet of the junction between the building drain and the building sewer (IPC 708.3.5).
64. **d.** An automatic clothes washer must drain through an air break to prevent backflow (IPC 802.4).
65. **d.** A vent terminal cannot be within 10 feet horizontally of any door, window that can be opened, or air intake in a building (IPC 904.5).
66. **d.** Crown vents are prohibited because they can easily plug up with debris (IPC 1002.3).
67. **b.** An air admittance valve must be installed a minimum of 4 inches above the horizontal drain being vented (IPC 917.4).
68. **b.** The vertical distance from a fixture outlet to the weir of the trap cannot be greater than 24 inches (IPC).
69. **a.** A leader conducts water on the outside of a building (IPC 202).
70. **c.** Slip joints can be on the trap inlet, the trap outlet, and within the trap seal (IPC 1002.2).
71. **d.** A showerhead can have a maximum flow of 2.5 GPM (IPC 604.4).
72. **a.** A water closet with a flushometer valve requires the water supply pipe to be a minimum diameter of 1 inch (IPC 604.5).
73. **c.** A grid strainer can have slots that are no wider than $\frac{1}{2}$ inch (IPC 304.2).
74. **b.** Piping, other than cast iron and steel, that passes through drilled or notched holes in framing members must be protected by shield plates if it is within $1\frac{1}{2}$ inches of the nearest edge of the framing member (IPC 305.8).
75. **a.** The maximum spacing for pipe supports on $\frac{3}{4}$-inch copper tube is 6 feet (IPC 308.5).
76. **a.** A compartment for a water closet must be a minimum of 30 inches wide (IPC 405.3.1).
77. **b.** The minimum size of water supply pipe to a wall hydrant is $\frac{1}{2}$ inch (IPC 604.5).
78. **c.** One psig of pressure will create 2.3 feet of head; 1 foot of head equals 0.433 psig (IPC 604.1).
79. **c.** A little-used floor drain must have a trap primer (IPC 1002.4).
80. **c.** When sizing and designing a roof drain system, any wall that diverts water onto the roof must be sized as part of the roof, but only 50% of its area is included in the calculation (IPC 1106.4).
81. **c.** The flushing rate for a urinal is 1 GPF (IPC Table 604.4).
82. **a.** A valve is required on the cold water inlet of a water heater (IPC 503.1).
83. **b.** Every building that is equipped with plumbing fixtures on a single lot is allowed to share a common sewer before connection to the public sewer (IPC 701.3).
84. **b.** CPVC is not on the list of approved materials for aboveground drainage and vent systems (IPC Table 702.1).
85. **b.** Because of its high concentration of organic waste, drainage from a kitchen sink is not considered gray water. The organic matter would accelerate bacteria growth in a gray-water holding tank if allowed into the tank (IPC C101.2 & C101.9).

ANSWERS AND EXPLANATIONS

86. **a.** Gray water needs to be disinfected when it is used for flushing fixtures (IPC C102.2).
87. **b.** A bell trap is not approved for floor drains because the depth of the trap seal is less than 2 inches, the seal is subject to a high rate of evaporation, and the trap seal area is easily clogged with debris (IPC 1002.2–1002.4).
88. **a.** The minimum slope allowed for a 6-inch horizontal drainpipe is $\frac{1}{8}$-inch per foot (IPC Table 704.1).
89. **d.** Solvent-welded PVC joints are permitted in all applications where PVC pipe is permitted (IPC 705.2.2).
90. **b.** A double tee wye can be used to drain back-to-back water closets if the horizontal developed length from the water closet to the connection with the double tee wye is 18 inches or greater (IPC 706.3).
91. **a.** In drainage piping, the position for a sanitary tee can be horizontal to vertical (IPC Table 706.3).
92. **a.** When a 2-inch drain serves a single fixture, a short sweep can be used in the horizontal-to-horizontal position (IPC Table 706.3).
93. **b.** A residential automatic clothes washer has a DFU of 2 (IPC Table 709.1).
94. **a.** The discharge pipe for a temperature and pressure relief valve must terminate within 6 inches of the floor (IPC 504.6).
95. **a.** A food preparation sink must drain through an air gap (IPC 802.1.1).
96. **c.** The primary purpose of a vent is to maintain the trap seals (IPC 901.2).
97. **c.** Any horizontal dry vent must rise vertically 6 inches above the flood-level rim of the highest trap or trapped fixture being vented (IPC 905.4).
98. **b.** Vitrified clay tile drainage piping must be joined with an elastomeric seal (IPC 705.15).
99. **b.** A full-opening valve is required in a drainage system after the last fixture draining into a gray-water recycling system (IPC C101.11.1).
100. **a.** A backflow preventer on a water supply pipe to a boiler must have a vent between the check valves (IPC 608.16.2).
101. **a.** At sea level, 43.3 psig of pressure is created by 100 feet of head (IPC 608.1).
102. **d.** A 4-inch diameter pipe that is 100 feet long can hold 65.25 gallons. $V = \frac{\pi r^2 h}{231}$
103. **c.** The minimum slope permitted for a $1\frac{1}{2}$-inch drain is $\frac{1}{4}$-inch per foot (IPC Table 704.1).
104. **c.** In a public bathroom, urinals must be spaced 30 inches apart (IPC 405.3.1).
105. **b.** There must be 21 inches of clearance in front of a water closet (IPC 405.3.1).
106. **c.** A clearance of 30 inches is required between the centerline of a water closet and any wall that may be on either side of the water closet (IPC 405.3.1).
107. **a.** An overflow built into a lavatory is optional (IPC 416.1 Commentary).
108. **a.** No plumbing system can be installed in an elevator shaft (IPC 301.6).
109. **a.** A building trap is permitted only when the Authority Having Jurisdiction has determined its necessity (IPC 1002.6).
110. **c.** A roof or floor truss can be cut only with written approval from a registered design professional (IPC 307.4).
111. **b.** Copper tubing with a $1\frac{1}{2}$-inch diameter that is horizontally installed must have hangers installed every 10 feet (IPC Table 308.5).
112. **b.** Each drain outlet of a laundry tray must have a minimum diameter of $1\frac{1}{2}$ inches (IPC 415.2).
113. **a.** The minimum allowable width of a shower floor is 30 inches (IPC 417.4).

ANSWERS AND EXPLANATIONS

114. a. Water closets with flushometer tanks must flush no more than 1.6 GPF (IPC Table 604.4).

115. b. The maximum outlet water temperature to a potable hot water distribution system is 140°F (IPC 501.6).

116. d. The temperature setting of a temperature and pressure relief valve installed in a water heater can be no more than 210°F (IPC 504.5).

117. c. A water closet with a flush valve requires a 1-inch water supply (IPC Table 604.5).

118. a. A water hammer arrestor is required where quick-closing valves are used (IPC 604.9).

119. a. The spout of a faucet must be a minimum of 1 inch above the flood-level rim of the fixture (IPC 608.15.1).

120. c. A barometric loop is a vacuum breaker (IPC 608.13.4).

121. b. A hospital requires two water services (IPC 609.2).

122. b. Manholes are required as cleanouts when a building sewer is 8 inches in diameter (IPC 708.3.2).

123. d. A drinking fountain has a DFU of 0.5 (IPC Table 709.1).

124. a. A 2-inch horizontal fixture branch has a DFU capacity of 6 (IPC 710.1[2]).

125. b. Wastewater can be no hotter than 140°F (IPC 701.7).

126. d. In a combination drain and vent system, which is a variation of a wet vent, a 3-inch floor drain could be installed an unlimited distance from a conventional vent (IPC 912.1–912.3).

127. c. The constant used to find the length of a 45° offset is 1.414.

128. c. In a combination drain and vent system, a kitchen sink, with a DFU of 2, would require the drain, starting from the vertical portion, to be 2 inches in diameter (IPC Table 912.3).

129. a. An air admittance valve must be installed at least 4 inches above the drain being served by the air admittance valve (IPC 917.4).

130. a. One vent through the roof is required in a vent system served by air admittance valves (IPC 917.7).

131. a. A grease interceptor requires a flow control on the inlet pipe (IPC 1003.3.4.1).

132. d. A secondary roof drain must drain where the discharge will be noticed (IPC 1107.2).

133. c. An approved material for building storm sewer pipe is acrylonitrile butadiene styrene (ABS) (IPC Table 1102.4).

134. d. A storm sewer can drain directly to a waterway, receive the discharge of a sump pump, and be laid side by side with a sanitary sewer (IPC 1101.2, 1111.1, and 703.3).

135. b. The minimum number of roof drains required for a roof that is 10,000 square feet or less is 2 (IPC 1110.4).

136. c. The minimum pipe diameter for a subsoil drain is 4 inches (IPC 1111.1).

137. a. Both a check valve and a gate valve are required on the discharge pipe of a sump pump (IPC 1113.1.4).

138. b. One of the materials that can be used for medical vacuum piping is galvanized steel pipe (IPC 1202.1).

139. a. Recycled gray water that is used for flushing fixtures must be dyed blue or green (IPC C102.4).

140. d. The operating vacuum in a vacuum drainage system is 16 to 20 inches of mercury (IPC G101.3 Commentary).

141. a. Where a storm drain connects to a combination sewer, the storm drain must be trapped (IPC 1103.1).

142. a. A grease interceptor with a flow-through rating of 10 GPM, must have a grease retention capacity of 10 pounds (IPC Table 1003.3.4.1).

ANSWERS AND EXPLANATIONS

143. a. The trap weir of a urinal can be an unlimited distance from a vent (IPC 906.1 Exception).

144. b. A bathtub cannot be vented using a combination drain and vent system (IPC 912.1).

145. b. A vent stack carries only air and sewer gas (IPC 202).

146. b. A high heel tee installed as the first fitting at the outlet of a water closet is not an acceptable way to connect a vent (IPC 706.4).

147. a. In 6-inch drainage pipes, the minimum cleanout size is 4 inches (IPC 708.7).

148. a. The minimum slope allowed in a 4-inch drain is $\frac{1}{8}$-inch per foot (IPC Table 704.1).

149. a. The maximum horizontal hanger spacing for 10-foot lengths of cast-iron pipe is 10 feet (IPC Table 308.5).

150. a. Polluted water is not considered a hazard to public health (IPC 202).

Chapter 1 Review

1. b. IPC (IPC 917.1). Although there is no language in the UPC specifically banning air admittance valves, all references to vents are those that terminate above the roof or those that connect to vents that terminate above the roof. There is no section that allows air admittance valves or sets any regulations for them.

2. The UPC only allows a 2-inch drain for a shower (UPC Table 7–3); the IPC allows a $1\frac{1}{2}$-inch drain (IPC Table 709.2).

3. d. The IPC disqualifies the use of mechanical valves, which are closed by spring pressure, in the definition of an air admittance valve (IPC Section 202 Definitions: Air Admittance Valve). The UPC does not allow any type of air intake valve in a vent system (UPC Chapter 9 Introduction).

4. b. Through observation and research, it has been found that frost closure is prevalent in vents that are smaller than 3 inches in diameter when subjected to temperatures colder than 0°F for a period of 24 hours or more. When the temperature is above 0°F, the warm air rising out of the vent prevents frost closure; but when the temperature is below 0°F, the warm, moist air rising up a vent cools very quickly leaving the moisture deposited as a buildup of frost in the vent extending above the roof (IPC 904.2).

5. c. DWV stands for drainage, waste, and vent. Those are the applications DWV copper tubing can be used for due to relatively thin wall thickness (IPC Table 702.1, UPC 701.1.4).

6. a. Often expressed as National Pipe Thread, Nominal Pipe Thread, and National Pipe Thread Tapered, NPT is really a code. The *N* stands for USA Standard; the *P* means pipe, as expected; and the *T* signifies that the threads are tapered (IPC Commentary for 605.10.3).

7. c. The UPC requires a 2-inch drain for a kitchen sink even though the trap may be $1\frac{1}{2}$ inches (UPC Table 7–3).

8. b. The drainage fixture unit, or DFU, as a measure of drainage loads was first established in the 1940s in Dr. Roy B. Hunter's Report BMS65, "Methods of Estimating Loads in Plumbing Systems."

ANSWERS AND EXPLANATIONS

9. **a.** In the UPC, the separate air gap is required to keep any backflow from the drainage system from getting into the dishwasher. This provides protection even for very old dishwashers that do not have integral backflow protection.
10. **a.** A vent stack is a vertical pipe that connects to the base of a soil stack or to the horizontal drain downstream of a soil stack. Its function is to provide air circulation for the drainage system and to relieve any excess pressures developed in the system. It carries no waste, only air and sewer gas (IPC 202, UPC 203.0).
11. **a.** In both codes, the DFU for a lavatory is 1; therefore, a $1\frac{1}{4}$-inch fixture branch is permitted (IPC Table 709.1, UPC Table 7–3).
12. **b.** Chlorinated polyvinyl chloride is a material used for supply pipes (IPC Table 605.3, UPC Table 6–4).
13. **b.** Underwriters Laboratories (UL) is a third-party testing agency.
14. **d.** None of the above. No state has adopted both codes as its official plumbing code. There are states in which different jurisdictions use parts or all of each code.
15. **b.** The UPC is generally used in the nation's western states.

Chapter 2 Review

1. **a.** With the exception of plastic piping materials, testing of drainage systems can be done with air pressure not exceeding 5 psig (IPC 312.1 and 312.3).
2. **b.** The same testing procedures are used for storm drains and for sanitary drains (IPC 312.8, UPC 1109.2).
3. **b.** For test pressures between 10 psig and 100 psig a gauge with 1 psi increments must be used (IPC 312.1.1, UPC 319.2).
4. **b.** If overexcavation occurs, sand or gravel may be backfilled in the trench in 6-inch layers and compacted to support the pipe (IPC 306.2.1).
5. **b.** Every plumbing fixture using water in its function is required to be connected to the building water supply system (IPC 301.4).
6. **a.** Operating pressures of at least 50 psig must be expected (IPC 312.5).
7. **b.** Neither fixtures nor piping can interfere with the action of any door or window (IPC 310.2).
8. **a.** Workers must be provided with toilet facilities. This is primarily for their comfort and health, and secondarily to allow them to remain at the job site rather than having to leave to find a bathroom (IPC 311.1).
9. **a.** The minimum depth is for protection from physical damage. If the local code requires a deeper burial, it is for frost protection (IPC 305.6).
10. **b.** Any pipe passing through a foundation needs a sleeve or arch (IPC 305.5). An exception to this rule, according to the UPC only, is that if the hole is drilled or bored, a sleeve is not required (UPC 313.10.1).
11. **c.** $\frac{1}{2}$ inch (IPC 304.2, UPC 313.12.1).
12. **c.** Flammable materials that could cause a fire or explosion are not approved (IPC 302.1, UPC 303.0).
13. **b.** When it is 12 inches or more above the pipe (IPC 306.3).
14. **6 (IPC 305.6).** UPC 313.6 simply states that buried pipes must be protected from freezing and the Authority Having Jurisdiction will determine the burial depth required.
15. **6 feet.** Smaller diameter pipes have less rigidity and require closer supports (IPC Table 308.5, UPC 314.1).

ANSWERS AND EXPLANATIONS

16. **b.** The UPC describes the permit holder as the permittee (IPC 312.1, UPC 103.5.4.2).
17. **b.** Plastic pipe cannot be air tested (IPC 312.5, UPC 723.0).
18. Overexcavate at least twice the diameter of the pipe and backfill up to the burial depth with gravel, crushed stone, or concrete (IPC 306.2.3).
19. **b.** When in service, the uppermost 10 feet will never have 10 feet of head pressure on it because the water column cannot go any higher (IPC 312.2).
20. **b.** Leakage from any piping system could corrode the parts of an elevator and threaten its safe operation (IPC 301.6).
21. **a.** Throughout the plumbing trade, pipe sizes have been traditionally expressed by inside diameter, or ID. The code rule augments that tradition by making it a regulation (IPC 301.5).
22. **a.** This creates consistency and ensures that the products used will meet the use for which they are intended (IPC 303.1).
23. **b.** Concrete contains limestone which is alkaline and corrosive to pipes (IPC 305.1, UPC 313.2)
24. Industrial waste must be made innocuous (harmless) to the sewer system or the sewage treatment plant (IPC 302.2).
25. **b.** This is referred to as a "grandfather clause." Existing piping may remain and it is assumed that it was installed in accordance with the code, if any, that existed at the time of installation (UPC 301.1.4).
26. **b.** Townhouses are considered to be separate individual residences and must be provided with their own water and sewer services (UPC 308.1).
27. **b.** Any drilling or tapping will weaken the pipe and any such connection would not follow the flow pattern of a drainage fitting (UPC 311.2).
28. For the UPC, 12 inches (UPC 313.4). For the IPC, it is determined by the local code official (IPC 305.6.1).
29. For the IPC, false; the depth must be $1\frac{1}{2}$ inches (IPC 305.8). For the UPC, true; the depth is 1 inch (UPC 313.9).
30. **d.** 3 (IPC table 311.1).
31. **b.** 1 inch WC (IPC 312.4).
32. **b.** The trench cannot go below the 45-degree bearing plane of the footing (IPC 307.5).
33. **a.** Inspectors are not responsible for supplying testing equipment (IPC 312.1, UPC 103.5.4.2).
34. **a.** Regular testing ensures proper operation of the backflow preventer (IPC 312.9.1).
35. **b.** Medical waste carries dangerous pathogens, drugs, and radioactivity that cannot be released into the public sewer system (IPC 302.1, UPC 303.0).

Chapter 3 Review

1. $1\frac{1}{2}$ in.
   $\frac{5}{8} + \frac{7}{8}$
   $\frac{5+7}{8} = \frac{12}{8} = 1\frac{4}{8} = 1\frac{1}{2}$
2. $\frac{7}{8}$ in.
   $\frac{3}{4} + \frac{1}{8}$
   $\frac{6}{8} + \frac{1}{8} = \frac{7}{8}$
3. $\frac{1}{8}$ in.
   $\frac{3}{4} - \frac{5}{8}$
   $\frac{6-5}{8} = \frac{1}{8}$
4. 19.625 sq. in.
   $A = \pi r^2$
   $A = 3.14 \times (2.5 \times 2.5)$
   $A = 19.625$ sq. in.
5. 52.85 gal.
   $V = \frac{\pi r^2 h}{231}$
   $V = \frac{3.14 \times (9 \times 9) \times 48}{231}$
   $V = \frac{12208.32}{231}$
   $V = 52.85$ gal.

ANSWERS AND EXPLANATIONS

6. $7\frac{1}{4}$ in.
$L = 8 - (\frac{3}{8} + \frac{3}{8})$
$L = 8 - \frac{6}{8}$
$L = 7\frac{8}{8} - \frac{6}{8}$
$L = 7\frac{2}{8} = 7\frac{1}{4}$ in.

7. 13.86 gal.
First, convert 85 ft. to inches so that both of the factors are expressed using the same measurement.
$85 \times 12 = 1{,}020$
$V = \frac{\pi r^2 h}{231}$
$V = \frac{3.14 \times (1 \times 1) \times 1{,}020}{231}$
$V = \frac{3{,}202.8}{231}$
$V = 13.86$ gal.

8. $23\frac{7}{16}$ in.
Convert the feet to inches so one unit of measurement is used.
1 foot 6 inches = 18 inches
L = Offset × 1.414 − (fitting allowance × 2)
$L = 18 \times 1.414 - (1 \times 2)$
$L = 25.452 - 2$
$L = 23.452$
Convert .452 to sixteenths by multiplying by 16.
$0.452 \times 16 = 7.232$
Round 7.232 to 7. That is the number of sixteenths we have ($\frac{7}{16}$). Put the $\frac{7}{16}$ with the 23 in. and the result is $23\frac{7}{16}$ in.

9. $41\frac{1}{8}$ in.
Convert the feet to inches so only one unit of measurement is used.
$4 \times 12 = 48$
$48 \quad \rightarrow \quad 47\frac{8}{8}$
$\underline{-6\frac{7}{8}} \quad \rightarrow \quad \underline{-6\frac{7}{8}}$
$\qquad\qquad\qquad 41\frac{1}{8}$

10. $10\frac{3}{16}$ in.
$C = \pi d$
$C = 3.14 \times 3\frac{1}{4}$
Convert the inches to a decimal by dividing the numerator (top number) by the bottom number (denominator):
$1 \div 4 = 0.25$
$C = 3.14 \times 3.25$
$C = 10.205$
Convert the decimal to sixteenths by multiplying by 16:
$0.205 \times 16 = 3.28$
Round 3.28 to 3. This is the number of sixteenths we have: $\frac{3}{16}$.
Put the $\frac{3}{16}$ back with the 10: $10\frac{3}{16}$ in.

11. $17\frac{3}{8}$ in.
The pipe will drop $\frac{1}{4}$ in. for every foot along the 18 feet.
It will drop 18 × 0.25, which equals 4.5 in.
The pipe is already $12\frac{7}{8}$ in. down.
Add the drop to the upstream distance:
$D = 12\frac{7}{8} + 4.5$
Convert $12\frac{7}{8}$ to a decimal.
$7 \div 8 = 0.875$
Solve for D:
$D = 12.875 + 4.5$
$D = 17.375$
Convert 0.375 to eighths by multiplying by 8: $8 \times 0.375 = 3$
3 is the number of eighths.
Put the $\frac{3}{8}$ back with the 17.
$D = 17\frac{3}{8}$ in.

ANSWERS AND EXPLANATIONS

12. $15\frac{7}{8}$ in.
L = Offset × 1.414 − (fitting allowance × 2)
First convert the fraction to a decimal:
9 ÷ 16 = 0.5625
L = 12 × 1.414 − (0.5625 × 2)
L = 16.968 − 1.125
L = 15.843
Decide how precise the measurement must be and multiply 0.843 by the denominator. Use eighths for this problem:
0.843 × 8 = 6.744
Round 6.744 to 7. That's 7 eighths ($\frac{7}{8}$).
L = $15\frac{7}{8}$ in.

13. 1 ft. $2\frac{3}{8}$ in.
2 ft. $\frac{3}{4}$ in. minus 3 ft. $3\frac{1}{8}$ in.
Denominators must be the same to subtract fractions:
2 ft. $\frac{6}{8}$ in. − 3 ft. $3\frac{1}{8}$ in.

$39\frac{1}{8}$ → $38\frac{9}{8}$
$-24\frac{6}{8}$ → $-24\frac{6}{8}$
$14\frac{3}{8}$ in.

$14\frac{3}{8}$ in. or 1 ft. $2\frac{3}{8}$ in.

14. $5\frac{1}{2}$ cups
V = $\frac{\pi r^2 h}{231}$ ÷ 2
V = $\frac{3.14 \times 1^2 \times 51}{231}$ ÷ 2
V = $\frac{160.14}{231}$ ÷ 2
V = 0.6932467 ÷ 2
V = 0.3466233 gal.
Round this off to 0.347 gal. Since this is a fractional part of 1 gallon, multiply by 16 to get the number of cups of antifreeze needed. There are 16 cups in a gallon:
0.347 × 16 = 5.552 cups.
Multiply 0.552 by 8 to get the number of ounces since there are 8 ounces in a cup:
0.552 × 8 = 4.416
Round to 4 ounces. Put that with the 5 cups and the answer is 5 cups, 4 ounces, or $5\frac{1}{2}$ cups.

15. 3 ft. 4 in., 40 in.
Convert all to inches:
1 ft. 8 in. = 20 in.
Add 20 in. plus 20 in. = 40 in. or 3 ft. 4 in.

16. 32
1 in. = $\frac{8}{8}$
$\frac{8}{8}$ × 4 → $\frac{8}{8}$ × $\frac{4}{1}$ = $\frac{32}{1}$

17. 791.44 gal.
There are 231 cubic inches in 1 gallon.
Divide 182,822 by 231
$\frac{182,822}{231}$ = 791.44 gal.

18. $19\frac{5}{16}$ in. and $19\frac{3}{16}$ in.
This requires two complete calculations. For the $\frac{3}{4}$-in. pipe:
L = Offset × 1.414 − (fitting allowance × 2)
L = Offset × 1.414 − (0.25 × 2)
L = 14 × 1.414 − 0.5
L = 19.796 − 0.5
L = 19.296
Convert .296 to sixteenths by multiplying by 16:
0.296 × 16 = 4.736
Round 4.736 to 5. That's the number of sixteenths we have. Put it back with the 19 in. to yield $19\frac{5}{16}$ in. or 1 ft. $7\frac{5}{16}$ in. for the $\frac{3}{4}$-in. pipe.
For the 1-in. pipe:
L = Offset × 1.414 − (fitting allowance × 2)
L = Offset × 1.414 − (0.3125 × 2)
L = 14 × 1.414 − 0.625
L = 19.796 − 0.625
L = 19.171
0.171 × 16 = 2.736
Round to 3:
$19\frac{3}{16}$ in. or 1 ft. $7\frac{3}{16}$ in. for the 1-in. pipe.

ANSWERS AND EXPLANATIONS

19. $1\frac{9}{16}$ in.
First, find the center of the joist:
$5.5 \div 2 = 2.75$
If the hole is drilled in the center, half of the hole will be on either side of the centerline. Divide the hole size by 2.
Convert $2\frac{3}{8}$ to decimal form by dividing 3 by 8, which is 2.375.
$2.375 \div 2 = 1.1875$
$2.75 - 1.1875 = 1.5625$ or $1\frac{9}{16}$ in.
The edge of the hole will be $1\frac{9}{16}$ in. from the edge of the 2 × 6.

20. 1,386 cu. in.
Since there are 231 cu. in. in a gallon,
231×6 gal. $= 1,386$ cu. in.

21. $31\frac{3}{8}$ in. × 22 in.
$C = \pi d$
$C = 3.14 \times 10$
$C = 31.4$
$C = 31\frac{3}{8}$ in.
The insulation blanket must be $31\frac{3}{8}$ in. wide by 22 in. high.

22.   6 ft. $10\frac{3}{4}$ in.
     7 ft. $14\frac{5}{4}$ in.
  − 1 ft. $4\frac{2}{4}$ in.
     6 ft. $10\frac{3}{4}$ in.

23. 2.2 gal.
$0.406 \times 12 = 4.872$
$V = \frac{\pi r^2 h}{231} \div 2$
$V = \frac{3.14 \times 6^2 \times 9}{231} \div 2$
$V = \frac{3.14 \times 36 \times 9}{231} \div 2$
$V = 4.4042 \div 2$
$V = 2.2021$ rounded to 2.2 gal.

24. 0.73 gal.
$V = \frac{\pi r^2 h}{231} \div 2$
$V = \frac{3.14 \times 1^2 \times (36 \times 18)}{231}$
$V = \frac{169.56}{231}$
$V = 0.73$ gal.

25. $8\frac{5}{8}$ in.
$C = \pi d$
$C = 3.14 \times (2.25 + 0.25 + 0.25)$
$C = 3.14 \times 2.75$
$C = 8.635$
Convert to inches with a fraction of an inch.
$0.635 \times 16 = 10.16$ sixteenths.
Round to $\frac{10}{16}$; reduce to $\frac{5}{8}$.
Bring the 8 inches back to give $8\frac{5}{8}$ in.

26. 3.36 gal.
$V = \frac{\pi r^2 h}{231}$
Convert 330 ft. to 3,960 in.
$V = \frac{3.14 \times .25^2 \times 3,960}{231}$
$V = \frac{3.14 \times 0.0625 \times 3,960}{231}$
$V = \frac{777.15}{231}$
$V = 3.36$ gal.

27. 16 ft. 5 in.
Divide 1,575 ft. by 8 to get the number of inches vertically in pitch:
$\frac{1,575}{8} = 196.875$ inches
Divide 196.875 inches by 12 to reduce the answer to feet and inches:
$\frac{196.875}{12} = 16.406$ feet
Multiply 0.406 by 12 to get the number of inches after the 16 ft.:
$0.406 \times 12 = 4.875$
Round to 5.
16 ft. 5 in.

28. $1\frac{3}{16}$ in.
All denominators must be the same:
$\frac{12}{16} + \frac{2}{16} + \frac{5}{16} = \frac{19}{16}$
$\frac{19}{16}$ reduces to $1\frac{3}{16}$ in.

29. Yes

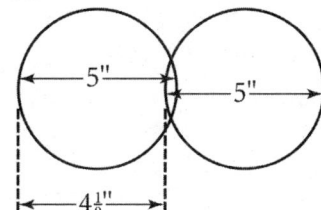

ANSWERS AND EXPLANATIONS

30. $5\frac{5}{8}$ in.
$L = 8 - 2\frac{3}{8}$
$L = 7\frac{8}{8} - 2\frac{3}{8}$
$L = 5\frac{5}{8}$ in.

31. $10\frac{1}{2}$ in.
$L = 12 - (2 \times .75)$
$L = 12 - 1.5$
$L = 10.5$ or $10\frac{1}{2}$ in.

32. 452.16 sq. in.
$A = \pi r^2$
Convert the 2 ft. to 24 in.
$A = 3.14 \times 12^2$
$A = 3.14 \times 144$
$A = 452.16$ sq. in.

33. 587.22 gal.
$V = \frac{\pi r^2 h}{231} \div 2$
Convert the 50 ft. to inches:
$V = \frac{3.14 \times (12^2) \times 600}{231} \div 2$
$V = \frac{3.14 \times 271296}{231} \div 2$
$V = 1174.44 \div 2$
$V = 587.22$ gal.

34. $7\frac{7}{16}$ in.
$V = \frac{\pi r^2 h}{231}$
$3 = \frac{3.14 \times r^2 \times 16}{231}$
$3 = \frac{50.24 \times r^2}{231}$
$231 \times 3 = 50.24 \times r^2$
$693 = 50.24 \times r^2$
$693 \div 50.24 = r^2$
$13.79 = r^2$
$\sqrt{13.79} = r$
3.71 in. $= r$
$D = r \times 2$
$D = 7.42$ Multiply 0.42 by 16 to convert to a fraction with 16 as the denominator:
$0.42 \times 16 = 6.72$
Round to 7. That is the number of sixteenths.
Put $\frac{7}{16}$ back with the 7, $D = 7\frac{7}{16}$ in.

35. 11 in.
Convert the feet to in.:
2 ft. = 24 in.
$L = 35 - 24$
$L = 11$ in.

Chapter 4 Review

1. **2 inches.** The volume of water that modern automatic washing machines can pump out, along with suds pressure, demand that a 2-inch drain be installed (IPC 406.3, UPC Table 7–3).

2. **a.** Slip joints are intended as an easy access to a drainage system and must, therefore, be accessible to permit maintenance of the system (IPC 405.8).

3. **b.** The outlet of a water closet ranges from $1\frac{7}{8}$ inches to about $2\frac{1}{2}$ inches. The IPC ignores the reduction in pipe size from 3 inches to 4 inches on this basis (IPC 420.4).

4. **b.** Public water closets seat are required to be open front to reduce personal contact with seat surfaces (IPC 420.3).

5. **b.** To protect patients and their caregivers, all piping in mental health facilities must be inaccessible to the patients (IPC 405.6).

6. **d.** All water closets are limited to 1.6 gallons per flush for economy of water usage (IPC 420.1, UPC 402.0).

7. **a.** Studies have determined the numbers of fixtures adequate for specific numbers of persons for particular building uses (IPC Table 403.1, UPC Table 4–1).

ANSWERS AND EXPLANATIONS

8. **b. for the IPC; d. for the UPC.** Studies have determined the numbers of fixtures adequate for specific numbers of persons for particular building uses. The range of numbers selected to be on the respective tables reflect the decisions made by the different code-writing bodies after they are presented with the results of studies (IPC Table 403.1, UPC Table 4–1).
9. **a.** The unisex water closet is considered an additional fixture for each men's and women's bathroom (IPC 403.1.1).
10. **c.** If the occupant load is less than 15, it has been determined that demand for the facilities will be low enough to allow individual usage (IPC 403.2).
11. **b.** All condensate must be kept out of a sterilizer. Because condensate is water at a lower temperature than the steam in a sterilizer, it has a greater risk of carrying pathogens (IPC 422.9.2, 422.9.3).
12. **a.** A water closet can be fitted with a hose and a washing arm to facilitate the washing of bedpans into the water closet; a clinical sink is a bedpan washer (IPC 422.3).
13. **b.** A service sink is intended for custodial use to keep a facility clean. If it was used as a clinical sink (bedpan washer), pathogenic waste might come in contact with sanitation materials (IPC 422.7).
14. **b.** The pump on a whirlpool tub must be above the weir of the trap of the tub so all water in the whirlpool piping and the pump drain as the tub is drained (IPC 421.2).
15. **a.** Backflow from the sanitary sewer into an ice-maker could have severe health consequences. Backflow prevention for icemakers is commonly done with an indirect waste (IPC 422.3).
16. **b.** Whirlpool circulation piping must drain after each use to discourage bacteria growth (IPC 421.3).
17. **a.** There is a greater risk of the presence of pathogens where therapy is performed and fixtures must be able to withstand the rinse temperatures required for sanitizing (IPC422.4).
18. **b.** The lavatory must be in the same room with the water closet to allow for private personal hygiene (IPC 405.3.2).
19. **b.** A closet flange must also be attached to the structure to create a solid mounting point for a water closet (IPC 405.4).
20. **b.** Thirty inches provides a minimum "elbow room" dimension that cannot be reduced (IPC 405.3.1, UPC 407.6).
21. **b.** Trough urinals do not have all surfaces rinsed when they are flushed. Therefore, they are banned (IPC 401.2, UPC 405.2).
22. **c.** Copper sheets thinner than 12 oz. per square foot would corrode relatively quickly and would be easy to puncture or tear during shaping and construction (IPC 402.3).
23. **b.** Studies have determined the numbers of fixtures adequate for specific numbers of persons for particular building uses (IPC Table 403.1, UPC Table 4–1).
24. **b.** Travel for customers and store employees is limited by code to facilitate individual comfort (IPC 403.4.2).
25. **b.** All points of contact between a plumbing fixture and the walls or floor must have a watertight seal. This is usually accomplished with silicone caulk, plaster of paris, or grout. Water must be kept out of the area between fixtures and walls or floors since cleaning is nearly impossible and bacteria growth would be likely in those areas (IPC 405.5).

ANSWERS AND EXPLANATIONS

26. **b.** In the past, overflows were required. It has been found that overflow passageways are a haven for bacteria growth because of lack of use. Both the IPC and UPC have now left the decision of whether to have an overflow up to the manufacturer of the fixture (IPC 416.1, UPC 403.0).
27. **b.** "Sinks shall be provided with waste outlets a minimum of 1.5 inches in diameter" (IPC 418.2).
28. **d.** "Water closet seats shall be sized for the water closet bowl type" (IPC 420.3).
29. **a.** A whirlpool pump is an item that will need maintenance and may have to be removed for repair or replacement. Clear access is vital (IPC 421.5).
30. **c.** The volume of water that modern automatic washing machines can pump out, along with suds pressure, demand that a 2-inch drain be installed (IPC 406.3, UPC Table 7–3).

Chapter 5 Review

1. **a.** Flushometer valves are mechanically opened and they do not all have handles if they are automatically controlled. Flushometer valves close when water pressure moves from one side of a diaphragm to the other, causing the diaphragm to flex and close the valve (IPC 425.2).
2. **d.** The flushometer has a closed, pressurized tank inside the water closet tank. Therefore, only the operating elements of the tank are seen and the water is inside the inner, closed tank (IPC 425.3).
3. **a.** Siphonage is essential for all waste and water to be drawn out of a water closet bow (IPC 425.1).
4. **b. for IPC; a. for UPC.** Sometimes, a group of urinals are automatically flushed where usage is very high. The UPC allows one flushing device for a group; the IPC only allows individual control (IPC 425.1.1, UPC 410.2).
5. **b.** Backflow prevention is essential and since transfer valves are commonly used to control handheld, flexible showers, the possibility of the water outlet being submerged is great (IPC 424.8).
6. **d.** Scalding will occur very quickly at temperatures higher than 120°F (IPC Table 424.4).
7. **a.** Because the outlet of a ball cock is submerged in a water closet tank, which is open to the atmosphere, a high risk of contaminated backflow could occur if no means of backflow prevention were present (IPC 425.3.1).
8. **b.** Any material that is acceptable as drainage piping can be used (UPC 404.3).
9. **b.** Water conservation is important with high-usage fixtures such as a public lavatory (IPC 604.4).
10. **b.** The water heater control cannot be trusted to provide 120°F water for a public restroom. A mixing valve is required (UPC 413.1).
11. **b.** All showers are required to have antiscald characteristics (IPC Table 424.4, UPC 415.5).
12. **b.** Hot and cold control positions are part of the plumbing code (UPC 415.0).
13. **a.** All materials that are permitted for above-ground drainage are permitted for use as in a continuous waste (IPC 424.1.2, UPC 404.3).
14. **a.** The only fixtures that do not require a strainer are water closets and siphon action wash down or blowout urinals (UPC 404.1).

ANSWERS AND EXPLANATIONS

15. **a.** The tailpiece is sized to carry the drainage load from a fixture. If the strainer had a smaller open area than the tailpiece, a restriction to the free flow of drainage from the fixture would occur (UPC 404.1).
16. **c.** Many urinals retain small quantities of water in the upper inlet area of the fixture. The requirement for the vacuum breaker in a flushometer to be 6 inches above the inlet of the urinal minimizes the chance of retained water splashing up into a flushometer and causing a contaminated backflow (UPC 409.0).
17. **b.** There is a high risk of backflow with an outlet submerged in a baptistery, so backflow protection is required (UPC 603.4.5).
18. **b.** A handheld shower can be immersed in water and must be provided with backflow protection (IPC 424.2).
19. **b.** Both tubs and showers are limited to a maximum hot water temperature of 120°F (IPC table 424.4, UPC 415.5).
20. **b.** A transfer valve is not required for the operation of a handheld shower; however, if one is used, it must be equipped with a vacuum breaker to prevent backflow (IPC 424.4).
21. **b.** As with all water closets, the amount of gallons per flush for a flushometer valve water closet is limited to 1.6 GPF (IPC 420.1, UPC 402.0).
22. **b.** Installing the wye branch before the trap provides a trap for the dishwasher (IPC 409.3, UPC Figure 4-3c).
23. **a. for the IPC.** Any flushing device is only permitted to flush one fixture (IPC 425.1.1). **d. for the UPC.** A single-flush tank is permitted for multiple urinals (UPC 410.2).
24. **c.** A vacuum breaker prevents back siphonage through a submerged handheld shower (IPC 424.8).
25. **a.** A vacuum breaker will prevent back siphonage through the outlet that is below the water line (UPC 407.8 and Figure 4-14).
26. **a. for the IPC.** A public lavatory must be supplied with tempered water (IPC 416.5). Tempered water is defined as having a temperature between 85°F and 110°F (IPC 202). **c. for the UPC.** Limiting water temperature is important to prevent scalding of those unable to move their hands out of the water stream quickly. At 120°F, third-degree burns will occur after $2\frac{1}{2}$ minutes for children, and 5 minutes for adults (UPC 413.1).
27. **c.** An emergency eyewash is used infrequently and does not require a large quantity of water. Building design should take into account the occasional water and possible chemical contamination that could occur when the fixture is used (IPC 411.2).
28. **c.** The $1\frac{1}{2}$-inch size reflects the same size requirement for the trap for a kitchen sink (IPC 413.2).
29. **a.** Because the outlet nozzle is below the flood-level rim of the bidet, it must be protected from backflow due to the high risk of contamination (IPC 408.2).
30. **a.** Water conservation measures have saved billions of gallons of water. Old unrestricted showerheads could deliver 7 gallons per minute or more depending on the pressure (IPC Table 604.4).

Chapter 6 Review

1. **b.** Actually, the discharge pipe must end within 6 inches of the floor in order to protect anyone nearby from being scalded in the event of a discharge from the relief valve (IPC 504.6, UPC 505.6).

ANSWERS AND EXPLANATIONS

2. **d.** All the provisions of choices **a**, **b**, and **c** can be found in the stated sections of code. Safe installations are paramount, and manufacturers are obligated to provide instructions that ensure a safe water heater installation (IPC 502.1, UPC 501.0).

3. **d.** The passageway can be no longer than 20 feet. The passageway can be as low as 30 inches. The platform must be a minimum of 30 inches wide by 30 inches deep. The only correct answer in the selections is choice **d** because continuous flooring is necessary for the safety of anyone servicing the water heater and to facilitate moving a water heater in and out of the installation location (IPC 502.3, UPC Figure 5–9).

4. **b.** A water heater pan is installed when damage can occur due to tank leakage. It must be drained indirectly, and in addition, the indirect drain must be located where drainage from the termination of the pipe will be noticed so repairs can readily be made (IPC 504.7.1, UPC).

5. **c.** Most T & P relief valves are rated for 150 psi or 125 psi. In the event a vessel has a lower working pressure rating than 150 psi, the relief valve must release at a pressure no higher than the working pressure of the vessel to prevent the possibility of a pressure explosion (IPC 504.5, UPC 505.4, 505.5).

6. **d.** The discharge pipe cannot be reduced in diameter at any point to ensure that the discharge pipe can carry the full water or steam discharge rate of the T & P relief valve (IPC 504.6, UPC 505.6).

7. **a.** Water vapor takes up approximately 1,600 times the volume, or space, as the same amount of liquid water. Water that has been superheated, that is, water under pressure heated above the atmospheric boiling point of 212°F without boiling, releases an incredible amount of energy when a tank rupture causes a sudden change of state in the water to water vapor in the form of steam. The force of steam pressure is the same thing used to catapult airplanes off aircraft carriers, and that steam pressure is under control; a tank rupture is not (IPC 504.4, UPC 505.4, 505.5).

8. **c.** When the ignition portions of a water heater are at least 18 inches above floor level, the chance of igniting vapors that collect at floor level, such as gasoline or solvents, is minimized. Normally, a water heater is installed on an 18-inch high platform to ensure that all the ignition sources on the water heater are at least 18 inches above the floor (IPC 502.1, UPC 508.14).

9. **b.** The T & P relief valve must be installed in the top 6 inches of the tank because the hottest water will be found at the top of the tank due to convection during a heating cycle (hot water rising, cooler water falling) and stratification (hot water staying at the top of the tank and cooler water settling to the bottom) during standby time (IPC 504.4.1, UPC 505.4, 505.5).

10. **c.** The relief valve must be self-closing to prevent damage from excessive release and to return the system to normal operation (IPC 504.4, UPC 505.4, 505.5).

11. **b.** A maximum water temperature of 140°F protects users from excessively hot water while being hot enough to kill bacteria that might grow (IPC 501.6).

ANSWERS AND EXPLANATIONS

12. **b.** A $\frac{3}{4}$-inch inlet size for a water heater drain valve is considered sufficient to minimize plugging of the drain valve with sediment (IPC 501.3 Commentary).
13. **b.** The water temperature for a tub or shower can be no more than 120°F (IPC 424.3, UPC 418.0).
14. **b.** Only water heaters that are installed in a location where flammable vapors are likely to occur must be 18 inches off the floor (IPC 502.1, UPC 508.14).
15. **a.** Automatic temperature control is required for all water heaters because manual controls would not be as dependable (IPC 501.8).
16. **b.** Relief valve discharge piping can be made of any material acceptable as water distribution piping (IPC 504.6, 605.4).
17. **b.** Water heaters can be installed in attics as long as the provisions particular to attic installations are followed (IPC 502.3, UPC 508.4).
18. **a.** A full-port valve will provide unrestricted water flow in the hot water distribution system (IPC 503.1, 606.1).
19. **b.** A valve is required only on the cold water inlet of a water heater. Once the cold water inlet is closed, there is no flow into the hot distribution system (IPC 503.1, 606.1).
20. **b.** In the event a backflow preventer eliminates the ability of heated water to expand back into the cold water supply, a hot water expansion tank must be provided (IPC 607.3, UPC 608.3).
21. **6.** The discharge pipe must end within 6 inches of the floor in order to protect anyone nearby from being scalded in the event of a discharge from the relief valve (IPC 504.6, UPC 505.6).
22. **Damage.** A water heater pan is installed when damage can occur due to tank leakage (IPC 504.7; UPC 508.4).
23. $1\frac{1}{2}$. A depth of $1\frac{1}{2}$ inches will accommodate the $\frac{3}{4}$-inch drain required and some room for water accumulation (IPC 504.7.1).
24. **15.** Efficient insulation saves energy (IPC 505.1).
25. **b.** The drain on a drain pan is not intended to receive the discharge from a relief valve, which could discharge faster than a gravity drain pan drain can accommodate (IPC 504.7 Commentary, UPC 508.5).
26. **b.** Seismic supports are required in earthquake-prone areas for the safety of the population. Falling water heaters can hurt people and broken gas supplies can cause fires (IPC 502.4, UPC 50/8.2).
27. **a.** A water heater will be damaged if it is fired with no water in it (IPC 504.1).
28. **b.** The T & P relief valve must be installed within the top 6 inches of the body of the water heater. This ensures that the temperature probe will sense the actual temperature in the top part of the tank (IPC504.4.1, UPC 505.4, 505.5).
29. **b.** Water would be retained in a relief valve discharge pipe that was piped in an upward direction. This would immediately cause a restriction to flow and could cause corrosion of the relief valve and discharge pipe or freeze. All those conditions would restrict the discharge outlet (IPC 504.6).
30. **b.** The discharge from a relief valve must be visible (IPC 504.6).

Chapter 7 Review

1. **d.** Incorrectly installed gas system piping can result in a failed inspection, appliance malfunction, or an explosion.
2. **b.** Coverage of gas piping systems begins at the point of delivery (UPC 1201.1).

ANSWERS AND EXPLANATIONS

3. **c.** Once a gas piping system has been installed, the pipes must be inspected.
4. **b.** The Authority Having Jurisdiction does the official gas piping inspection.
5. **b.** There are two inspections of a gas piping system.
6. **b.** Gas piping installed in walls but not yet concealed is referred to as rough piping.
7. **d.** The first inspection determines the correct sizing of piping, use of the correct material, and that the code requirements are met.
8. **c.** If gas piping is not installed in accordance with the relevant codes, the Authority Having Jurisdiction can disconnect the gas supply.
9. **c.** Before a gas system can be installed, a sketch or plan must be prepared for approval by the Authority Having Jurisdiction.
10. **c.** Schedule 40 metallic pipe must be used for gas piping (IPC 403.4.2).
11. **c.** Before they are installed in gas piping systems, pipes, valves, and fittings must be free of foreign material (NFPA 54:5.6.1.2).
12. **c.** If gas contains more than 0.3 grains of hydrogen sulfide per 100 standard cubic feet of gas, copper or brass cannot be used for the gas piping (IPC 403.4.3).
13. **b.** Threads are an accepted way of joining gas piping for installations (UPC 1208).
14. **c.** Appliance malfunction can result from insufficient gas pressure in a system (UPC 1215, 1216).
15. **a.** A gas pressure regulator should be installed on a gas piping system when there is too much gas pressure (UPC 1208.7.1; NFPA 54-12:5.8.1).
16. **d.** The inspector is responsible for verifying that overpressure protection is installed (UPC 1208.7.1).
17. **b.** In order to prevent gas buildup, all gas conduits terminating in buildings must not be sealed (UPC 1210.1.6.2).
18. **d.** The termination of gas conduits in buildings must be accessible (UPC 1210.1.6.2).
19. **d.** Gas piping systems must be purged if they are industrial, commercial, or installed in public facilities (UPC 1213.6.1.1–1213.6.1.4).
20. **b.** When gas lines are installed under grade or slab, they must be in a conduit (UPC 1210.1.6.2).

Chapter 8 Review

1. **b.** Sometimes the antifreeze solutions in solar systems can be toxic (IPC 601).
2. **a.** One must be careful to ensure that electrical grounding is present to protect yourself and others (IPC 601.3).
3. **d.** Any building intended for human habitation is required to have a potable water supply (IPC 602.1).
4. **c.** Whether a water supply or drainage system is public or private, any building inhabited by people must be connected to those systems. Our culture has dictated that these are considered basic necessities (IPC 602.1).
5. **b.** A system can only function properly if it is sized to accommodate peak demands (IPC 602.3.2).
6. **d.** A committee has judged 18 inches to be a sufficient height to keep a water pump out of collected water or sewage backup that may occur in a basement (IPC 602.3.5.1).

ANSWERS AND EXPLANATIONS

7. **c.** Even if the demand of a building could be accommodated with a smaller pipe size, $\frac{3}{4}$ inch is the minimum allowed because usually a water service pipe is in service for many decades and the water demand of the building may increase over time. In addition, corrosion could reduce the nominal size of the water service pipe (IPC 603.1).

8. **b.** Generally, leakage from a sewer main would be expected to drain downward. That is why a water service pipe that is on a ledge 12 inches above the sewer main is acceptable (IPC 603.2).

9. **d.** Copper is an acceptable material for a water service pipe; 1 inch is an acceptable size since $\frac{3}{4}$ inch is the minimum size allowed; and a water service is safe lying 5 feet away from a septic tank since it is expected that a leak would be evident in the septic tank or the water service, and if the water service did leak, under normal circumstances water would flow out of the leak rather than be drawn into it (IPC Table 603.2.1).

10. **c.** The provision for 100% sizing to the demand of the building ensures that all fixtures will have sufficient pressure at all times (IPC 604.1).

11. **a.** For sizing purposes, a normal demand rate is 4 GPM for a bathtub faucet (IPC Table 604.3).

12. **a.** The blowout water closet with a flushometer valve has a flow rate of 35 GPM (IPC Table 604.3).

13. **b.** This reduction of the last 30 inches of a supply pipe allows for the flexible supply tubes that are commonly used for fixtures (IPC 604.5).

14. **a.** Quick-closing valves create a high potential for water hammer arrestors due to the immediate stoppage of water flow by these types of valves. Water hammer arrestors provide a "cushion" upon which the water can decelerate (IPC 604.9).

15. **c.** A booster pump must be installed to satisfy the requirement that the distribution system have an adequate flow to all fixtures at all times (IPC 604.7).

16. **a.** Choices **b**, **c**, and **d** all include "air chamber" as one of the accepted types of water hammer arrestor. Air chambers only work for a short time because there is no division between the water and the air in the chamber. The result is that the air in the chamber is absorbed by the water and there is no longer a chamber of compressed air left to act as a water hammer arrestor (IPC 604.9).

17. **c.** Lead must be limited in plumbing components to minimize human exposure to this metal (IPC 605.2).

18. **d.** Many materials are acceptable for a water service pipe. Tolerance for the water it will carry, the pressure of the water, and the corrosivity of the soil in which the pipe is lain determine the types of materials that are suitable for a water service pipe (IPC Table 605.3).

19. **a.** In order to provide full flow to all fixtures, a full port valve is required on the cold water inlet of the water heater (IPC 606.1).

20. **c.** A $1\frac{1}{2}$-inch drain will empty the tank in a reasonable time (IPC 606.5.7)

21. **c.** The working pressure of a water storage tank cannot be exceeded. A pressure relief valve ensures this (IPC 606.5.10).

22. **b.** It would take an unreasonable amount of time for hot water to reach a hot water outlet that is more than 100 feet from the hot water supply source (IPC 607.2).

ANSWERS AND EXPLANATIONS

23. **b.** Insulation diminishes heat loss in a hot water recirculation pipe (IPC 607.2.1).
24. **d.** The main purpose for pressure-reducing valves lies in the name. They are made of bronze and brass and a check valve is integral in a pressure-reducing valve. When a pressure-reducing valve is installed on the water main for a building, it is usually necessary to add a thermal expansion tank to the piping near the water heater to prevent the temperature and pressure relief valve on the water heater from relieving constantly due to thermal expansion (IPC 607.3.1).
25. **c.** At 111°F it takes a minimum of 5 minutes to develop first-degree burns and 7 minutes to begin to develop second- and third-degree burns. The 110°F limit makes bathing relatively safe for everyone (IPC 607.1).
26. **a.** Backflow preventers preclude backflow by not allowing a vacuum condition (siphon) from drawing water that has entered the environment back into the water supply system (IPC 608.1).
27. **b.** Valves equipped with a waste (integral drain) are at risk of allowing water outside the water main from being drawn into the water supply system (IPC 608.7).
28. **b.** Only new pipe or pipe that has been previously used to carry potable water can be used for potable water. If other materials are conducted through a pipe, they can react with the interior coating of the pipe or leave deposits that might contaminate potable water (IPC 608.10).
29. **a.** Hospitals must have a continuous supply of potable water. Therefore, two water services must be installed to ensure, as well as possible, a continuous supply of potable water (IPC 609.2).
30. **b.** The water heater thermostat is intended to control the temperature inside the water heater, not the temperature of the hot water supply. The approximate nature of the temperature range of a water heater thermostat, combined with stratified temperatures within a water heater, does not yield temperature control that is accurate enough to ensure protection of those using the hot water supply (IPC 607.1).
31. **b.** The pressure of a relief valve can be no more than the rated working pressure of the tank in which it is installed (IPC 606.5.10).
32. **a.** The 3-inch overflow will ensure that the tank will not spill over the top (IPC Table 606.5.4).
33. **b.** (IPC Commentary 604.1) P (pressure) = 0.433 × h (height in feet, or head)
    P = 0.433 × h
    P = 0.433 × 10
    P = 4.33 psi
34. **a.** It only makes sense that a water supply pump must be protected from freezing. By making this a provision of the plumbing code, the responsibility falls to the installer (IPC 602.3.5.1).
35. **b.** Only pipe that will readily conduct electricity can be used for electrical grounding (IPC 601.3).

Chapter 9 Review

1. **b.** All sewage must be made to be harmless to the sewage disposal system, humans, and the environment before being introduced into the sewage system (IPC 701.4).
2. **c.** Detergent is not considered to be harmful to the system's operation (IPC 701.5).

ANSWERS AND EXPLANATIONS

3. **d.** By its very nature, steam is at or close to 212°F depending on altitude. Only sewage that is at or below 140°F is permitted in a sewage disposal system (IPC 701.7).

4. **d.** The possibility of sewage leaking onto any food preparation area is a real possibility if drainage pipes are overhead. Therefore, none are allowed over these areas (IPC 701.9).

5. **b.** Vitrified clay tile pipe may be used as a building sewer, but not as a building drain. Proper pipe support and joint integrity would prove difficult in an aboveground application (IPC Table 702.2).

6. **b.** No use of bituminized fiber pipe is permitted in the plumbing codes due to its tendency to disintegrate (IPC Table 702.3).

7. **a.** Lead pipe is quite inert and is applicable to chemical waste (IPC Commentary 702.6).

8. **a.** Solvent welded joints are permitted below ground. Therefore, they can be used in a building sewer (IPC Table 702.3).

9. **b.** Concrete pipe is only permitted to be used outside the building as a building sewer (IPC Table 702.2).

10. **a.** No primer is needed for ABS as it is for PVC. Only ABS cement is needed to connect ABS pipe and fittings in a one-step process (IPC 705.2.2).

11. **b.** Purple primer is required so the Authority Having Jurisdiction can see the purple and is assured that primer has been used in the joint (IPC 705.8.2).

12. **a.** The elastomeric O-ring is the only method approved for joining concrete pipe (IPC 705.6).

13. **d.** All the methods shown are time-tested and approved (IPC 705.5).

14. **b.** Elastomeric couplings are the only approved connecting methods for glass pipe (IPC 705.11).

15. **a.** All-purpose or ABS-PVC cement are not approved for the solvent welding of PVC. Only PVC cement is permitted (IPC 705.14.2).

16. **b.** Schedule 40 PVC is not thick enough for threading. In PVC, only schedule 80 or thicker can be threaded (IPC 705.14.3).

17. **b.** The only approved position for a tee wye in a drainage system is for the branch to be horizontal and the run to be vertical (IPC Table 706.3).

18. **d.** Since a wye has superior flow characteristics, it can be used in all the positions (IPC Table 706.3).

19. **a.** The heel inlet fittings would have backsplash occur in them and the 4 × 3 coupling would be considered a reduction downstream in a drainage pipe, which is not permitted (IPC 706.4).

20. **b.** Cleanouts are not needed at every change in direction in a building drain (IPC 708.3.3).

21. **b.** Manholes are also required at any connection point in the system, such as a wye (IPC 708.3.2).

22. **d.** All of these scenarios warrant the installation of a cleanout (IPC 7.8.3.2–708.3.3).

23. **b.** Smaller drains require cleanouts; 8-inch and larger sewers need manholes for proper access (IPC 708.3.2).

24. **c.** A bathroom group consisting of a water closet (3 DFUs), a lavatory (1 DFU), and a bathtub or shower (2 DFUs) is counted as 5 DFUs. The rationale is that even though the drainage fixture units of the individual fixtures add up to 6 DFUs, all of the fixtures will not be used at the same time, thus justifying the reduction to 5 DFUs (IPC Table 709.1).

ANSWERS AND EXPLANATIONS

25. **a.** Many flat-bottomed fixtures drain at approximately the same rate because gravity affects the drainage water over a flat surface in the same way no matter what fixture it is in (IPC Table 709.1).
26. **b.** The sizing is easily found by reading the IPC Table 710.1(2).
27. **d.** One bathroom group equals 5 DFUs. The laundry tray is 2 DFUs. The kitchen sink also has a DFU value of 2. The water closet is 3 DFUs and the lavatory has 1 DFU (IPC Table 709.1).
28. **d.** A building subdrain can receive the same types of waste as the building drain (IPC 712.1).
29. **d.** Twenty-four inches allows enough room for the sewage pump and collection of sewage (IPC 712.3.2).
30. **a.** Only automatic sewage ejectors are allowed. Any type of manual control would be less dependable (IPC 712.1).
31. **c.** The local vent is only used on bedpan washers with a connection for a local vent. Local vents allow foul-smelling or pathogen-carrying gasses to move out of the building (IPC 713.9).
32. **a.** The backwater valve will ensure there is no backflow of sewage from a shallow sewer. If the manhole is higher than the lowest fixture in a building, the fixture would become the low point and in the event of a sewer blockage, all sewage from the sewer main would flow out of the lowest fixture (IPC 715.1).
33. **a.** Only a full-opening valve can ensure proper flow of the sewer under normal conditions (IPC 715.5).
34. **a.** IPC Table 710.0(1) clearly shows the 6 DFU limit for a 2-inch vertical stack.
35. **a.** Blockages are more likely to occur at the base of a stack, making the base of a stack a reasonable location for a cleanout (IPC 708.3.4).

Chapter 10 Review

1. **a.** It's true that steam cannot be introduced into a sanitary drainage system, but condensate that is at a temperature lower than 140°F can (IPC 803.1 and 802.1.5, UPC 810.0).
2. **d.** Choice a, outdoors, is acceptable for swimming pools that are not in a building or that have access to a building drain (IPC 802.1.4, UPC 813.0).
3. **b.** The potable water supply is not at risk from the release of water from a boiler pressure relief valve. Therefore, an air break is sufficient (IPC 802.1.5, UPC 812.0).
4. **b. for the IPC; a. for the UPC.** The IPC takes into consideration that domestic dishwashing machines have an integral air gap (IPC 802.1.6). The UPC does not make that assumption (UPC 807.4).
5. **c.** An air gap or air break is required to prevent the possibility of sewage backing up into the dishwasher (IPC 802.1.7, UPC 807.4).
6. **b. for the IPC; c. for the UPC.** Longer pipes would allow an area of bacteria growth that would result in odors. The difference in length between the two codes is a result of the committee approach to code provision approval. Neither answer is wrong (IPC 802.2, UPC 803.0).
7. **d. for the IPC.** A vent is never required on an indirect waste (IPC 802.2). **b. for the UPC** (UPC 803.0).

ANSWERS AND EXPLANATIONS

8. **a.** This provision gives backflow protection to the refrigeration equipment while allowing the equipment to be installed on the floor without the necessity of raising all the equipment off the floor to create the free air distance for an air gap (IPC 802.1.2, UPC 801.2.2).

9. **a.** An air break is formed when the drain pipe is not directly connected but the termination of the indirect pipe (in this case the washing machine hose) is below the overflow level of the receiver (IPC 802.1.2, UPC 804.1).

10. **c.** Corrosive waste would compromise the drainage system and present a hazard to the environment (IPC 803.2, UPC 811).

11. **c.** This question only applies to the UPC. No vent is required for an indirect waste in the IPC (UPC 803.0).

12. **a.** Chemical wastes require treatment before they can enter a sanitary drainage system (IPC 803.0, UPC 811).

13. **d.** Indirect waste piping materials must be consistent with standard drainage piping material regulations (IPC 804.1).

14. **a.** A standpipe is treated as a fixture in the codes so it needs an individual trap (IPC 802.4, UPC 804.0).

15. **a.** An air gap is an upgrade from an air break. The discharge outlet of an air gap is above the flood-level rim of the receptor. The discharge outlet from an air break is below the flood-level rim of the receptor. Effluent in the drainage system cannot back up into any portion of an indirect waste that ends with an air gap (IPC 802.2.2).

16. **a.** This guarantees that the dishwasher drainage is trapped (IPC 802.1.6, UPC 807.4).

17. **b. for IPC; d. for UPC.** Any reasonable gap will prevent back-siphonage. The two codes simply took a different approach to be sure that an air gap is created (IPC 802.2.1, UPC 801.1).

18. **b.** Since a floor drain is below floor level, backflow is a possibility. The backwater valve is an extra system element to help prevent a backflow (IPC 802.1.2).

19. **c.** An indirect waste drain is required when food is handled in the fixture (IPC 802.1, UPC 801.2.3).

20. **a. for IPC; b. for UPC.** UPC 813.0 simply requires an indirect waste; it does not specify an air gap. IPC 802.1.4 specifies an air gap.

21. **b.** There is no limitation on the length of a drain from a refrigeration coil (UPC 801.2.1).

22. **b.** An air gap must be used rather than an air break to protect the potable water supply (IPC 802.1.3, UPC 801.4).

23. **a.** The air gap protects the equipment and the instruments being sterilized from any possible contamination through the drain (IPC 802.1, UPC 806.0).

24. **b. for IPC; a. for UPC.** Again, this is simply a difference in approach between the codes (IPC 802.1.5, UPC 807.2).

25. **a. for UPC; b. for IPC.** The IPC insists that any fixture that is not required to be drained indirectly must be directly connected to the sanitary drainage system (IPC 802.1). The UPC allows drinking fountain drains to be connected either directly or indirectly (UPC 809.0).

26. **c.** Any fixture that is not required to drain indirectly must be directly connected to the sanitary drainage system (IPC 802.1, UPC 801.2).

ANSWERS AND EXPLANATIONS

27. d. Choices **a**, **b**, and **c** are provided by an air gap (IPC 801.2, UPC 801.0).

28. b. Drains can be combined and must only be trapped if they exceed the length stated in the respective code (IPC 802.2, UPC 803.0).

29. a. It is not critical that an automatic clothes washer have an air gap since it is not a sterile or food-handling fixture (IPC 802.1, UPC 807.10).

30. a. Innocuous wastes are the only type that will flow in an indirect waste allowing the same materials to be used as those used in a sanitary drainage system (IPC 804.1).

Chapter 11 Review

1. **2 inches.** The larger size accommodates both water flow and air circulation (IPC table 912.3).

2. At least to above the drain outlet of the fixture being served. Generally, the high point is just below the underside of the countertop (IPC 913.2).

3. **c.** A traditional vent is preferred, but the combination system can be used when standard venting is not practical (IPC 912).

4. **b.** As long as the trap seals are maintained, no sewer gas can enter the building through a trap (IPC 901.2).

5. **d.** The minimum trap seal allowed is 2 inches. The 1-inch WC pressure allowed leaves a 1-inch trap seal if it is attained (IPC 901.2).

6. **a.** S-traps are prohibited because they self-siphon (IPC 901.2.1).

7. **a.** A vent flashing must be able to withstand decades of weather and building settlement (IPC 902.2).

8. **a.** This provides for a standard system with similar materials (IPC 902.1).

9. **b.** A closed vent can allow pressures to rise above the 1-inch WC limit (IPC 904.2).

10. **d. for IPC; a. for UPC.** Traditionally vents are installed to pitch upward from the trap to the vent termination. This is what the UPC requires (UPC 905.0). The IPC allows for pitch in either direction as long as no portion of the vent can collect water (IPC 905.2).

11. **a.** Crown vents are not permitted because they can easily be plugged by debris in the drain (IPC 906.3).

12. **c.** S-traps can be self-siphoning and are not allowed for any application (IPC 906.2, UPC Chapter 9 Introduction).

13. **c.** Seven feet is considered high enough to carry rising sewer gas away from the occupants of the roof (IPC 904.1).

14. $1\frac{1}{4}$ **inches.** The smallest vent size allowed is $1\frac{1}{4}$ inches (IPC 916.1, UPC 904.0).

15. **2 inches.** The minimum size vent for a drain can be one-half the diameter of the drain served (IPC 916.1, UPC 904.0).

16. 12 (IPC Table 912.3)

17. **Unlimited.** The operation of a water closet depends on siphonage, therefore no limit is placed on the distance between a water closet and a vent (IPC 906.1, UPC 905.2).

18. **c.** The minimum trap seal allowed is 2 inches. The 1-inch WC pressure allowed leaves a 1-inch trap seal if it is attained (IPC 901.2, UPC Chapter 9 Introduction).

19. 5 (IPC Table 906.1)
20. 6 (IPC Table 906.1)
21. 8 (IPC Table 906.1)
22. 12 (IPC Table 906.1)
23. 16 (IPC Table 906.1)

24. **b.** It is unlikely that a floor drain in a large open building can be located near a vent. That is why a combination drain and vent system is used for a floor drain in such an installation (IPC 912.1, UPC 910).

ANSWERS AND EXPLANATIONS

25. **10 for IPC; 5 for UPC.** The relief vent helps to keep pressures balanced in a drainage and vent system. Less than a 1-inch WC pressure or vacuum will be maintained whether it is installed every 5 branch intervals or every 10 branch intervals (IPC 914.1, UPC 907.0).

26. **b.** Only floor drains, sinks, lavatories, and drinking fountains may be drained with a combination drain and vent system. No food waste grinders or clinical sinks (bedpan washers) may be drained utilizing a combination system (IPC 912.1, UPC 910.0).

27. **a.** Having the vent connection between the last and second-to-last fixture on a horizontal drain provides an individual vent for the last fixture and a connection point in a nonturbulent portion of the horizontal drain (IPC 911.1).

28. **a.** The waste stack is sized according to the total load on the stack. With the stack remaining the same diameter all the way to the top termination, adequate venting in the stack is ensured (IPC 910.4).

29. **a.** Only very small stacks ($1\frac{1}{2}$-inch and 2-inch) have any limitation on the DFU load discharged into a waste stack at one branch interval (IPC Table 910.4).

30. **a.** A stack vent, which is simply a continuation of the waste stack above branch connections to a termination, is needed to keep pressure in the waste stack below 1-inch WC (IPC 910.3).

31. **b.** Only floor drains, sinks, lavatories, and drinking fountains may be drained with a combination drain and vent system (IPC 912.1, UPC 910.0).

32. **a. for IPC (IPC 917.3); b. for UPC**, which has no provision in code allowing them.

33. **a.** The air admittance valve must be 4 inches above the drain it serves (IPC 917.4).

34. **a.** (IPC 917.3) **b. for UPC.** No air admittance valves are permitted under the UPC.

35. **a.** Condensate drains must drain indirectly into a trapped fixture (IPC 901.4).

Chapter 12 Review

1. **b.** (IPC Ch. 10 Purpose, UPC Ch. 10 Introduction)

2. **A trap that is manufactured as part of a fixture and is not removable** (IPC 1002.1).

3. **a.** A grease interceptor can be a fixture trap as long as the vertical distance from the fixture outlet to the inlet of the grease interceptor is not greater than 30 inches and the total developed length of the drain is no longer than 60 inches (IPC 1002.1).

4. **d.** One trap can serve a combination fixture such as a three-compartment sink (IPC 1002.1, UPC 1001.2).

5. **a.** If the height were greater than this, a siphon could develop that might break the trap seal (IPC 1002.1, UPC 1001.4).

6. **c.** Three compartments in a combination fixture is the most that can be served by a single trap (IPC 1002.1, UPC 1001.2).

7. **b.** Some special-use drum traps are permitted, such as those that have screens for hair in hair salons or those designed to recover precious metals as in a dental office (IPC 1002.2, UPC 1004.0).

8. **b.** A minimum of 2 inches ensures a secure trap seal. Some large-diameter (3-inch and 4-inch) traps have a deeper seal to allow for a smooth bend in the larger pipe (IPC 1004.4, UPC 1005.0).

9. **b.** A slip joint is also allowed on the trap outlet (IPC 1002.2, UPC 1003.2).

ANSWERS AND EXPLANATIONS

10. c. A crown vent would be located close to the point where wastewater would be moving in an upward direction out of a trap, thereby throwing debris up into the crown vent, which cannot be washed out (IPC 1002.3, UPC 1004.0).
11. c. Some special-use drum traps are permitted, such as those that have screens for hair in hair salons or those designed to recover precious metals (IPC 1002.3, UPC 1004.0).
12. a. Floor drains that normally have no water flowing into them must be primed (IPC 1002.4, UPC 1007.0).
13. d. If a trap were larger than the drain pipe it drains into, it would be a reduction in drain size, which is prohibited (IPC 1002.5, UPC 1003.3).
14. d. Only when there are unique air pressure issues in a sewer can a building trap be used. It must be specified by the Authority Having Jurisdiction (IPC 1002.6, UPC 1008.0).
15. c. The relief vent on a building trap is a fresh air intake on the inlet side (IPC 1002.6, UPC 1008.0).
16. a. If a trap were not level, the trap seal would be less than the required 2 inches (IPC 1002.7).
17. a. A vitrified clay trap requires the extra support of being encased in concrete when used underground (IPC 1002.9).
18. b. A grease interceptor must be installed when the discharge contains grease (IPC 1003.1, UPC 1009.1).
19. a. Grease interceptor manufacturers use accepted practices to size their interceptors (IPC 1003.2).
20. d. Fixtures that have grease-laden waste require a grease interceptor (IPC 1003.3.1, UPC 1009.0).
21. c. Food waste would quickly fill a grease interceptor and render it useless (IPC 1003.3.2).
22. b. Individual dwellings do not require a grease trap (IPC 1003.3.3).
23. b. The flow control must be installed on the inlet to prevent overburdening the grease interceptor (IPC 1003.3.4.2).
24. d. If a facility has oil-bearing waste, that waste must first pass through an interceptor (IPC 1003.4, UPC 1009.0).
25. a. Separators send the separated material to a separate approved receptacle (IPC 1003.4.1).
26. c. The capacity of a grease interceptor determines the GPM rate that can flow through the interceptor (IPC Table 1003.4.1).
27. a. The UPC uses a slightly different approach to grease interceptor sizing than the IPC, using GPM flow and DFU value rather than pounds of grease and GPM (UPC Table 10–2).
28. a. Drum-type traps are approved as interceptors for solids to prevent these solids from clogging the drain (IPC 1002.3, UPC 1010.0).
29. b. Only when there are unique air pressure issues in a sewer can a building trap be used. It must be specified by the Authority Having Jurisdiction (IPC 1002.1 and 1002.6, UPC 1008.0).
30. b. A water closet has an integral trap and cannot have another trap installed in the soil pipe as this would be a double trap, which is prohibited (IPC 1002.1, UPC 1001.0).

Chapter 13 Review

1. a. If a floor drain was connected to the storm drainage, it would be possible for sanitary waste to enter the storm drainage system, which does not get treated (IPC 1104.3).

ANSWERS AND EXPLANATIONS

2. **d.** Storm water is relatively clean water and lands on streets and lawns anyway. Secondary roof drains purposely discharge onto open surfaces so the drainage will be noticed (IPC 1101.2).

3. **a.** A sanitary drainage system would become overloaded during a storm (IPC 1101.3).

4. **a.** The potential for leakage and building damage is at least as high in a conductor as it is in a sanitary drainage system (IPC 1101.4, UPC 1109.2).

5. **a.** As with sanitary drainage systems, conductors must not be reduced in size or restriction to flow will occur (IPC 1101.5).

6. **a.** Water flows by gravity in a storm drainage system just as it does in a sanitary drainage system (IPC 1101.6).

7. **c.** The roof must be sturdy enough to support the full potential depth of water that could occur if the primary roof drains do not function and the secondary roof drain system must take over (IPC 1101.7).

8. **c.** Water and debris flow in a similar manner in a storm drainage system and in a sanitary drainage system (IPC 1101.8).

9. **a.** Once again, water and debris flow in a similar manner in a storm drainage system and in a sanitary drainage system (IPC 1102.1).

10. Only when it connects to a combination sewer. This keeps sewer gas from entering the storm drainage system (IPC 1103.1).

11. **a.** No reduction or increase is allowed. A reduction would restrict flow and an increase would only slow the velocity of flow in the trap, creating the opportunity for debris to settle in the trap (IPC 1103.3, UPC 1103.3).

12. **b.** Only storm water may drain into conductors (IPC 1104.1).

13. **a.** This height has been adopted to keep most debris out of the roof drains (IPC 1105.1, UPC 1105.2).

14. **a.** The increase in size allows for a collection area (IPC 1105.2, UPC 1105.3).

15. **b.** This is the same slope used in a sanitary drainage system (IPC 1106.3).

16. **b.** Rain falling at an angle can strike a wall above the roof and flow onto the roof. Only half of the wall area is considered in anticipation of the rain falling at an angle no greater than 45 degrees (IPC 1106.4, UPC 1106.4).

17. **c. for IPC; b. for UPC.** For the IPC, the secondary roof drain system must be independent of the primary system to ensure its operation if the primary system fails (IPC 1107.2). The UPC allows the secondary roof drain to be connected at a point below any horizontal portion of the primary drain below the roof. However, when this connection is made, the conductor must be sized for double the rainfall rate (UPC 1101.11.2.2.2).

18. **a.** The sizing strategy is the same because the rainfall rate for the location remains the same, and the flow of the primary system cannot be included because a complete failure of the primary system must be assumed (IPC 1107.3).

19. **d.** Any of the three means can be used to allow for an entrance point for water in the soil (IPC 1111.1).

20. **a.** Groundwater does not generate sewer gas so it does not require a gastight cover (IPC 1111.1).

21. **b.** A smaller discharge pipe would cause a restriction that would reduce the output capacity of the pump (IPC 1113.1.4).

ANSWERS AND EXPLANATIONS

22. **a.** If there were no check valve, some of the same water contained in the discharge pipe would return to the sump through the pump and would have to be pumped out again. This would reduce the gallons-per-hour delivery of the system (IPC 1113.1.4).
23. **a. for IPC; d. for UPC.** The dimension was chosen by the individual code writers. The reason for a width dimension being chosen is to allow room for a sump pump to be installed in the sump. Common sump pumps will fit into either size sump pit (IPC 1113.1.2, UPC 1101.5.3).
24. **c. for IPC; b. for UPC.** A depth is specified to create an area for water to collect to prevent short-cycling of the sump pump (IPC 1113.1.2, UPC 1101.5.3).
25. **d.** Rainfall data can be obtained from any approved local weather data source (IPC 1106.1, UPC 1106.0).
26. **a.** When a floor drain is used as an area drain in an application such as this, it is not considered to be a floor drain (IPC 1104.3 Commentary).
27. **a.** The same materials that are permitted in sanitary drainage systems are approved for use in a storm drainage system (IPC 1102.1, UPC 1105.1.2).
28. **c. for IPC; b. for UPC.** The IPC specifies a 4-inch minimum and the UPC specifies a 3-inch minimum size for subdrains (IPC 1111.1, UPC 1101.5.1).
29. **a.** Both codes require a backwater valve to be installed in a storm drain that could be subject to backflow. Imagine if the storm sewer termination became inundated. Water at the level of the flood area would flow back into the building up to the level of the inundated area without a backwater valve (IPC 1111.1, UPC 1101.5.5).
30. **a.** If ponding occurs on a roof, then roof drains are required. If roof drains are required, then a secondary roof drain system is required as a backup (IPC 1107.1, UPC 1101.11.2.2).

Chapter 14 Review

1. **OXY, MED, OXY/MED, ACR/OXY, ACR/MED.** These markings are essential so the inspector(s) of a system are assured that approved materials have been used in the installation (IPC 1201.1). The UPC does not require these markings. Regular L or K copper is acceptable as long as it has been cleaned and sealed to protect it from contamination.
2. **Copper.** Copper is resistant to physical damage and will not react with the gasses conducted through it (IPC 1202.1, UPC 1316.3).
3. **Nitrogen.** Flowing nitrogen through a tube that is being brazed forces all the air out of the tube. Oxidation is prevented from forming on the inside of the pipe since there is no oxygen present (IPC 1202.1, UPC 1319.0).
4. **$1\frac{1}{2}$.** A successful pressure test well above the system's operating pressure ensures a leak-proof system (IPC 1203.1, UPC 1327.7).
5. **Redundancy.** If one pump fails, the other will operate to ensure continuous functioning of the system (UPC 1326.0).
6. **Third-party certification.** If a medical gas system fails, individuals who use the gas and who handle the gas can be put at high risk. Third-party certification ensures that a system is correctly functioning when it is put into service (UPC 1328.0).

ANSWERS AND EXPLANATIONS

7. **Nitrogen.** The presence of a mostly nitrogen atmosphere around the fittings prevents oxidation (IPC 1202.1, UPC 1317.2).
8. **All fixtures and fittings must be vandal-proof.** This is often accomplished by putting all the fixtures inside a locked cabinet (UPC 1304.0).
9. **Stainless steel.** Stainless steel is corrosion resistant and resistant to physical damage (IPC 1202.1, UPC 1316.3).
10. **Ball.** Ball valves are dependable and because of their quarter-turn operation from the on to off positions, it is obvious to operators whether the valve is on or off (UPC 1320.0).
11. **a.** Brazing is the acceptable method of making copper-to-copper joints due to the strength of the joints (IPC 1202.1, UPC 1319.0).
12. **c.** The IPC refers medical vacuum exhausts to the International Mechanical Code. The UPC requires that the termination be outdoors, turned down, and screened. In addition, the termination must be at least 10 feet away from any opening in the building to include doors, windows, or air intakes, and the termination must be at a different level than any air intake (IPC 1202.1 exception 2, UPC 1326.2).
13. **b.** A vacuum system by its nature becomes contaminated with all varieties of bodily liquids. It can never be converted for the delivery of any medical gas (IPC 1202.1).
14. **b.** All medical gas systems must be designed by an engineer, architect, or other certified designer. The plans, after approval by the Authority Having Jurisdiction, must be strictly adhered to by the installer. Any unplanned change must be approved in writing before it can be done (UPC 1312.1).
15. **b.** No medical gas system can be installed in an elevator shaft (UPC 13.18.2).

Chapter 15 Review

1. **d.** Because subsurface irrigation water will not be seen, there is no need to color it (IPC C103.5).
2. **a.** Coloring identifies the recycled water as nonpotable (IPC C102.4).
3. **c.** Gray water is drained from showers, tubs, laundry trays, and clothes washers. No black water (water containing human waste) or kitchen waste can be included (IPC C101.2).
4. **b.** Filtering must occur before gray water enters the reservoir (IPC C101.11).
5. **b.** The potable water supply must be protected, which demands a reduced pressure principle backflow preventer (IPC C102.3).
6. **c.** The retention time for an irrigation system is limited to 24 hours since no disinfection is required for this application (IPC C103.1).
7. **a.** This is still a gravity drainage system just like one that terminates in a sanitary drainage system. The same rules apply as for standard drainage materials (IPC C101.5).
8. **d.** Each individual uses 25 gallons per day for showers, bathtubs, and lavatories, and another 15 gallons per day for laundry (IPC C103.6).
9. **d.** Black letters on a yellow background is the color code for nonpotable water (IPC C103.1.1 and Table 608.8).
10. **a.** Gray water must be disinfected when used for fixture flushing. (IPC C102.2).
11. **b.** The smallest allowable reservoir tank is 50 gallons (IPC C102.1).
12. **d.** Because the gray water for a fixture-flushing gray-water system is disinfected, it can retain water for up to 72 hours (IPC C102.1).

ANSWERS AND EXPLANATIONS

13. **a.** An indirect drain allows the reservoir to be safely drained with no possibility of a backflow from the sanitary drainage system (IPC C101.13).
14. **d.** There are numerous approved methods of disinfection (IPC C102.2).
15. **a.** The overflow pipe on a gray-water reservoir must be at least the same size as the inflow pipe (IPC C101.12).
16. **a.** When the overflow is connected to the sanitary drainage system indirectly, there is no chance for the sanitary drainage system to backflow into the reservoir (IPC C101.12).
17. **a.** A full-opening valve must be installed before the gray water enters the filter before the reservoir to accommodate necessary servicing of filters and other equipment (IPC C101.11.1).
18. **c.** The reservoir must be a gastight container. Access openings must be provided for inspection and cleaning of the container (IPC C101.10).
19. **c.** (IPC C03.6 and Table C103.6)
    $C = A \times B \times 3$; where A is the number of occupants; B represents flow per person; and 3 days is the maximum retention time.
    $C = 4 \times 40 \times 3$
    $C = 480$ gallons
20. **a.** No makeup water is necessary because an irrigation system does not supply a fixture that requires a certain amount of water (IPC C103.3).

Chapter 16 Review

1. **a.** Due to very low water consumption, the water supply fixture unit is only 1 (IPC G101.2.4).
2. **d.** Testing to 19 inches of mercury ensures a vacuum-tight system (IPC G101.3).
3. **a.** A discharge valve is opened to flush along with a water valve to rinse and refill the bowl (IPC G101.2.1).
4. **c.** The manufacturer will specify or suggest materials that must be used. For the vacuum drainage pipe, any material that is on the manufacturer's list and is also on the IPC list for drainage piping found in Table 702.1 may be used (IPC G101.2.6).
5. **a.** Variation in vacuum will be experienced during normal operations such as flushing, vacuum pumping, and emptying of the receiver tank by a sewage ejector pump (IPC G101.3).
6. **a.** Although the IPC does not specify sizes, the engineered design of these systems almost always specifies smaller drainage pipe sizes in order to maintain velocity through the system (IPC G101.2.1).
7. **c.** Only the water closet has a special design. All other fixtures drain to a receiver with an interface valve that automatically empties the receiver (IPC G101.2.4).
8. **b.** Sources of fresh water are facing increasing demand. There is high interest in all water-saving technologies (IPC G101.2.4).
9. **a.** All standard gravity-type fixtures are required to have traps and cleanouts according to the drainage provisions of the IPC (IPC G101.2.5).
10. **b.** Due to the specialized nature of these systems, the authorities must be given an opportunity to review, research, and approve them before installation begins (UPC G101.2.1).

Posttest 1

1. **b.** DWV copper tubing can only be used for drainage, waste, and vent applications, due to its relatively thin wall thickness (IPC Table 702.1, UPC 701.1.4).

ANSWERS AND EXPLANATIONS

2. **d.** The abbreviation NPT, commonly expressed as National Pipe Thread, Nominal Pipe Thread, and National Pipe Thread Tapered, is actually a code. The N stands for USA Standard, the P means pipe, as expected, and the T signifies that the threads are tapered (IPC 605.10.3).

3. **a.** For test pressures between 10 and 100 psig, a gauge with 1-psi increments must be used (IPC 312.1.1, UPC 319.2).

4. **c.** Flammable materials that could cause a fire or explosion, such as naphtha, are not approved for a building drainage system (IPC 302.1, UPC 303.0).

5. **c.** A plumbing fixture is defined as a receptacle that discharges liquid-borne solid waste to a system (IPC Commentary for 2012).

6. **a.** The limestone in concrete is alkaline, so it is corrosive to pipes (IPC 305.1, UPC 313.2).

7. **b.** To determine the area of a circle, you multiply π (3.14) by the radius squared (2.25 × 2.25 = 5.0625), giving you 3.14 × 5.0625, which equals 15.896 square inches (math for plumbers, not code specific).

8. **d.** The pipe will drop $\frac{1}{4}$ inch for every foot along the 20 feet, so it will drop a total of 5 inches (0.25 × 20). Because the pipe is already $14\frac{1}{2}$ inches down, you add the drop to the upstream distance (14.5 + 5), which equals 19.5 or $19\frac{1}{2}$ inches (math for plumbers, not code specific).

9. **b.** The formula for volume in gallons is $\pi r^2 h$ divided by 231 cubic inches (1 gallon of water has 231 cubic inches in it). Divide 275 feet by 12 inches per foot to get h: 3,300 inches. We therefore have 3.14 × (0.5 inch × 0.5 inch) × 3,300 inches, which equals 2,590.5 cubic inches. When you divide 2,590.5 cubic inches by 231 cubic inches, you get 11.214 gallons (math for plumbers, not code specific).

10. **a.** Condensate is water at a lower temperature than the steam in a sterilizer, so it poses a greater risk of carrying pathogens. Therefore, all condensate must be kept out of a sterilizer (IPC 422.9.2, 422.9.3).

11. **c.** Studies have determined certain numbers of fixtures for certain numbers of persons to be suitable for particular building uses (IPC Table 403.1, UPC Table 4–1).

12. **b.** All materials permitted for aboveground drainage, such as PVC, are also permitted for use as a continuous waste (IPC 424.1.2, UPC 404.3).

13. **c.** If the fixture backs up, 6 inches is high enough for the vent not to be flooded with the backup (IPC 905.4).

14. **a.** Hot and cold water piping can be bundled together as long as the hot water piping is insulated (IPC 308.9).

15. **d.** Water vapor takes up approximately 1,600 times the volume, or space, that the same amount of liquid water takes up (IPC 504.4, UPC 505.4, 505.5).

16. **d.** It is generally assumed that service sinks in small occupancy retail stores are rarely, if ever, used (IPC 403.1).

17. **a.** In order to protect anyone nearby from being scalded in the event of a discharge from the relief valve, the discharge pipe must end within 6 inches of the floor (IPC 504.6, UPC 505.6).

18. **b.** A drain from a swimming pool filter backwash can also be drained outdoors or through an air break into a receptor that is trapped and that is directly connected to a building drain (IPC 802.1.4, UPC 813.0).

19. **b.** To prevent the possibility of sewage backing up into the dishwasher, an air gap or air break is required (IPC 802.1.7, UPC 807.4).

ANSWERS AND EXPLANATIONS

20. **c.** Gray water is produced by bathtubs, showers, lavatories, clothes washers, and laundry tubs (UPC 209, IPC Chapter 13 Code 2012).
21. **a.** In order to guarantee that the dishwasher drainage is trapped, the drain hose must be connected to a wye tailpiece on the inlet side of a trap (IPC 802.1.6, UPC 807.4).
22. **a.** Equipment and fixtures utilized for the storage, preparation, and handling of food must discharge through an indirect waste pipe by means of an air gap (IPC 802.1, UPC 801.2).
23. **b.** The use of gray water promotes significant savings, which is particularly important in areas of the United States in which there is a shortage of potable water (IPC Chapter 13 Code 2012).
24. **a.** The minimum trap seal allowed is 2 inches. The 1-inch WC pressure allowed leaves a 1-inch trap seal if it is attained (IPC 901.2, UPC Chapter 9 Introduction).
25. **a.** This situation is considered highly unsanitary (IPC 403.3.6).
26. **d.** Slip joints are allowed on trap inlets and trap outlets, as well as within the trap seal (IPC 1002.2, UPC 1003.2).
27. **c.** When installing a vent, it should not be located close to the point where wastewater would be moving upward out of a trap where it will enter the vent and clog it (IPC 1002.3, UPC 1004.0).
28. **a.** The extra available space is in compliance with the Americans with Disabilities Act (IPC 405.3.1).
29. **d.** A building trap can only be used when there are unique air pressure issues in a sewer, so the use of building traps is dictated by the local jurisdiction (IPC 1002.6, UPC 1008.0).
30. **b.** An oil separator is needed only when the facility has oil-bearing waste (IPC 1003.4, UPC 1009.0).
31. **c.** No reduction or increase in size is allowed when working with a storm water trap (IPC 1103.3, UPC 1103.3).
32. **c.** The most important duty of a plumber is to protect people's health. Preventing backflow is crucial for sanitary conditions (IPC 405.1).
33. **d.** You can obtain rainfall data from any approved local weather source (IPC 1106.1, UPC 1106.0).
34. **b.** Nitrogen is used both to prevent oxidation inside the pipe and to force all other air out of the tube (IPC 1202.1, UPC 1319.0).
35. **c.** Approved copper tubing or stainless steel piping can be used in a medical vacuum system because they are corrosion resistant and resistant to physical damage (UPC 1302.0).
36. **d.** Termination must be outdoors, turned down, screened, at least 10 feet away from any opening in the building, and at a different level than any air intake (IPC 1202.1 exception 2, UPC 1326.2).
37. **c.** Any building inhabited by people must be connected to a public or private water supply and drainage system (IPC 602.1).
38. **d.** It has been determined that 15 inches is adequate for the average person (IPC 405.3.1, UPC 402.5).
39. **b.** The water distribution system design criteria for a bathtub is 4 GPM (IPC Table 604.3).
40. **a.** In this situation, you would need to install a booster pump to satisfy the requirement that the distribution system must have an adequate flow to all fixtures at all times (IPC 604.7).
41. **b.** To be able to empty the tank in a reasonable time, a $1\frac{1}{2}$-inch drainpipe is required (IPC 606.5.7).
42. **b.** Because a hospital must have a continuous supply of potable water, at least two water services must be installed in order to ensure a continuous supply of potable water (IPC 609.2).

ANSWERS AND EXPLANATIONS

43. b. A waste outlet is not required (IPC 411.2, UPC 416.5).

44. a. Exposed drainage piping should never be installed over a food preparation or storage area due to the danger of leaks (IPC 701.9).

45. b. Because solvent welded joints are permitted below ground, it is appropriate to use them in a building sewer (IPC Table 702.3).

46. d. The only approved method for connecting glass pipe is the use of elastomeric couplings (IPC 705.11).

47. b. Backsplash would occur in a heal inlet fitting in this situation. A side-inlet quarter bend should be installed instead (IPC 706.4).

48. b. Although a bathroom group with a 1.6 GPF water closet has a combined mathematical DFU of 6, it is actually considered to have a DFU of 5 because the fixtures will not all be used simultaneously (IPC Table 709.1).

49. d. Any type of waste that the building drain can receive can also be received by the building subdrain (IPC 721.1).

50. b. 30 inches × 30 inches is the minimum size for a shower stall for use by the average person (IPC 417.4).

51. b. A pipe with a diameter between $2\frac{1}{2}$ inches and 6 inches must be identified with a color field that is 8 inches long (IPC Table 608.8.3).

52. a. The water supply fixture unit is only 1, due to very low water consumption (IPC G101.2.4).

53. c. Vacuum pumps maintain an operating range of 16 to 20 inches of mercury vacuum, and are subjected to a vacuum test of 19 inches of mercury (IPC G101.3).

54. c. To protect underground water pipes, they must be buried at least 12 inches below grade (IPC 305.6).

55. a. In general, plumbing systems cannot be located in elevator shafts, but there is an exception for drains, sumps, and sump pumps, which may be needed to remove water from the elevator shaft in the event of flooding (IPC 301.6).

56. c. Rodents such as rats can squeeze into small openings. Therefore, to prevent them from getting into the water system and spreading disease, the openings on strainer plates on drain inlets cannot be any larger than $\frac{1}{2}$ inch in the smallest dimension (IPC 304.2, UPC 313.12.1).

57. c. Copper tubing that is $1\frac{1}{2}$ inches or smaller being hung horizontally must be supported at least every 6 feet. (IPC Table 308.5, UPC Figure 3–29).

58. a. Adapting plastic pipe to other materials requires the use of proper adapters (IPC 605.23.2, UPC 316.2.3).

59. a. In drain pipes that are 4 inches or larger in diameter, the force of the flow of drainage can push apart couplings unless they are sufficiently restrained (IPC 308.7.1).

60. d. If openings other than the highest opening are not tightly closed, they may allow water out, which will prevent you from locating leaks (IPC 312.2, UPC 712.2).

61. a. The minimum pressure for testing water supply piping is the actual pressure that the pipes will be subjected to during normal use (IPC 312.5, UPC 609.4).

62. c. The purpose of testing the sewer is to determine how well water will flow from the drains in the building, into the building sewer system, and then out into the public sewer (IPC 312.7, UPC 723.0).

63. b. All public water closet seats must be open front in order to reduce personal contact with seat surfaces (IPC 420.3).

ANSWERS AND EXPLANATIONS

64. b. In this scenario, the IPC requires 1 water closet per 50 people, for a total of 5 water closets. The UPC requires 2 water closets for the first 50 people, and then 1 water closet for each additional 50 people, for a total of 6 water closets (IPC Table 403.1, UPC Table 4–1).

65. c. The circulation piping in a whirlpool tub must drain after each use in order to discourage the growth of bacteria (IPC 422.4).

66. b. The water temperature used in therapy can be 180°F or higher (IPC 422.2).

67. c. To create a solid mounting point for the water closet, the closet flange must be attached to both the base and the structure below (IPC 405.4).

68. b. In a covered mall, toilet facilities must be located no further than one floor above or below the space for which such facilities are required (IPC 403.4.2).

69. b. 12 inches is large enough to remove, replace, or repair the pump. Note, however, that if the pump is located more than 2 feet from the access opening, then the opening should not be less than 18 inches square (IPC 421.5; UPC 409.6).

70. b. Because a whirlpool bath pump will need maintenance, and may need to be removed for repair or replacement, it must be accessible (IPC 421.5).

71. d. Specialty fixtures can be made of soapstone, chemical stoneware, corrosion-resistant steel, or any other material specified by the code (IPC 402.2, UPC 406.3).

72. c. In order to ensure sufficient room to clean the fixtures, water closets cannot be placed any closer than 30 inches from center to center (IPC 405.3.1, UPC 407.6).

73. b. These devices must be accessible, as they have to be periodically inspected, serviced, maintained, and repaired (IPC 422.5).

74. b. 110°F is hot enough for sanitary conditions, but not so hot as to cause scalding (IPC 408.3, UPC 410.3).

75. b. Six inches is low enough so as not to cause injury to a person but still high enough not to allow the discharge to hit the floor and back up into the discharge line (IPC 504.6).

76. b. Both the IPC and UPC require that a floor drain be at least 2 inches in diameter for standard use (IPC 412.3, UPC Table 7–3).

77. d. The tank must be installed in a galvanized pan with a 1-inch drain that terminates outside of the building (IPC 504.7, UPC 508.4).

78. a. A clinical sink must include a flushing rim to cleanse the interior surface (IPC 422.6).

79. c. The flushometer has a single closed tank inside the water closet tank. The water is inside that inner tank, so it cannot be seen. Only the operating elements of the tank, including the actuator level, are visible (IPC 425.3).

80. b. A mixing valve is used to mix the hot and cold water to control the water temperature and prevent burns (UPC 413.1).

81. a. The plumbing code requires the "hot on the left, cold on the right" standard for both uniformity and safety (UPC 415.0).

82. c. The strainer and the tailpiece must be equally sized in order to allow the free flow of drainage (UPC 404.1).

83. a. The GPF limit on all water closets is 1.6 GPF (IPC 420.1, UPC 402.0).

84. b. As part of the modern water conservation movement, showerheads are now limited to a flow rate of 2.5 GPM (IPC Table 604.4).

ANSWERS AND EXPLANATIONS

85. d. The distance of 5 feet ensures that there is no cross contamination. The bottom of the water service pipe, however, must not be less than 12 inches above the building sewer if the distance between the two pipes is less than 5 feet (IPC 603.2).

86. b. A water heater drain pan must be at least $1\frac{1}{2}$ inches deep (IPC 504.7.1).

87. b. The slope of pipes sized 3 inches to 6 inches is $\frac{1}{8}$ inch per foot because this allows the proper drainage of solid and liquid wastes (IPC 704.1, Table 704.1).

88. d. A flue collar is a portion of an appliance designed for the attachment of a draft hood, vent connector, or venting system (UPC 502.8).

89. b. Potable water is used as the heat transfer fluid in a direct connection solar heating system (IPC 601.2).

90. b. A blowout water closet with a flushometer valve actually has the highest flow rate for fixtures—35 GPM (IPC Table 604.3).

91. c. A full port valve is required on the cold inlet of a water heater in order to provide full flow to all fixtures (IPC 606.1).

92. c. A cleanout must be installed on a horizontal drain every 100 feet, otherwise it would be nearly impossible to clear if there is a blockage since the drain cleaning machine has only 100 feet of cable (IPC 708.3.1).

93. b. Because a frozen water pump can easily crack or break, water pumps must be protected from freezing regardless of when they are expected to be used (IPC 602.3.5).

94. c. An individual water supply can come from a drilled well, driven well, dug well, bored well, spring, stream, or cistern. Surface bodies of water, such as lakes and ponds, cannot be used as an individual water supply unless they are properly treated (IPC 602.3.1).

95. a. The septic or treatment tank must be at least 25 feet from the cistern to prevent contamination (IPC Figure 603.2.1).

96. b. The minimum flow rate for a residential sink is 2.5 GPM, and the minimum flow pressure is 8 psi (IPC Table 604.3).

97. b. The building's water distribution system must be designed for the minimum pressure available in order to ensure that the water needs of each fixture are always met (IPC 604.6).

98. c. Backflow must be avoided because it can lead to the contamination of the potable water supply (IPC 608.1, UPC 204.0).

99. d. It must not rise within 2 inches of the invert of the gravity drain inlet into the pump, as the effluent may start backing up into the drain line (IPC 712.3.4).

100. d. A solvent-cement joint needs 24 hours to fully set before you apply working pressure; otherwise, you could damage the connection (IPC 705.2.2).

101. c. for the IPC; d. for the UPC. The IPC groups the water closet, lavatory, and bathtub/shower together as a bathroom group with a DFU of 5. It also states that the clothes washer and kitchen sink each have a DFU of 2. When you add 5 + 2 + 2, you get 9. The UPC identifies each fixture individually, with a DFU of 3 for the clothes washer, 2 for the kitchen sink, 3 for the water closet, 1 for the lavatory, and 2 for the bathtub/shower. When you add 3 + 2 + 3 + 1 + 2, you get 11 (IPC Table 709.1, UPC Table 7–3).

102. b. The discharge opening should not be less than $1\frac{1}{4}$ inches or else it may get clogged and need servicing (IPC 712.4.2).

ANSWERS AND EXPLANATIONS

103. c. for the IPC; d. for the UPC. The IPC and UPC differ on this issue. According to the IPC, a 2-inch trap has a drainage fixture unit value of 3. For the UPC, it has a drainage fixture unit alue of 4 (IPC Table 709.2, UPC 702.0).

104. d. An air break is "a piping arrangement in which a drain from a fixture, appliance, or device discharges indirectly into another fixture, receptacle, or interceptor at a point below the flood-level rim and above the trap seal" (IPC 802.4, UPC 804.1).

105. a. The UPC states that all "indirect waste vent pipes shall terminate through the roof separate from any other venting" (UPC 803.0).

106. b. A backwater valve is necessary because a floor drain is below floor level, meaning backflow is a significant possibility (IPC 802.1.2).

107. b. This type of connection is prohibited because cross contamination can occur (IPC 713.8).

108. c. The high point is generally just below the underside of the countertop (IPC 913.2).

109. a. Self-siphoning is a problem with S-traps (IPC 901.2.1).

110. a. for the IPC; b. for the UPC. The IPC allows for pitch in either direction, as long as no portion of the vent can collect water. The UPC, however, requires that the vent be installed to pitch upward from the trap to the vent termination. (IPC 905.2, UPC 905.0).

111. c. Crown vents are vents within 2 pipe diameters of the trap weir. They are not permitted because they can easily become plugged with debris in the drain (IPC 906.3).

112. b. According to IPC Table 906.1, a trap weir that is 11 inches in diameter can be 5 feet from a vent (IPC Table 906.1).

113. b. for the IPC; a. for the UPC. The IPC and UPC differ on this matter. The IPC requires relief vents every 10 branch intervals. The UPC requires relief vents every 5 branch intervals (IPC 914.1, UPC 907.0).

114. b. The stack must remain the same diameter all the way to the top termination in order ensure adequate venting (IPC 910.4).

115. d. Because the average person's height is less than 6 feet, the lowest height should be 7 feet (IPC 903.1).

116. d. Both the IPC and the UPC state that a wet vent can only connect fixtures on the same floor. Although the UPC does not specify where those fixtures would be located, the IPC limits wet venting to bathrooms on the same floor (IPC 909.1, UPC 908.0).

117. c. Any distance greater than 3 feet will prevent the sewer gases from entering the air intake of the building (IPC 903.5).

118. c. Approved flashing is used to seal holes in roofs (IPC 903.3).

119. b. A food waste grinder can only connect to a grease interceptor if it first drains to a solids interceptor (IPC 1003.3.2).

120. b. At 6 inches above the flood level rim, backup can be avoided from the highest fixture (IPC 913.3).

121. d. The UPC specifies that a grease interceptor with a flow capacity of 150 GPM can handle 216 DFUs (UPC Table 10–2).

122. a. Both the IPC and the UPC require that fixture traps be self-cleaning (IPC 1002.2, UPC 1003.1).

123. d. Any type of business where oily and flammable liquid wastes are produced must have an oil separator installed (IPC 1003.4, UPC 1017.1).

124. b. Storm runoff can also run to lawns and streets, where it lands anyway (IPC 1101.2).

ANSWERS AND EXPLANATIONS

125. d. Because water and debris flow in a similar manner in storm systems and sanitary systems, cleanout requirements are the same for both (IPC 1101.8).

126. a. The maximum slope for a horizontal storm drain is the same as for a sanitary drainage system—$\frac{1}{2}$ inch per foot (IPC 1106.3).

127. b. Because groundwater does not generate sewer gas, a gastight cover is not required (IPC 1111.1).

128. d. If a smaller discharge pipe were used, it would cause a restriction that would reduce the output capacity of the pump (IPC 1113.1.4).

129. b. for the IPC; a. for the UPC. The IPC and UPC differ on this issue. The IPC requires a minimum diameter of 4 inches. The UPC requires a minimum diameter of 3 inches (IPC 1111.1, UPC 1101.5.1).

130. b. Annealed copper is also known as soft copper; drawn copper is also known as hard copper (IPC 1102.4 Commentary).

131. b. The vent stack starts at the base of the soil stack and joins the stack vent just before it goes through the roof (IPC 913.3).

132. b. Medical gas piping is not allowed in kitchens or elevator shafts (UPC 1318.2).

133. d. All fixtures, except for the water closet, drain to a receiver with an interface valve that automatically empties the receiver (IPC G101.2.4).

134. a. The presiding authority must be given an opportunity to review, research, and approve a vacuum drainage system before installation begins (UPC G101.2.1).

135. c. The UPC states that "firestop systems for piping must have an F rating of at least one hour, but not less than the required fire-resistance rating of the assembly being penetrated" (UPC 1505.3).

136. d. The closet ring is the gasket between the base of the water closet bowl and the closet flange (UPC 203.0).

137. b. A 2-inch liquid seal is sufficient to prevent a fixture trap from siphoning (IPC 1002.4).

138. b. The full-opening valve on the cold water inlet of a water heater provides for full flow through the hot water system and a means for shutting off the complete hot water distribution system (UPC 605.0–605.7, IPC 606.1).

139. d. Horizontal hanger spacing for steel piping must be 12 feet apart (IPC Table 308.5, UPC Table 3–2 and Figure 3–28).

140. a. Debris, such as leaves and dead birds, can get into the system and cause a blockage (IPC 1101.8).

141. d. All fixtures installed in a building intended for human habitation must be connected to the building water distribution system (IPC 301.4, UPC 601.0).

142. b. for IPC, d. for UPC. The IPC and UPC differ on this matter. For the IPC, a vent stack is required when there are 5 or more branch intervals. The UPC, however, requires a vent stack when there are 10 branch intervals (IPC 903.2, UPC 907.1).

143. b. The purpose of a percolation test is to see how quickly gray water will be absorbed into the soil (IPC 1303.7).

144. d. Frost closure can be prevalent at or below 0°F (IPC 904.2).

145. b. A 1-inch water column is the greatest pressure that a vent and drainage system can apply to a trap seal (IPC 901.2, UPC 1002.1).

146. b. Laundry trays, floor drains, sinks, lavatories, and drinking fountains are the fixtures intended for combination drain and vent systems (IPC 912.1, UPC 910.0, Appendix B1).

147. d. The vertical pipe can be no longer that 8 feet in a combination drain and vent system (IPC 912.2, UPC 910.0, Appendix B1).

148. a. A water supply pipe in the same trench as a sewer pipe must be a minimum of 12 inches higher than the sewer pipe (IPC 603.2, UPC 609.2.2).

ANSWERS AND EXPLANATIONS

149. a. Potable water supply systems must be protected with an air gap (IPC 802.1.3, UPC 801.4).

150. b. Strainers on roof drains must extend 4 inches above the roof level (IPC 1105.1, UPC 1105.2).

Posttest 2

1. b. A DFU is a drainage fixture unit, which is used when measuring drainage loads. It was first established in the 1940s in Dr. Roy B. Hunter's Report BMS65, "Methods of Estimating Loads in Plumbing Systems" (unit of measure, not code specific).

2. c. Both the IPC and UPC define a bathroom group as a water closet, a lavatory, a bathtub or shower, and possibly a bidet in one bathroom (IPC and UPC definitions).

3. d. The openings in strainer plates on drain inlets must be no larger than $\frac{1}{2}$ inch (IPC 304.2, UPC 313.12.1).

4. d. Horizontal copper tubing that is $1\frac{1}{4}$ inch or less must be supported at least every 6 feet (IPC Table 308.5, UPC 314.1).

5. d. The inspector does not provide any testing equipment (IPC 312.1, UPC 103.5.4.2).

6. c. When you change $\frac{1}{2}$ inch into eighths of an inch, you get $\frac{4}{8}$ inch. Next, you add the numerators (7 + 4), but leave the denominator (8) the same, giving you $\frac{11}{8}$ inch, or $1\frac{3}{8}$ inch (math for plumbers, not code specific).

7. c. The formula for determining the volume of a cylinder in gallons is $\pi r^2 h$ divided by 231 cubic inches (1 gallon of water has 231 cubic inches in it). Since the water heater measures 20 inches across, the radius is 10 inches (half the diameter). 10 × 10 = 100; 100 × 3.14 = 314; 314 × 60 = 18,840 cubic inches; and 18,840 cubic inches ÷ 231 cubic inches = 81.558 gallons (math for plumbers, not code specific).

8. c. The formula for determining the length is offset × 1.414 − (fitting allowance × 2), which equals 16 × 1.414 − (0.75 × 2). As you solve the equation you get 22.624 − 1.5, which equals 21.124, or approximately $21\frac{1}{8}$ inches (math for plumbers, not code specific).

9. c. Considering the volume of water that a typical automatic washing machine can pump out, along with suds pressure demand, a 2-inch drain should be installed (IPC 406.3, UPC Table 7–3).

10. b. Although overflows were required in the past, it has been found that overflow passageways are a haven for bacteria growth because of lack of use. Both the IPC and UPC have now left the decision of having an overflow to the manufacturer of the fixture (IPC 416.1, UPC 403.0).

11. d. Because not all surfaces in a trough urinal are rinsed when it is flushed, trough urinals are banned (IPC 401.2, UPC 405.2).

12. b. Showers and bathtubs are both limited to a maximum hot water temperature of 120°F (IPC Table 424.4, UPC 415.5).

13. c. The passageway can be no longer than 20 feet, and may be as low as 30 inches (IPC 502.3, UPC Figure 5–9).

14. c. Because of convection and stratification, the hottest water in the tank will be in the top 6 inches of the tank; therefore, the temperature and pressure relief valve must be installed toward the top of the body (IPC 504.4.1, UPC 505.4, 505.5).

15. b. A combination fixture is two or more different types of fixtures manufactured as a single unit (IPC and UPC definitions).

16. b. In earthquake-prone areas, seismic supports are required for the safety of the population (IPC 502.4, UPC 50, 8.2).

ANSWERS AND EXPLANATIONS

17. **b.** The outlet pipe from a T & P relief valve must be at least the same diameter as the T & P valve outlet. The discharge pipe cannot be reduced in diameter at any point to ensure that the discharge pipe can carry the full water or steam discharge rate of the temperature and pressure relief valve (IPC 504.6, UPC 505.6).

18. **b.** The potable water supply is not at risk from the release of water from a boiler pressure relief valve (IPC 802.1.5, UPC 812.0).

19. **d.** Allowing the use of an air break gives backflow protection to the refrigeration equipment while allowing the equipment to be installed on the floor without the necessity of raising all the equipment off the floor to create the free air distance for an air gap (IPC 802.1.2, UPC 801.2.2).

20. **a.** An air gap must be used rather than an air break to protect the potable water supply (UPC 801.4, IPC 802.1.3).

21. **d.** The purposes of an air gap include preventing backflow into the indirect waste, preventing contamination of food in a food prep sink, and protecting the potable water supply (IPC 801.2, UPC 801.0).

22. **b.** The minimum size vent for a drain is one-half the diameter of the drain served, so for a 4-inch soil stack, the vent can be no smaller than 2 inches (IPC 916.1, UPC 904.0).

23. **b.** Only floor drains, sinks, lavatories, and drinking fountains may be drained with a combination drain and vent system (IPC 912.1, UPC 910.0).

24. **c.** The trap is specifically designed to stop sewer gas from backing up the pipe and out the drain (IPC Chapter 10 Purpose, UPC Chapter 10 Introduction).

25. **a.** A single trap can serve up to three compartments on a combination fixture (IPC 1002.1, UPC 1001.2).

26. **b.** The top rim of a toilet or sink, the flood-level rim, is where the water overflows (IPC and UPC definitions).

27. **a.** The relief vent on a building trap is a fresh air intake on the inlet side (IPC 1002.6, UPC 1008.0).

28. **c.** The purpose of the grease interceptor is to trap grease before it enters the wastewater disposal system (IPC 1003.1, UPC 1009.1).

29. **a.** A sink in a facility where plaster casts are regularly made would need an interceptor to trap the solid residue that might wash down the drain. Drum traps are approved for this purpose (IPC 1002.3, UPC 1010.0).

30. **b.** A standard of 4 inches has been adopted for strainers on roof drains in order to keep most debris out (IPC 1105.1, UPC 1105.2).

31. **c.** A solid piece of waste $\frac{1}{2}$ inch or less will not cause blockage in a plumbing waste drainage system (IPC 304.2).

32. **c.** Roof drains are required when ponding occurs on a roof, and if roof drains are required, a secondary roof drainage system is also necessary (IPC 1107.1, UPC 1101.11.2.2).

33. **a.** This height is sufficient to allow privacy, and the distance from the floor is sufficient to prevent splatter (IPC 405.3).

34. **b.** Brazing, not the use of flare fittings, is the accepted method of making copper-to-copper joints (IPC 1202.1, UPC 1319.0).

35. **d.** A branch is any part of the piping system other than a main, riser, or stack (IPC 202, UPC 203.0).

36. **b.** In order to function properly, a system must be sized to accommodate peak demands (IPC 602.3.2).

37. **d.** Most residential plumbing fixtures are not equipped to handle water pressures greater than 80 psi (IPC 604.8, UPC 608.2).

ANSWERS AND EXPLANATIONS

38. c. Because there is no division between the water and the air in air chambers, they only work for a short time, so they are not considered an acceptable form of water hammer arrestor (IPC 604.9).

39. b. Hot water temperature maintenance is required for pipes longer than 100 feet (IPC 607.2).

40. d. Carbon steel pipe material corrodes very rapidly (IPC Table 605.4, UPC Table 604.1).

41. d. A 3-inch overflow is required to ensure that the tank will not spill over the top (IPC Table 605.4).

42. c. Cross connection on potable water supplies has caused epidemics around the world (health concerns).

43. a. Although vitrified clay tile would be inappropriate for aboveground use, it can safely be used for a building sewer (IPC Table 702.2).

44. c. The reason that purple primer is used on PVC solvent weld joints is to make it easy for the inspecting authority to visually verify that primer has been used in the joint (IPC 705.8.2).

45. b. In PVC, only schedule 80 or thicker can be threaded. Schedule 40 PVC is too thin (IPC 705.14.3).

46. a. Manholes are also required at changes of direction (IPC 708.3.2).

47. d. Identification is the only sure way to know the contents of the pipes (IPC 608.8, UPC 10.2).

48. c. A sump pit must be at least 24 inches deep to allow enough room for the sewage pump and collection of sewage (IPC 712.3.2).

49. a. Gray water includes all water drained from bathtubs, lavatories, showers, clothes washers, and laundry trays (IPC C101.2).

50. b. Gray water must be disinfected when used for fixture flushing (IPC C102.2).

51. d. Testing a vacuum drainage system to 19 inches of mercury assures a vacuum-tight system (IPC G101.3).

52. d. Depending on the type of industrial waste, there may be products that can block a sewer pipe, cause corrosion, cause fire or explosion, poison life forms, or leak radioactive materials. For this reason, industrial waste must be treated or neutralized before it enters the public sewer system (IPC 302.1–302.2, UPC 303.0).

53. c. The distance between a faucet outlet and the flood-level rim of a sink is an example of an air gap, which is considered the most effective means of preventing backflow (IPC 608.15.1, UPC 603.3.1).

54. c. Freezing will not allow the mechanisms in the backflow protector to function properly (IPC 608.14.1, UPC 603.4.7).

55. b. A metal collar fits around the base of the pipe at the floor and completely covers the opening, making it impossible for a rodent to get through (IPC 304.4, UPC 313.12.3).

56. a. Cast-iron pipe is sturdy, but heavy, so it must be supported at least every 15 feet when run vertically (IPC Table 308.5, UPC Figure 3–23).

57. b. Pipes must never be embedded in a concrete or block wall because any slight shifting or settling of the wall could cause a pipe or pipe connection to burst (UPC 313.2).

58. c. The walls and surfaces in a public restroom must be smooth, hard, and nonabsorbent. This helps to prevent the growth of germs and bacteria, and makes sanitation much easier (IPC 310.3).

59. c. Both the IPC and the UPC require that a gauge with increments of 1 psi or less must be used to test a pressure between 10 psi and 100 psi (IPC 312.1.1, UPC 319.2).

ANSWERS AND EXPLANATIONS

60. **d.** This valve, made specifically for fire sprinkler systems, can withstand the very high pressures that occur when firefighters connect their hoses to the automatic sprinkler system (IPC 608.16.4, UPC 603.3.8).
61. **a.** Impurities in the fuel and combustion air can cause condensate to be acidic and corrosive to many materials (IPC 314.1).
62. **d.** A condensate disposal system can include components made of cast iron, galvanized steel, copper, cross-linked polyethylene, polybutylene, polyethylene, ABS, CPVC, or PVC pipe or tubing (IPC 314.2.2).
63. **b.** The unisex water closet is considered an additional fixture for each men's and women's restroom (IPC 403.1.1).
64. **d.** This distance is the length of the average cable used to clear blockages (IPC 708.3.2, UPC 707.4).
65. **b.** Both the IPC and the UPC require 1 water closet for every 10 males living in a dormitory (IPC Table 403.1, UPC Table 4–1).
66. **a.** All points of contact between a plumbing fixture and the walls or floor must have a watertight seal in order to facilitate cleaning and prevent bacterial growth (IPC 405.5).
67. **d.** The seat on a water closet must be smooth, nonabsorbent, and designed to fit the bowl. It must also be designed to facilitate cleaning to prevent bacterial growth (IPC 420.3).
68. **b.** Water closets are prohibited from having concealed or hidden traps (IPC 401.2, UPC 405.0).
69. **c.** In order to be used for a shower pan, a section of sheet lead must weigh at least 4 pounds per square foot and be coated with an asphalt paint or other approved coating (IPC 402.4).
70. **a.** Restaurant and nightclub owners need not incorporate separate toilet facilities in their buildings for employees and patrons (IPC 403.5).
71. **d.** Closet bolts are required to be made of brass because of its strength and resistance to corrosion (IPC 405.4.1).
72. **d.** Slip joints may be installed on trap outlets, on trap inlets, and within trap seals (IPC 405.8).
73. **b.** The accepted materials for bathtubs include enameled cast iron, porcelain-enameled formed steel or plastic, and a variety of plastic materials specified in the code (IPC 407.1, UPC 405.3).
74. **a.** Plastic bathtubs must be impregnated with fire-retardant chemicals to reduce the likelihood that they will ignite in the event of a fire (IPC 407.1).
75. **d.** Pop-up stoppers, crossbars, and strainers are all acceptable devices for restricting the opening of the waste outlet of a lavatory (IPC 416.3).
76. **d.** The walls of a shower must rise at least 6 feet above the floor of the room in order to prevent water damage from the spray (IPC 417.4.1).
77. **a.** The trap seal depth and water surface area for the siphon washbowl are the smallest of all the water closet bowl designs (IPC 420.1).
78. **b.** 12 inches has been determined to be a safe height to install the water service above the building sewer so as to prevent cross contamination in case of a break in the sewer (IPC 603.2, UPC 609.2).
79. **d.** Under the UPC, if a material can be used for drainage piping, it can also be used for an exposed tubular trap (UPC 404.3).
80. **b.** This is the approved method of getting rid of waste under these circumstances (IPC 712.1, UPC 710.2).
81. **b.** Because a handheld shower can be immersed in water, it must be designed with backflow protection (IPC 424.2).

ANSWERS AND EXPLANATIONS

82. **b.** A food waste grinder attached to a residential sink must be attached to a drain that is at least $1\frac{1}{2}$ inches in diameter—the same requirement for the trap size of a kitchen sink (IPC 413.2).
83. **d.** Placing the ignition portions at least 18 inches above the floor significantly reduces the likelihood of igniting floor-level vapors such as gasoline or solvents (IPC 502.1, UPC 508.14).
84. **b.** All water heaters must have automatic temperature controls (IPC 501.8).
85. **a.** Only the inlet pipe on a hot water heater requires a valve because once the inlet pipe is turned off, there is no flow into the hot distribution system (IPC 503.1 and 606.1).
86. **b.** Hot water storage tanks must be designed and insulated to lose no more than 15 BTUs per hour in order to save energy (IPC 505.1).
87. **a.** All water heaters must be certified by an approved third-party agency such as AGA or UL (IPC 505.1, UPC 501.0).
88. **d.** A permit is generally required to remove, replace, or install a water heater (UPC 503.0).
89. **b.** The minimum distance between a water service pipe and a septic tank is 5 feet (IPC 603.2).
90. **a.** It is a plumbing waste receptor and as such must be vented to remove all gases that may be produced by the effluent that enters it (IPC 1003.9, UPC 1014.2.2).
91. **a.** Any device that incorporates the use of a stop-and-waste feature is prohibited if the waste opening is underground or in any other location where wastewater or waterborne contaminants could enter by reversal of flow (IPC 608.7).
92. **c.** A $2\frac{1}{2}$-inch overflow will ensure that the tank will not spill over the top (IPC Table 606.5.4).
93. **d.** A water pump in a basement must be elevated at least 18 inches in order to protect it from damage in the event of a flood (IPC 602.3.5.1).
94. **b.** The absorption field must be at least 50 feet from the cistern in order to prevent contamination (IPC Figure 603.2.1).
95. **c.** The vent prevents the water from the trap seal from being siphoned and thereby broken (IPC 901.2, UPC 902.1).
96. **d.** In the event of insufficient pressure from the public main, an approved gravity tank, pressure booster system, or pressure booster pump must be used to increase the water pressure in a building. These types of devices are commonly used in multistory buildings (IPC 606.5.1, UPC 608.1).
97. **d.** An irrigation system can be protected from backflow by an atmospheric vacuum breaker, a pressure-type vacuum breaker, or a reduced pressure principle backflow preventer (IPC 608.16.5, UPC 603.4.6).
98. **c.** The number of fixture units connected to a drainage system is the yardstick used to determine the size of the vents attached (IPC 906.1, UPC 904.1).
99. **c.** The code states that every time a building drain changes direction at an angle greater than 45 degrees, a cleanout must be installed (IPC 708.3.3).
100. **d.** The approved materials for sump pit construction include, but are not limited to, tile, concrete, plastic, and steel (IPC 712.3.2, UPC 710.8).
101. **a. for the IPC; b. for the UPC.** The IPC and UPC differ on this issue (IPC Table 702.4, UPC 701.2).
102. **b.** Once the steam condenses and cools to no more than 140°F, it can travel through a condensate drain to the drainage system (IPC 803.1 and 802.1.5, UPC 810.0).

ANSWERS AND EXPLANATIONS

103. c. ABS, copper, PVC, and cast iron are among the materials approved for indirect waste pipes, but concrete is not (IPC 804.1).

104. a. An air gap is an upgrade from an air break, so an air gap can be used in place of an air break, but an air break cannot necessarily be used in place of an air gap (IPC 802.2).

105. b. for the IPC; a. for the UPC. The IPC and UPC differ on this matter (IPC 802.1.4, UPC 813.0).

106. d. A drain from a refrigeration coil can be as long as you like—there is no maximum approved limit (UPC 801.2.1).

107. a. Although a traditional vent is preferred, the combination system can be used when standard venting is impractical (IPC 912).

108. b. Vent piping helps to maintain trap seals, which prevents sewer gas from entering the building through a trap (IPC 901.2).

109. a. The minimum weight requirement is in place because vent flashing must be able to withstand decades of weather and building settlement (IPC 902.2).

110. d. The vent shall terminate 7 feet above the roof, as that height is taller than the average person and the sewer gases should not be inhaled by anyone (IPC 903.1, UPC 906.3).

111. c. According to IPC Table 906.1, a trap weir that is 3 inches in diameter can be 8 feet from a vent (IPC Table 906.1).

112. b. There is no limit on the number of fixture units that can be discharged into one branch interval of a 3-inch waste stack as long as the load for the stack is not exceeded (IPC Table 910.4).

113. c. A wet vent is literally a vent that is wet because it also conveys drainage (IPC 909.1, UPC 908.0).

114. d. Horizontal and vertical offsets are prohibited with waste stack vents (IPC 910.2).

115. b. In case the fixture backs up due to a blockage in the drain line, this height is sufficient to prevent it from entering the venting system (IPC 905.5, UPC 905.3).

116. a. Neither code permits the use of traps that require moving parts to function (IPC 1002.3, UPC 1004.0).

117. c. Because a vitrified clay trap requires extra support, it must be encased in concrete when used underground (IPC 1002.9).

118. b. The waste outlets must not be more than 30 inches apart in order to avoid a pressure buildup in the waste piping (IPC 1002.1, UPC 1001.1).

119. b. It's the other way around. The interceptor retains waste, while the separator sends waste out to a storage tank (IPC 1003.4.1).

120. b. Fixture traps must have a liquid seal between 2 inches and 4 inches, unless the Authority Having Jurisdiction states otherwise (IPC 1002.4, UPC 1005.0).

121. a. Sand interceptors are used in commercial establishments to intercept heavy solids (IPC 1003.5, UPC 1016.0).

122. a. The runoff from a strong storm could quickly overload a sanitary drainage system (IPC 1101.3).

123. b. A distance longer than 24 inches will start to create a vacuum in the system (IPC 1002.1, UPC 1001.1).

124. c. The same materials are required for storm systems and sanitary systems (IPC 1102.1).

125. a. Gray water cannot be drained into conductors when connecting to a combination sewer (IPC 1104.1).

126. a. It must not be larger than the trap arm, as this will obstruct the flow from the fixture outlet (IPC 1002.5, UPC 1003.3).

127. a. A check valve is required because some of the same water contained in the discharge pipe would return to the sump through the pump, and would have to be pumped out again (IPC 1113.1.4).

ANSWERS AND EXPLANATIONS

128. c. for the IPC; b. for the UPC. The IPC and UPC differ on this issue, but common sump pumps will fit into either size pit (IPC 1113.1.2, UPC 1101.5.3).

129. b. This ensures that there is not a buildup of pressure in the system and that siphoning does not occur in the trap (IPC 1002.6, UPC 1008.0).

130. c. for the IPC; d. for the UPC. The IPC requires that the following markings appear on a medical gas system to indicate to the inspector that the materials are approved for use: OXY, MED, XY/MED, ACR/OXY, ACR/MED; the UPC does not require any of these markings (IPC 1201.1).

131. a. A medical vacuum system becomes contaminated with a wide variety of bodily liquids that would contaminate any gas passing through if it was converted to a medical gas system (IPC 1202.1).

132. d. Gas pipes must typically be buried at least 18 inches underground to protect them from the weight of heavy vehicle loads and similar pressures that could lead to rupture (UPC 1211.1).

133. c. With the increasing demand on fresh water sources in the United States, there is an increased interest in technologies that save water (IPC G101.2.4).

134. b. A trap primer adds water to a trap periodically to replenish the trap seal (IPC 1002.4, UPC 1007.1).

135. b. 30 inches is required for sufficient room in a shower (UPC 411.7, IPC 417.4).

136. b. In addition, when the water and sewer main are in the same trench, the sewer main must be made of materials that are permitted for aboveground use (UPC 609.2, IPC 603.2).

137. c. Floor drains require a 2-inch trap (IPC Table 709.1, UPC Table 7-3).

138. b. Food preparation creates a lot of grease, and this must not be allowed to enter the drainage system because it will create a blockage when the grease solidifies (IPC 1003.1–1003.10, UPC 1014.1–1017.2).

139. d. Unless otherwise specified, piping is expressed in nominal size, which is the same as inside diameter and standard size (IPC 301.5).

140. b. A trap seal must be a minimum of 2 inches deep (IPC 1002.4, UPC 1005.0).

141. a. Air admittance valves are permitted under the IPC (IPC 917.1).

142. a. A sewage ejector discharge must connect to the building drainage pipe with a wye fitting on the top of the pipe (IPC 712.3.5).

143. d. All of these locations produce runoff rainwater that must be drained (IPC 110.1, UPC 1101.2).

144. d. Automatic clothes washers must drain into the building drainage system (IPC 701.2, UPC 305.1–305.2).

145. c. A backwater valve is made to prevent backflow in drainage systems (IPC 1101.9, UPC 1101.5.4).

146. d. All of the above produce gray water that can be recycled (IPC 1301.7, UPC 1602.8.1).

147. b. ASME stands for American Society of Mechanical Engineers (IPC Chapter 13, UPC Chapter 14).

148. c. Nonpotable clear water waste must enter the drainage system through an air break into a trap (IPC 802.1.5, UPC 814.3).

149. b. for IPC, c. for UPC. For the IPC, a 2-inch vent can serve a drain carrying 20 drainage fixture units; for the UPC, the DFU number is 24 (IPC Table 916.1, UPC Figure 9-34).

150. b. Disinfected gray water can be used to flush urinals and water closets (IPC 1302, UPC 1604.10.2).

APPENDIX A
CORRELATION GUIDE

This correlation guide will help you as you go through the Answers and Explanations section of this book. Each answer explanation provided for the Pretest, Chapter Reviews, and Posttests contains a cross-reference to an IPC or UPC code. This guide breaks down the plumbing code reference numbers not only by codebook topic, but also by the chapters in this book. Simply locate the code in the chart on pages 286–288 and read across to find the relating chapter, or section within a chapter, in this book. The first number in each code reference relates to the corresponding chapter in either the IPC or UPC. For instance, if an answer provides a reference to IPC 301.3, this can be found in Chapter 3 of the IPC.

APPENDIX A

CHAPTER CONTENT	IPC CODE	UPC CODE
Different Plumbing Codes and Abbreviations (Chapter 1)	Applies to all code	Applies to all code
General Regions Served by the IPC and the UPC	Applies to all code	Applies to all code
Differences between the IPC and the UPC Abbreviations	Applies to all code	Applies to all code
General Regulations (Chapter 2)	301.1–314.2.4	301.0–319.4
General Principles of Plumbing Installation	301.1	301.0–301.2.1
Connections to Building Sanitary Drainage Systems and Water Supply Systems	301.3–301.4	304.0–305.3
Testing of New Plumbing Systems	312.1–312.9.2	319.0–319.4
Math for Plumbers (Chapter 3)	Not code specific	Not code specific
Basic Facts	Not code specific	Not code specific
Basic Calculations	Not code specific	Not code specific
Fixtures (Chapter 4)	401.4–423.2	401.0–407.7, 408.0–410.2, 411.1–413.0
Fixture Material	402.1–402.4	410.0
Fixture Requirements in Various Buildings	403.1–404.1	412.0, 412.1
Fixture Locations	404.1	407.6
Bathroom Design Relative to Fixtures	405.3.1	407.6
Approved Fixture Connections	405.4–405.9	404.2
Regulations for Specific Fixtures	422.1–423.2	408.0, 409.0, 411.1–411.5
Faucets and Fixture Fittings (Chapter 5)	424.1–427.1	407.8, 410.3, 413.1–418.0
Water Consumption	401.3, 424.1.1	402.0
Design	424.2	401.0, 403.0, 405.0
Temperature Control	424.3–424.8	403.1–418.0
Water Heaters (Chapter 6)	501.1–505.1	501.0–512.0
Location	501.4, 502.2, 502.3	505.1–505.1.2
Temperature Control	501.6, 501.8	501.0
Installation Requirements	502.1, 503.1, 503.2	506.0–512.0
Safety	504.1–504.7.2	505.4–505.5

(Continued)

APPENDIX A

CHAPTER CONTENT	IPC CODE	UPC CODE
Gas Piping Installations (Chapter 7)		
General Rules		1201.1, 318, 1213.13
Material		1208
Pressure		1208.7, 1210.1.6.2, 1213.6.1.1–1213.6.1.4
Sizing		1215–1216
Water Supply and Distribution (Chapter 8)	**601.1–613.1**	**601.0–610.10**
General Rules	601.1–601.4	601.0–603.0
Quantity	602.1–602.3.5.1	608.1–608.2
Water Service	603.1–603.2.1	609.1–609.3, 610.8
Distribution System	604.1–604.11	610.9
Sanitary Drainage (Chapter 9)	**701.1–715.5**	**701.0–723.0**
General Requirements	701.1–701.9	
Allowable Sewage	701.4–701.5	714.0–714.2
Piping Materials	702.1–702.6	701.0–701.2
Piping Joints	705.1–705.22, 707.1	705.0–705.3.3
Use of Fittings	706.1–706.4	706.0–706.4
Cleanouts	708.1–708.9	707.0–707.10
Manholes	708.3.6	719.6
Sizing	710.1–710.2	703.0–703.2
Sumps and Ejectors	712.1–712.4.2	710.1–710.13
Healthcare Drainage Piping	713.1–713.11.4	801.5, 806.0
Backwater Valves	715.1–715.4	710.1–710.6
Indirect and Special Wastes (Chapter 10)	**801.1–804.1**	**801.1–814.3**
Objective of Indirect Waste	801.1	801.0
Specific Waste (Special Wastes)	802.1–802.1.1, 802.1.3, 802.2, 802.2.1–802.2.2, 803.1–803.3	801.1–801.2.3, 801.4, 801.6, 808.0, 810.0–812.0, 814.0–814.3
Specific Fixtures	802.1.2, 802.1.4, 802.1.6, 802.1.7, 802.3–802.4	801.3, 801.5, 809.0, 813.0
Vents (Chapter 11)	**901.1–919.2**	**901.0–910.7**
Objective of Vents	901.2	901.0
Materials	902.1–902.3	903.0–903.4
Installation	903.1.1	904.0, 905.0–909.0
Vent Patterns	903.2–911.5	905.0–908.0
Combination Drain and Vent System	912.1–912.3	902.0, 910.0–910.7

(Continued)

APPENDIX A

CHAPTER CONTENT	IPC CODE	UPC CODE
Multistory buildings	914.1–916.4, 910.1–910.4	907.0–907.2
Air Admittance Valves	917.1–917.8	None
Traps, Interceptors, Separators (Chapter 12)	**1001.1–1004.1**	**1001.0–1017.2**
Trap Requirements	1002.1	1001.0–1002.4
Trap Configurations	1002.2–1002.10	1003.0–1004.0
Interceptors and Separators	1003.1–1003.10	1009.0–1017.2
Storm Drainage (Chapter 13)	**1101.1–1113.1.4**	**1101.1–1109.2**
General Requirements	1101.1–1101.9	1101.1
Roof Drains	1105.1–1105.3	1101.11.2.1–1109.2
Subsoil	1111.1	1101.5.1–1101.6
Building Subdrains	1112.1	1101.1
Pipes and Storage Systems (Chapter 14)	**1201.1–1203.1**	**1301.0–1328.4**
Nonflammable Medical Gasses	1202.1	1309.0
Medical Oxygen	1202.1	1310.6
Nonmedical Oxygen	1203.1	None
Vacuum System Piping	1202.1	1326.0–1327.9
Gray-Water Recycling (Chapter 15)	**C101.1–C103.11**	**None**
What Is Gray-Water Recycling?	C102.2	None
Connections	C101.9	None
Water Closets	C102.1–C102.6	None
Landscape Irrigation	C103.1–C103.11	None
Vacuum Drainage Systems (Chapter 16)	**G101.1–G101.4**	**None**
What Is a Vacuum Drainage System?	G101.2.1	None

APPENDIX B
PROJECT MANAGEMENT

You need a plan when you are the project manager. To a large extent, your plan will be based on the blueprints for the building you are working in. However, many other factors will affect your formula for the project. For example, you must consider the probable weather conditions that will occur during the period of construction. Temporary enclosures are desirable for some steps of construction to keep materials at a reasonable temperature. A side benefit is a bit of comfort for workers who are using temperature-sensitive materials.

The project manager must be thoroughly familiar with the codes that are used in the project's locality. That not only includes the plumbing code, but the building code, the fire code, the electrical code, and so on. Of course, no one knows all the codes by heart, but the project manager must be able to know when a point of construction begs a code-based decision. A knowledge of where to go for code answers is invaluable. Having a network of experts is as valuable a tool as any. The names and telephone numbers of the building owner, the Authority Having Jurisdiction as far as code enforcement is concerned, the general contractor, and the subcontractors are important facts that need to be gathered and kept in one place.

It is also important for the project manager to have direct knowledge of and the ability to readily contact the material suppliers for your projects. An intimate knowledge of the availability of ordered material for your crew and delivery dates of key equipment is a must. The architect needs to be on the list of contacts and the project manager cannot hesitate to contact any and all of the key players in the project as necessary. Prompt communication and explanation from all the individuals involved makes for the smooth progression of construction. One cannot assume anything. The project manager can only deal in hard facts and firm obligations. All too often, experienced builders fall prey to overconfidence and overwork resulting in poor decisions or no decisions, which destroys the proper sequence of a project.

If you are to be a good project manager you must be thoroughly knowledgeable in your field and you must be a masterful communicator. That does not mean you have to use all of the big words in *Webster's Dictionary*, it just means that you must stay in contact with all of the individuals important to the progress of the project in a professional and assertive manner. Only accept concrete, specific answers.

Your job is to complete the project as specified, in the time specified, and still turn a profit for your company. You must be on top of *all* aspects of the project at *all* times. Call for the rough-in and finish inspections at least three days ahead of time so your workers aren't held up by waiting for an inspection. Avoid micromanaging your workers. You should have people in the field that you can trust. Generally, they will do better work and more work if they are given the responsibility for its completion. As long as the job is meeting the time, quality, and cost schedules, let your workers do what they have trained for and love to do. Keep the materials flowing and the other contractors in your loop, and you will have a successful project.

APPENDIX C
PLUMBING ELEMENTS AND FACILITIES
as Directed by the Americans with Disabilities Act

601	General
602	Drinking Fountains
603	Toilet and Bathing Rooms
604	Water Closets and Toilet Compartments
605	Urinals
606	Lavatories and Sinks
607	Bathtubs
608	Shower Compartments
609	Grab Bars
610	Seats
611	Washing Machines and Clothes Dryers
612	Saunas and Steam Rooms

Source: Adapted from the ADA and ABA Accessibility Guidelines for Buildings and Facilities. www.access-board.gov

601 General

601.1 Scope. The provisions of Chapter 6 shall apply where required by Chapter 2 or where referenced by a requirement in this document.

602 Drinking Fountains

602.1 General. Drinking fountains shall comply with 307 and 602.

602.2 Clear Floor Space. Units shall have a clear floor or ground space complying with 305 positioned for a forward approach and centered on the unit. Knee and toe clearance complying with 306 shall be provided.

> **EXCEPTION:** A parallel approach complying with 305 shall be permitted at units for children's use where the spout is 30 inches (760 mm) maximum above the finish floor or ground and is 3½ inches (90 mm) maximum from the front edge of the unit, including bumpers.

602.3 Operable Parts. Operable parts shall comply with 309.

602.4 Spout Height. Spout outlets shall be 36 inches (915 mm) maximum above the finish floor or ground.

602.5 Spout Location. The spout shall be located 15 inches (380 mm) minimum from the vertical support and 5 inches (125 mm) maximum from the front edge of the unit, including bumpers.

602.6 Water Flow. The spout shall provide a flow of water 4 inches (100 mm) high minimum and shall be located 5 inches (125 mm) maximum from the front of the unit. The angle of the water stream shall be measured horizontally relative to the front face of the unit. Where spouts are located less than 3 inches (75 mm) of the front of the unit, the angle of the water stream shall be 30 degrees maximum. Where spouts are located between 3 inches (75 mm) and 5 inches (125 mm) maximum from the front of the unit, the angle of the water stream shall be 15 degrees maximum.

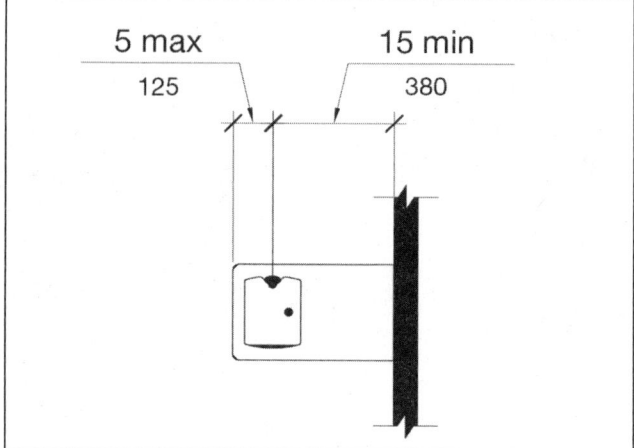

Figure 602.5 Drinking Fountain Spout Location

> **ADVISORY**
> **602.6 Water Flow.** The purpose of requiring the drinking fountain spout to produce a flow of water 4 inches (100 mm) high minimum is so that a cup can be inserted under the flow of water to provide a drink of water for an individual who, because of a disability, would otherwise be incapable of using the drinking fountain.

602.7 Drinking Fountains for Standing Persons. Spout outlets of drinking fountains for standing persons shall be 38 inches (965 mm) minimum and 43 inches (1090 mm) maximum above the finish floor or ground.

603 Toilet and Bathing Rooms

603.1 General. Toilet and bathing rooms shall comply with 603.

603.2 Clearances. Clearances shall comply with 603.2.

603.2.1 Turning Space. Turning space complying with 304 shall be provided within the room.

603.2.2 Overlap. Required clear floor spaces, clearance at fixtures, and turning space shall be permitted to overlap.

603.2.3 Door Swing. Doors shall not swing into the clear floor space or clearance required for any fixture. Doors shall be permitted to swing into the required turning space.

> EXCEPTIONS:
> 1. Doors to a toilet room or bathing room for a single occupant accessed only through a private office and not for common use or public use shall be permitted to swing into the clear floor space or clearance provided the swing of the door can be reversed to comply with 603.2.3.
> 2. Where the toilet room or bathing room is for individual use, and a clear floor space complying with 305.3 is provided within the room beyond the arc of the door swing, doors shall be permitted to swing into the clear floor space or clearance required for any fixture.

> **ADVISORY**
> **603.2.3 Door Swing Exception 1.** At the time the door is installed, and if the door swing is reversed in the future, the door must meet all the requirements specified in 404. Additionally, the door swing cannot reduce the required width of an accessible route. Also, avoid violating other building or life safety codes when the door swing is reversed.

603.3 Mirrors. Mirrors located above lavatories or countertops shall be installed with the bottom edge of the reflecting surface 40 inches (1015 mm) maximum above the finish floor or ground. Mirrors not located above lavatories or countertops shall be installed with the bottom edge of the reflecting surface 35 inches (890 mm) maximum above the finish floor or ground.

> **ADVISORY**
> **603.3 Mirrors.** A single full-length mirror can accommodate a greater number of people, including children. In order for mirrors to be usable by people who are ambulatory and people who use wheelchairs, the top edge of mirrors should be 74 inches (1880 mm) minimum from the floor or ground.

603.4 Coat Hooks and Shelves. Coat hooks shall be located within one of the reach ranges specified in 308. Shelves shall be located 40 inches (1,015 mm) minimum and 48 inches (1,220 mm) maximum above the finish floor.

604 Water Closets and Toilet Compartments

604.1 General. Water closets and toilet compartments shall comply with 604.2 through 604.8.

> EXCEPTION: Water closets and toilet compartments for children's use shall be permitted to comply with 604.9.

604.2 Location. The water closet shall be positioned with a wall or partition to the rear and to one side. The centerline of the water closet shall be 16 inches (405 mm) minimum to 18 inches (455 mm) maximum from the sidewall or partition, except that the water closet shall be 17 inches (430 mm) minimum and 19 inches (485 mm) maximum from the sidewall or partition in the ambulatory accessible toilet compartment specified in 604.8.2. Water closets shall be arranged for a left-hand or right-hand approach.

604.3 Clearance. Clearances around water closets and in toilet compartments shall comply with 604.3.

604.3.1 Size. Clearance around a water closet shall be 60 inches (1,525 mm) minimum measured perpendicular from the sidewall and 56 inches (1,420 mm) minimum measured perpendicular from the rear wall.

604.3.2 Overlap. The required clearance around the water closet shall be permitted to overlap the water closet, associated grab bars, dispensers, sanitary napkin disposal units, coat hooks, shelves, accessible routes, clear floor space and clearances required at other fixtures, and the turning space. No other fixtures or obstructions shall be located within the required water closet clearance.

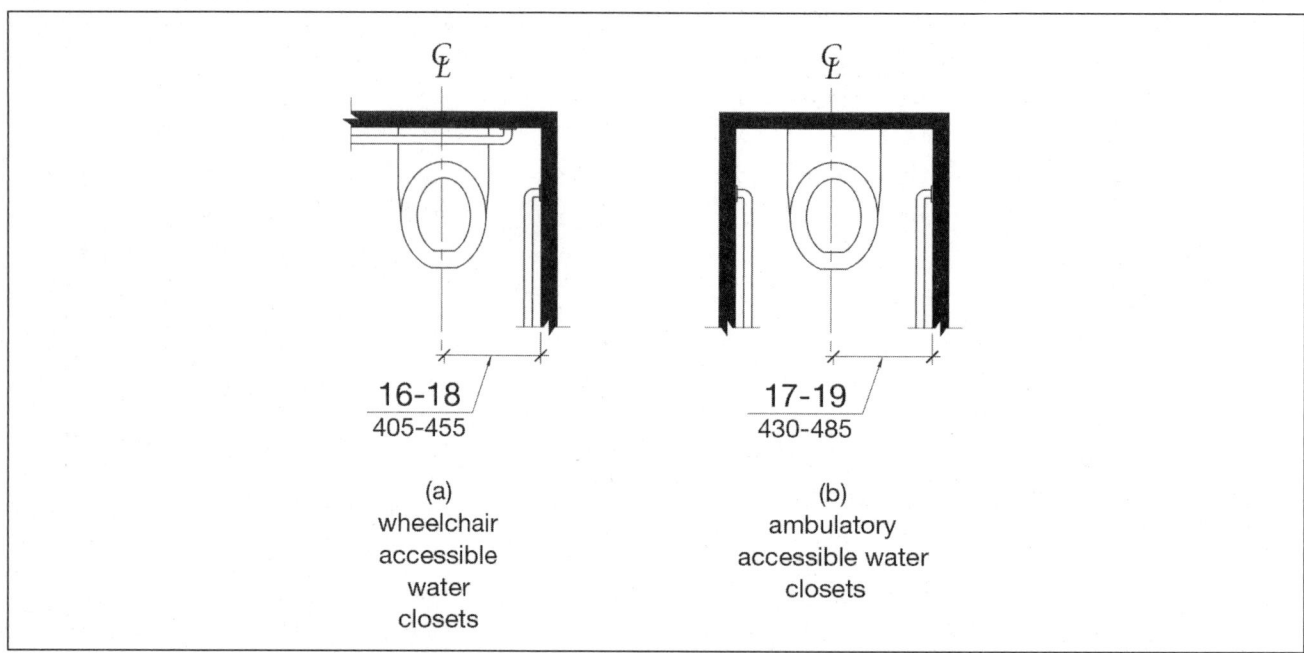

Figure 604.2 Water Closet Location

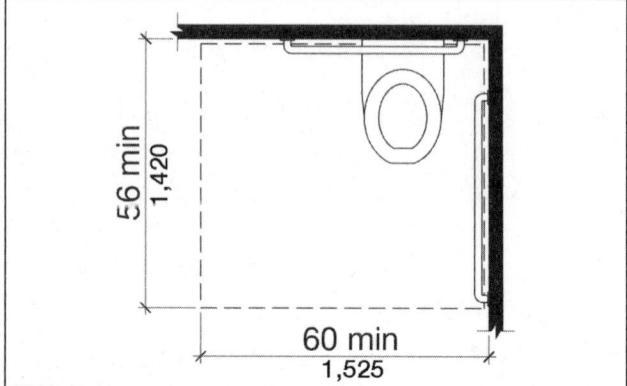

EXCEPTION: In residential dwelling units, a lavatory complying with 606 shall be permitted on the rear wall 18 inches (455 mm) minimum from the water closet centerline where the clearance at the water closet is 66 inches (1,675 mm) minimum measured perpendicular from the rear wall.

Figure 604.3.1 Size of Clearance at Water Closets

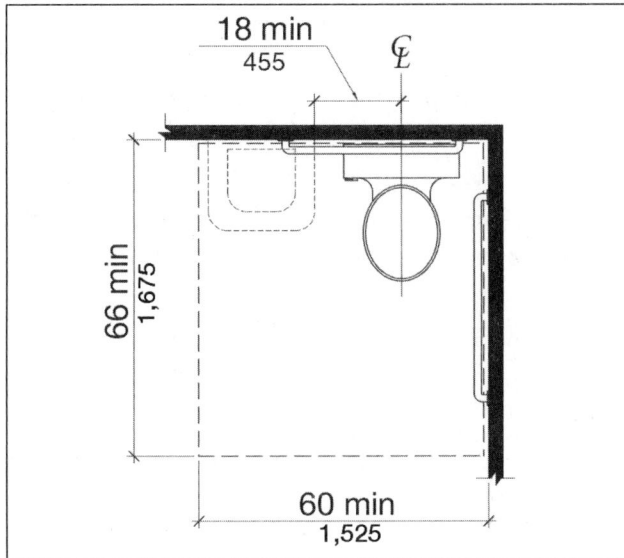

Figure 604.3.2 (Exception) Overlap of Water Closet Clearance in Residential Dwelling Units

> **ADVISORY**
> **604.3.2 Overlap.** When the door to the toilet room is placed directly in front of the water closet, the water closet cannot overlap the required maneuvering clearance for the door inside the room.

604.4 Seats. The seat height of a water closet above the finish floor shall be 17 inches (430 mm) minimum and 19 inches (485 mm) maximum measured to the top of the seat. Seats shall not be sprung to return to a lifted position.

EXCEPTIONS:
1. A water closet in a toilet room for a single occupant accessed only through a private office and not for common use or public use shall not be required to comply with 604.4.
2. In residential dwelling units, the height of water closets shall be permitted to be 15 inches (380 mm) minimum and 19 inches (485 mm) maximum above the finish floor measured to the top of the seat.

604.5 Grab Bars. Grab bars for water closets shall comply with 609. Grab bars shall be provided on the sidewall closest to the water closet and on the rear wall.

EXCEPTIONS:
1. Grab bars shall not be required to be installed in a toilet room for a single occupant accessed only through a private office and not for common use or public use provided that reinforcement has been installed in walls and located so as to permit the installation of grab bars complying with 604.5.
2. In residential dwelling units, grab bars shall not be required to be installed in toilet or bathrooms provided that reinforcement has been installed in walls and located so as to permit the installation of grab bars complying with 604.5.
3. In detention or correction facilities, grab bars shall not be required to be installed in housing or holding cells that are specially designed without protrusions for purposes of suicide prevention.

> **ADVISORY**
> **604.5 Grab Bars Exception 2.** Reinforcement must be sufficient to permit the installation of rear and side wall grab bars that fully meet all accessibility requirements including, but not limited to, required length, installation height, and structural strength.

604.5.1 Side Wall. The side wall grab bar shall be 42 inches (1,065 mm) long minimum, located 12 inches (305 mm) maximum from the rear wall and extending 54 inches (1,370 mm) minimum from the rear wall.

604.5.2 Rear Wall. The rear wall grab bar shall be 36 inches (915 mm) long minimum and extend from the centerline of the water closet 12 inches (305 mm) minimum on one side and 24 inches (610 mm) minimum on the other side.

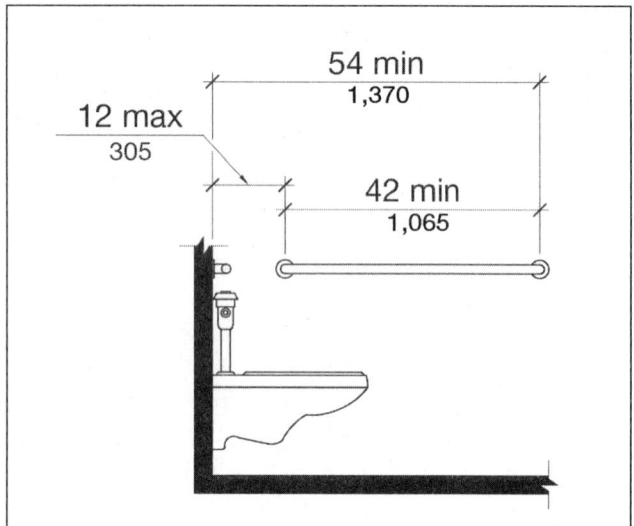

Figure 604.5.1   Side Wall Grab Bar at Water Closets

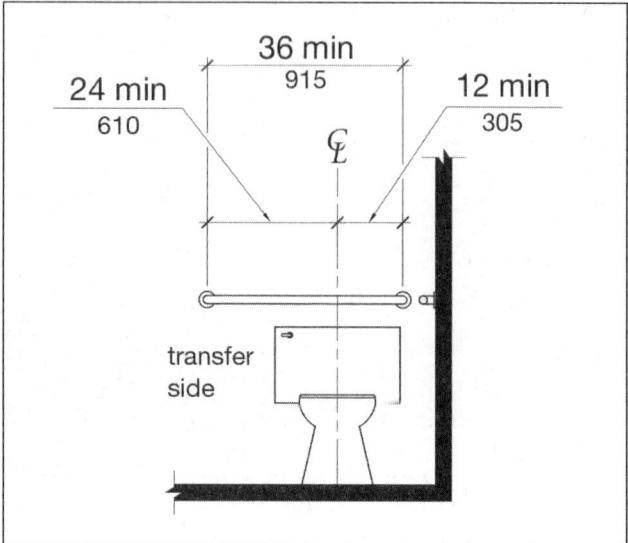

Figure 604.5.2   Rear Wall Grab Bar at Water Closets

EXCEPTIONS:
1. The rear grab bar shall be permitted to be 24 inches (610 mm) long minimum, centered on the water closet, where wall space does not permit a length of 36 inches (915 mm) minimum due to the location of a recessed fixture adjacent to the water closet.
2. Where an administrative authority requires flush controls for flush valves to be located in a position that conflicts with the location of the rear grab bar, then the rear grab bar shall be permitted to be split or shifted to the open side of the toilet area.

604.6 Flush Controls. Flush controls shall be hand operated or automatic. Hand operated flush controls shall comply with 309. Flush controls shall be located on the open side of the water closet except in ambulatory accessible compartments complying with 604.8.2.

> **ADVISORY**
>
> **604.6 Flush Controls.** If plumbing valves are located directly behind the toilet seat, flush valves and related plumbing can cause injury or imbalance when a person leans back against them. To prevent causing injury or imbalance, the plumbing can be located behind walls or to the side of the toilet; or if approved by the local authority having jurisdiction, provide a toilet seat lid.

604.7 Dispensers. Toilet paper dispensers shall comply with 309.4 and shall be 7 inches (180 mm) minimum and 9 inches (230 mm) maximum in front of the water closet measured to the centerline of the dispenser. The outlet of the dispenser shall be 15 inches (380 mm) minimum and 48 inches (1,220

APPENDIX C

mm) maximum above the finish floor and shall not be located behind grab bars. Dispensers shall not be of a type that controls delivery or that does not allow continuous paper flow.

> **ADVISORY**
> **604.7 Dispensers.** If toilet paper dispensers are installed above the side wall grab bar, the outlet of the toilet paper dispenser must be 48 inches (1,220 mm) maximum above the finish floor and the top of the gripping surface of the grab bar must be 33 inches (840 mm) minimum and 36 inches (915 mm) maximum above the finish floor.

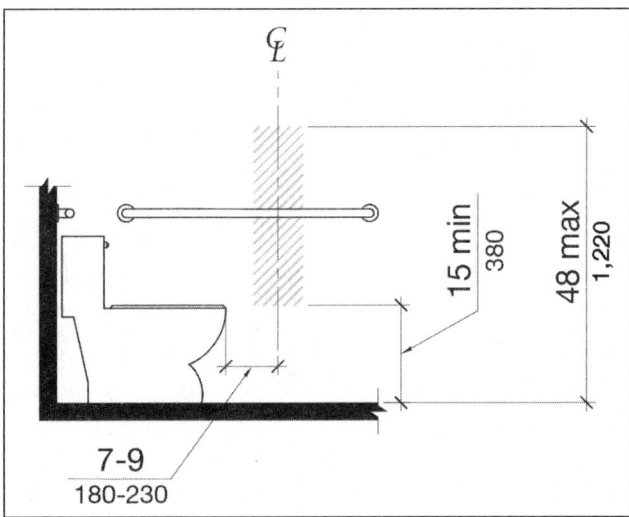

Figure 604.7 Dispenser Outlet Location

604.8 Toilet Compartments. Wheelchair accessible toilet compartments shall meet the requirements of 604.8.1 and 604.8.3. Compartments containing more than one plumbing fixture shall comply with 603. Ambulatory accessible compartments shall comply with 604.8.2 and 604.8.3.

604.8.1 Wheelchair Accessible Compartments. Wheelchair accessible compartments shall comply with 604.8.1.

604.8.1.1 Size. Wheelchair accessible compartments shall be 60 inches (1,525 mm) wide minimum measured perpendicular to the side wall, and 56 inches (1,420 mm) deep minimum for wall hung water closets and 59 inches (1,500 mm) deep minimum for floor-mounted water closets measured perpendicular to the rear wall. Wheelchair accessible compartments for children's use shall be 60 inches (1,525 mm) wide minimum measured perpendicular to the side wall, and 59 inches (1,500 mm) deep minimum for wall-hung and floor-mounted water closets measured perpendicular to the rear wall.

> **ADVISORY**
> **604.8.1.1 Size.** The minimum space required in toilet compartments is provided so that a person using a wheelchair can maneuver into position at the water closet. This space cannot be obstructed by baby changing tables or other fixtures or conveniences, except as specified at 604.3.2 (Overlap). If toilet compartments are to be used to house fixtures other than those associated with the water closet, they must be designed to exceed the minimum space requirements. Convenience fixtures such as baby changing tables must also be accessible to people with disabilities as well as to other users. Toilet compartments that are designed to meet, and not exceed, the minimum space requirements may not provide adequate space for maneuvering into position at a baby changing table.

APPENDIX C

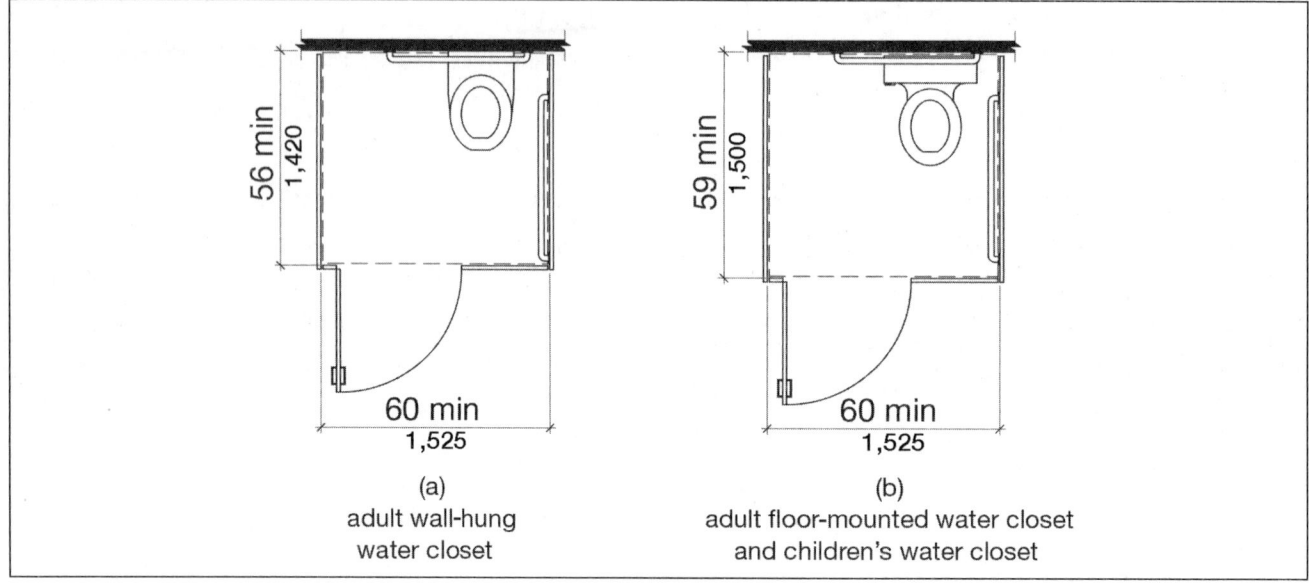

Figure 604.8.1.1 Size of Wheelchair Accessible Toilet Compartment

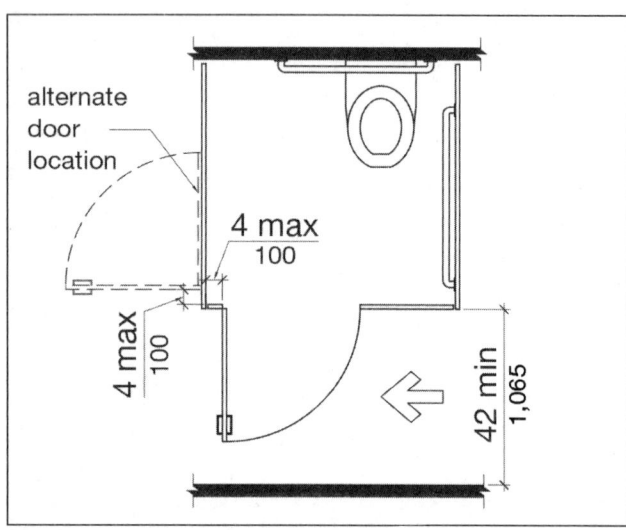

Figure 604.8.1.2 Wheelchair Accessible Toilet Compartment Doors

604.8.1.2 Doors. Toilet compartment doors, including door hardware, shall comply with 404 except that if the approach is to the latch side of the compartment door, clearance between the door side of the compartment and any obstruction shall be 42 inches (1,065 mm) minimum. Doors shall be located in the front partition or in the side wall or partition farthest from the water closet. Where located in the front partition, the door opening shall be 4 inches (100 mm) maximum from the side wall or partition farthest from the water closet. Where located in the side wall or partition, the door opening shall be 4 inches (100 mm) maximum from the front partition. The door shall be self-closing. A door pull complying with 404.2.7 shall be placed on both sides of the door near the latch. Toilet compartment doors shall not swing into the minimum required compartment area.

604.8.1.3 Approach. Compartments shall be arranged for left-hand or right-hand approach to the water closet.

604.8.1.4 Toe Clearance. The front partition and at least one side partition shall provide a toe clearance of 9 inches (230 mm) minimum above the finish floor and 6 inches (150 mm) deep minimum beyond the compartment-side face of the partition, exclusive of partition support members. Compartments for children's use shall provide a toe clearance of 12 inches (305 mm) minimum above the finish floor.

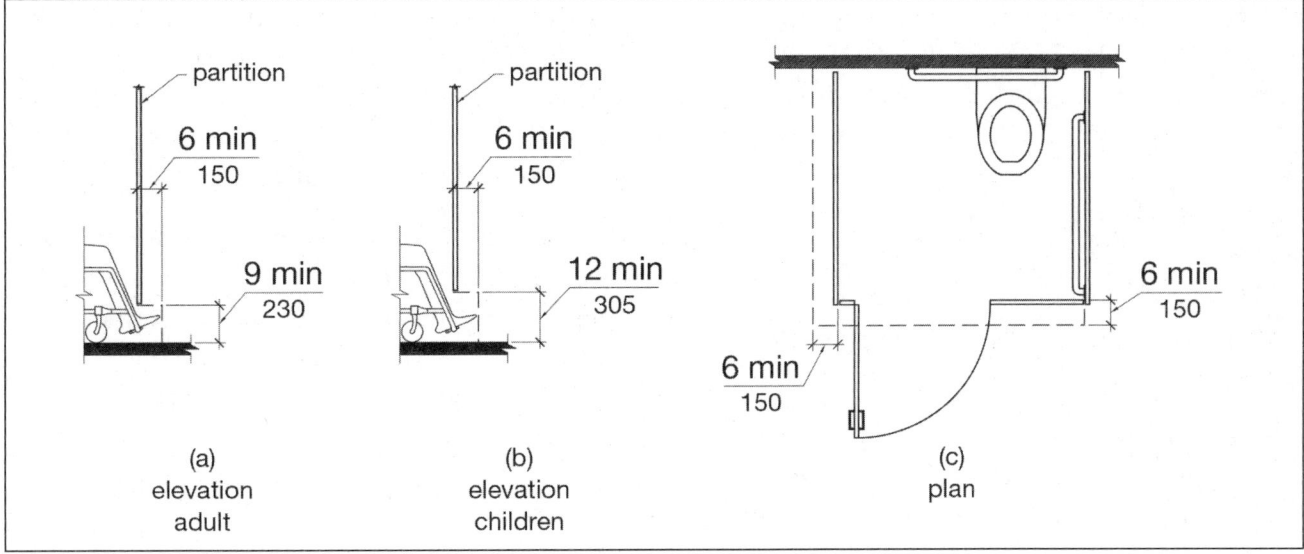

Figure 604.8.1.4 Wheelchair Accessible Toilet Compartment Toe Clearance

EXCEPTION: Toe clearance at the front partition is not required in a compartment greater than 62 inches (1,575 mm) deep with a wall-hung water closet or 65 inches (1,650 mm) deep with a floor-mounted water closet. Toe clearance at the side partition is not required in a compartment greater than 66 inches (1,675 mm) wide. Toe clearance at the front partition is not required in a compartment for children's use that is greater than 65 inches (1,650 mm) deep.

604.8.1.5 Grab Bars. Grab bars shall comply with 609. A side-wall grab bar complying with 604.5.1 shall be provided and shall be located on the wall closest to the water closet. In addition, a rear-wall grab bar complying with 604.5.2 shall be provided.

604.8.2 Ambulatory Accessible Compartments. Ambulatory accessible compartments shall comply with 604.8.2.

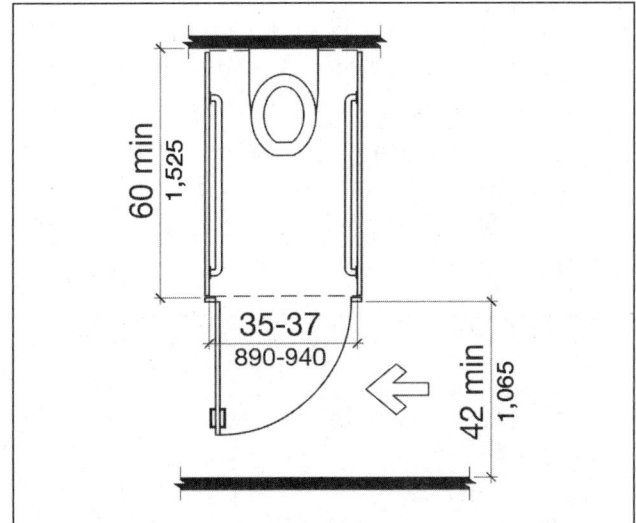

Figure 604.8.2 Ambulatory Accessible Toilet Compartment

604.8.2.1 Size. Ambulatory accessible compartments shall have a depth of 60 inches (1,525 mm) minimum and a width of 35 inches (890 mm) minimum and 37 inches (940 mm) maximum.

APPENDIX C

604.8.2.2 Doors. Toilet compartment doors, including door hardware, shall comply with 404, except that if the approach is to the latch side of the compartment door, clearance between the door side of the compartment and any obstruction shall be 42 inches (1,065 mm) minimum. The door shall be self-closing. A door pull complying with 404.2.7 shall be placed on both sides of the door near the latch. Toilet compartment doors shall not swing into the minimum required compartment area.

604.8.2.3 Grab Bars. Grab bars shall comply with 609. A side-wall grab bar complying with 604.5.1 shall be provided on both sides of the compartment.

604.8.3 Coat Hooks and Shelves. Coat hooks shall be located within one of the reach ranges specified in 308. Shelves shall be located 40 inches (1,015 mm) minimum and 48 inches (1,220 mm) maximum above the finish floor.

604.9 Water Closets and Toilet Compartments for Children's Use. Water closets and toilet compartments for children's use shall comply with 604.9.

ADVISORY

604.9 Water Closets and Toilet Compartments for Children's Use. The requirements in 604.9 are to be followed where the exception for children's water closets in 604.1 is used. The following table provides additional guidance in applying the specifications for water closets for children according to the age group served and reflects the differences in the size, stature, and reach ranges of children ages 3 through 12. The specifications chosen should correspond to the age of the primary user group. The specifications of one age group should be applied consistently in the installation of a water closet and related elements.

ADVISORY SPECIFICATIONS FOR WATER CLOSETS SERVING CHILDREN AGES 3 THROUGH 12

	AGES 3 AND 4	AGES 5 THROUGH 8	AGES 9 THROUGH 12
Water Closet Centerline	12 inches (305 mm)	12 to 15 inches (305 to 380 mm)	15 to 18 inches (380 to 455 mm)
Toilet Seat Height	11 to 12 inches (280 to 305 mm)	12 to 15 inches (305 to 380 mm)	15 to 17 inches (380 to 430 mm)
Grab Bar Height	18 to 20 inches (455 to 510 mm)	20 to 25 inches (510 to 635 mm)	25 to 27 inches (635 to 685 mm)
Dispenser Height	14 inches (355 mm)	14 to 17 inches (355 to 430 mm)	17 to 19 inches (430 to 485 mm)

604.9.1 Location. The water closet shall be located with a wall or partition to the rear and to one side. The centerline of the water closet shall be 12 inches (305 mm) minimum and 18 inches (455 mm) maximum from the side wall or partition, except that the water closet shall be 17 inches (430 mm) minimum and 19 inches (485 mm) maximum from the side wall or partition in the ambulatory accessible toilet compartment specified in 604.8.2. Compartments shall be arranged for left-hand or right-hand approach to the water closet.

604.9.2 Clearance. Clearance around a water closet shall comply with 604.3.

604.9.3 Height. The height of water closets shall be 11 inches (280 mm) minimum and 17 inches (430 mm) maximum measured to the top of the seat. Seats shall not be sprung to return to a lifted position.

604.9.4 Grab Bars. Grab bars for water closets shall comply with 604.5.

604.9.5 Flush Controls. Flush controls shall be hand operated or automatic. Hand operated flush controls shall comply with 309.2 and 309.4 and shall be installed 36 inches (915 mm) maximum above the finish floor. Flush controls shall be located on the open side of the water closet except in ambulatory accessible compartments complying with 604.8.2.

604.9.6 Dispensers. Toilet paper dispensers shall comply with 309.4 and shall be 7 inches (180 mm) minimum and 9 inches (230 mm) maximum in front of the water closet measured to the centerline of the dispenser. The outlet of the dispenser shall be 14 inches (355 mm) minimum and 19 inches (485 mm) maximum above the finish floor. There shall be a clearance of $1\frac{1}{2}$ inches (38 mm) minimum below the grab bar. Dispensers shall not be of a type that controls delivery or that does not allow continuous paper flow.

604.9.7 Toilet Compartments. Toilet compartments shall comply with 604.8.

605 Urinals

605.1 General. Urinals shall comply with 605.

> **ADVISORY**
> **605.1 General.** Stall-type urinals provide greater accessibility for a broader range of persons, including people of short stature.

605.2 Height and Depth. Urinals shall be the stall type or the wall-hung type with the rim 17 inches (430 mm) maximum above the finish floor or ground. Urinals shall be $13\frac{1}{2}$ inches (345 mm) deep minimum measured from the outer face of the urinal rim to the back of the fixture.

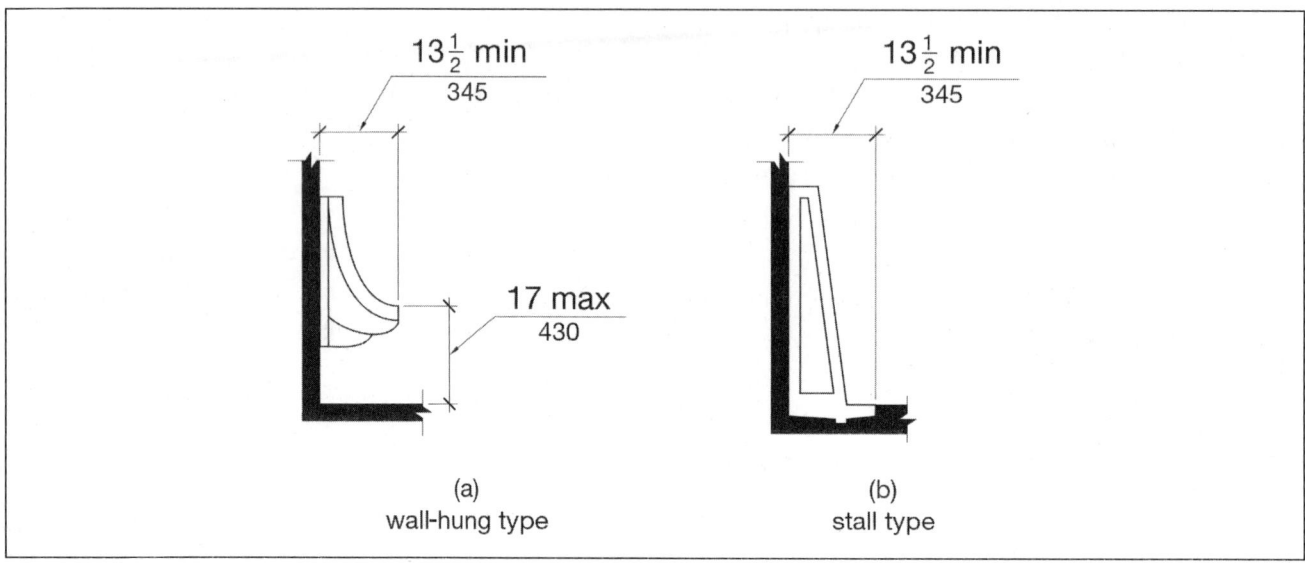

Figure 605.2 Height and Depth of Urinals

605.3 Clear Floor Space. A clear floor or ground space complying with 305 positioned for forward approach shall be provided.

605.4 Flush Controls. Flush controls shall be hand operated or automatic. Hand-operated flush controls shall comply with 309.

606 Lavatories and Sinks

606.1 General. Lavatories and sinks shall comply with 606.

> **ADVISORY**
> **606.1 General.** If soap and towel dispensers are provided, they must be located within the reach ranges specified in 308. Locate soap and towel dispensers so that they are conveniently usable by a person at the accessible lavatory.

606.2 Clear Floor Space. A clear floor space complying with 305, positioned for a forward approach, and knee and toe clearance complying with 306 shall be provided.

EXCEPTIONS:
1. A parallel approach complying with 305 shall be permitted to a kitchen sink in a space where a cook top or conventional range is not provided and to wet bars.
2. A lavatory in a toilet room or bathing facility for a single occupant accessed only through a private office and not for common use or public use shall not be required to provide knee and toe clearance complying with 306.
3. In residential dwelling units, cabinetry shall be permitted under lavatories and kitchen sinks provided that all of the following conditions are met:
   (a) the cabinetry can be removed without removal or replacement of the fixture;
   (b) the finish floor extends under the cabinetry; and
   (c) the walls behind and surrounding the cabinetry are finished.
4. A knee clearance of 24 inches (610 mm) minimum above the finish floor or ground shall be permitted at lavatories and sinks used primarily by children 6 through 12 years where the rim or counter surface is

31 inches (785 mm) maximum above the finish floor or ground.
5. A parallel approach complying with 305 shall be permitted to lavatories and sinks used primarily by children 5 years and younger.
6. The dip of the overflow shall not be considered in determining knee and toe clearances.
7. No more than one bowl of a multibowl sink shall be required to provide knee and toe clearance complying with 306.

606.3 Height. Lavatories and sinks shall be installed with the front of the higher of the rim or counter surface 34 inches (865 mm) maximum above the finish floor or ground.

EXCEPTIONS:
1. A lavatory in a toilet or bathing facility for a single occupant accessed only through a private office and not for common use or public use shall not be required to comply with 606.3.
2. In residential dwelling unit kitchens, sinks that are adjustable to variable heights, 29 inches (735 mm) minimum and 36 inches (915 mm) maximum, shall be permitted where rough-in plumbing permits connections of supply and drain pipes for sinks mounted at the height of 29 inches (735 mm).

606.4 Faucets. Controls for faucets shall comply with 309. Hand-operated metering faucets shall remain open for 10 seconds minimum.

606.5 Exposed Pipes and Surfaces. Water supply and drain pipes under lavatories and sinks shall be insulated or otherwise configured to protect against contact. There shall be no sharp or abrasive surfaces under lavatories and sinks.

607 Bathtubs

607.1 General. Bathtubs shall comply with 607.

607.2 Clearance. Clearance in front of bathtubs shall extend the length of the bathtub and shall be 30 inches (760 mm) wide minimum. A lavatory complying with 606 shall be permitted at the control end of the clearance. Where a permanent seat is provided at the head end of the bathtub, the clearance shall extend 12 inches (305 mm) minimum beyond the wall at the head end of the bathtub.

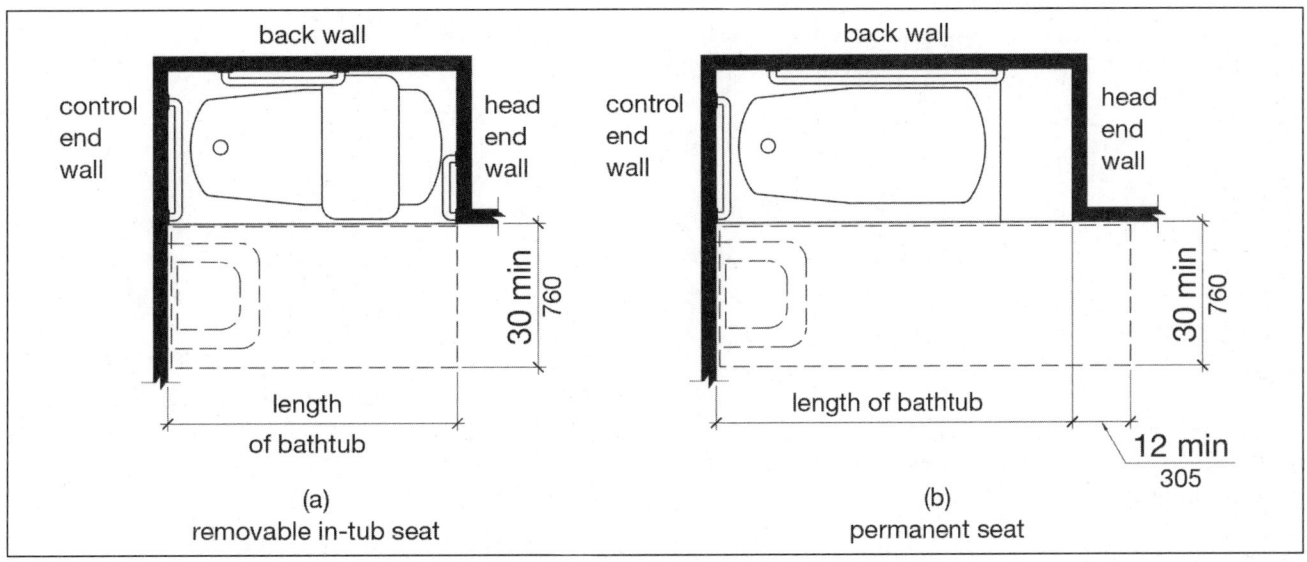

Figure 607.2 Clearance for Bathtubs

APPENDIX C

607.3 Seat. A permanent seat at the head end of the bathtub or a removable in-tub seat shall be provided. Seats shall comply with 610.

607.4 Grab Bars. Grab bars for bathtubs shall comply with 609 and shall be provided in accordance with 607.4.1 or 607.4.2.

EXCEPTIONS:
1. Grab bars shall not be required to be installed in a bathtub located in a bathing facility for a single occupant accessed only through a private office and not for common use or public use provided that reinforcement has been installed in walls and located so as to permit the installation of grab bars complying with 607.4.

2. In residential dwelling units, grab bars shall not be required to be installed in bathtubs located in bathing facilities provided that reinforcement has been installed in walls and located so as to permit the installation of grab bars complying with 607.4.

607.4.1 Bathtubs with Permanent Seats. For bathtubs with permanent seats, grab bars shall be provided in accordance with 607.4.1.

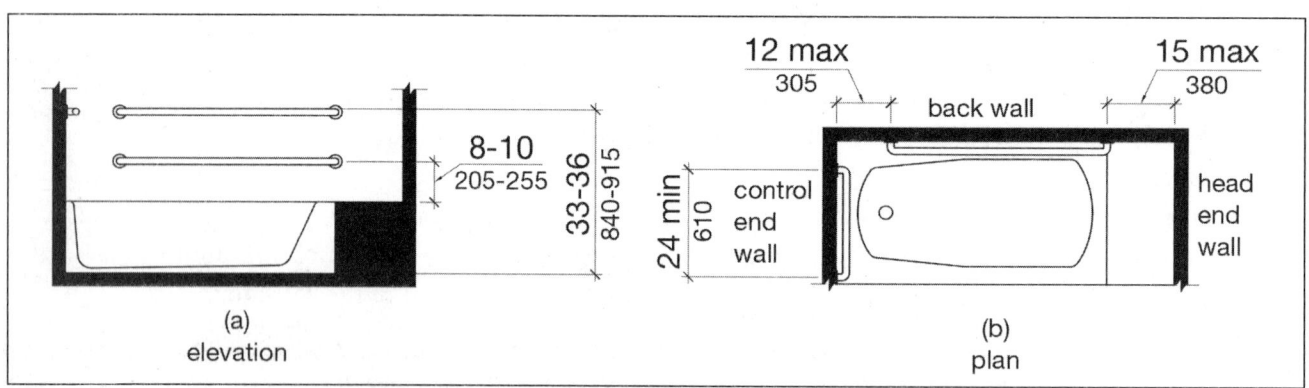

Figure 607.4.1 Grab Bars for Bathtubs with Permanent Seats

607.4.1.1 Back Wall. Two grab bars shall be installed on the back wall, one located in accordance with 609.4 and the other located 8 inches (205 mm) minimum and 10 inches (255 mm) maximum above the rim of the bathtub. Each grab bar shall be installed 15 inches (380 mm) maximum from the head end wall and 12 inches (305 mm) maximum from the control end wall.

607.4.1.2 Control End Wall. A grab bar 24 inches (610 mm) long minimum shall be installed on the control end wall at the front edge of the bathtub.

607.4.2 Bathtubs without Permanent Seats. For bathtubs without permanent seats, grab bars shall comply with 607.4.2.

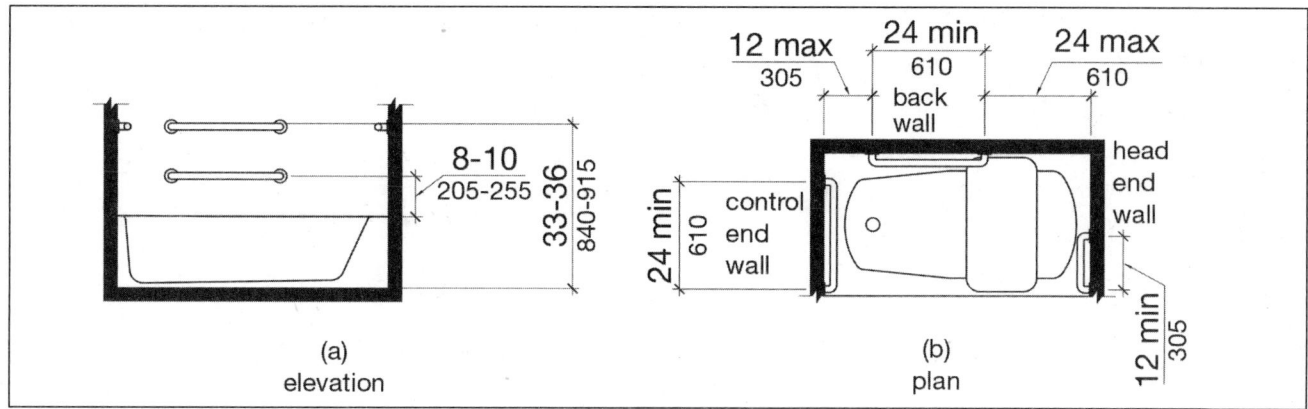

Figure 607.4.2 Grab Bars for Bathtubs with Removable In-Tub Seats

607.4.2.1 Back Wall. Two grab bars shall be installed on the back wall, one located in accordance with 609.4 and the other located 8 inches (205 mm) minimum and 10 inches (255 mm) maximum above the rim of the bathtub. Each grab bar shall be 24 inches (610 mm) long minimum and shall be installed 24 inches (610 mm) maximum from the head end wall and 12 inches (305 mm) maximum from the control end wall.

607.4.2.2 Control End Wall. A grab bar 24 inches (610 mm) long minimum shall be installed on the control end wall at the front edge of the bathtub.

607.4.2.3 Head End Wall. A grab bar 12 inches (305 mm) long minimum shall be installed on the head end wall at the front edge of the bathtub.

607.5 Controls. Controls, other than drain stoppers, shall be located on an end wall. Controls shall be between the bathtub rim and grab bar, and between the open side of the bathtub and the centerline of the width of the bathtub. Controls shall comply with 309.4.

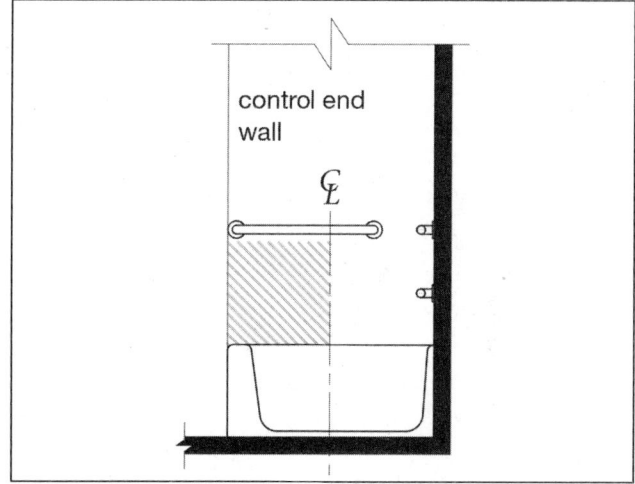

Figure 607.5 Bathtub Control Location

607.6 Shower Spray Unit and Water. A shower spray unit with a hose 59 inches (1,500 mm) long minimum that can be used both as a fixed-position shower head and as a handheld shower shall be provided. The shower spray unit shall have an on/off control with a nonpositive shut-off. If an adjustable-height shower head on a vertical bar is used, the bar shall be installed so as not to obstruct the use of grab bars. Bathtub shower spray units shall deliver water that is 120°F (49°C) maximum.

APPENDIX C

> **ADVISORY**
> **607.6 Shower Spray Unit and Water.** Ensure that handheld shower spray units are capable of delivering water pressure substantially equivalent to fixed shower heads.

607.7 Bathtub Enclosures. Enclosures for bathtubs shall not obstruct controls, faucets, shower and spray units or obstruct transfer from wheelchairs onto bathtub seats or into bathtubs. Enclosures on bathtubs shall not have tracks installed on the rim of the open face of the bathtub.

608 Shower Compartments

608.1 General. Shower compartments shall comply with 608.

> **ADVISORY**
> **608.1 General.** Shower stalls that are 60 inches (1,525 mm) wide and have no curb may increase the usability of a bathroom because the shower area provides additional maneuvering space.

608.2 Size and Clearances for Shower Compartments. Shower compartments shall have sizes and clearances complying with 608.2.

608.2.1 Transfer-Type Shower Compartments. Transfer-type shower compartments shall be 36 inches (915 mm) by 36 inches (915 mm) clear inside dimensions measured at the center points of opposing sides and shall have a 36 inch (915 mm) wide minimum entry on the face of the shower compartment. Clearance of 36 inches (915 mm) wide minimum by 48 inches (1,220 mm) long minimum measured from the control wall shall be provided.

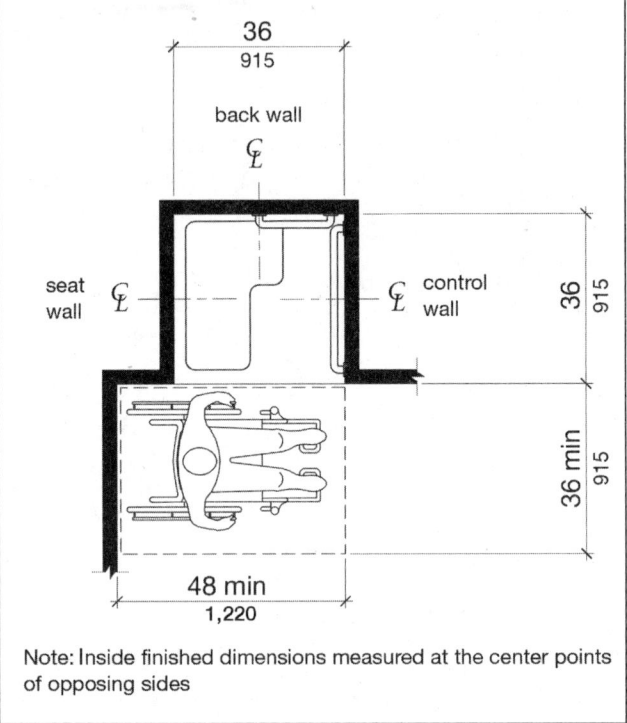

Figure 608.2.1 Transfer-Type Shower Compartment Size and Clearance

608.2.2 Standard Roll-In Type Shower Compartments. Standard roll-in type shower compartments shall be 30 inches (760 mm) wide minimum by 60 inches (1,525 mm) deep minimum clear inside dimensions measured at center points of opposing sides and shall have a 60 inches (1,525 mm) wide minimum entry on the face of the shower compartment.

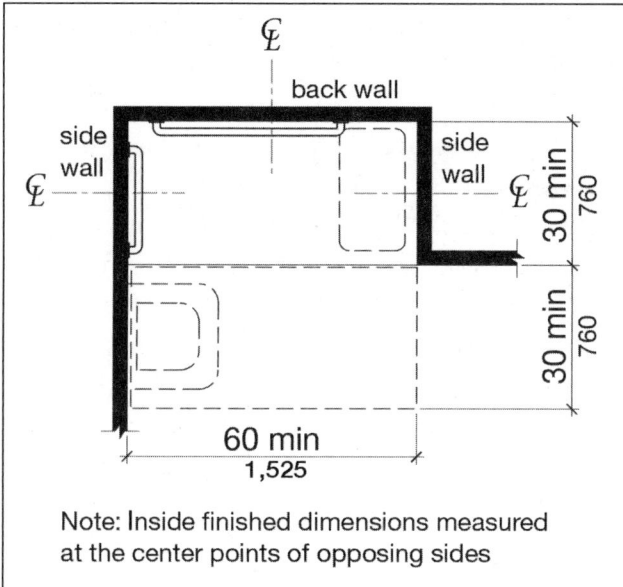

Figure 608.2.2 Standard Roll-In Type Shower Compartment Size and Clearance

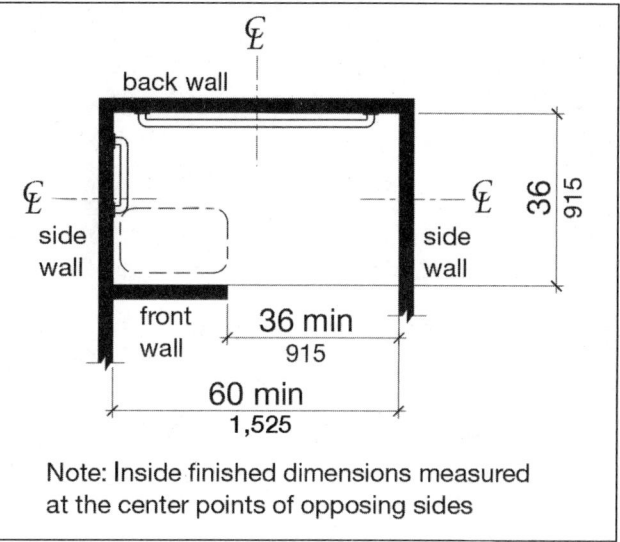

Figure 608.2.3 Alternate Roll-In Type Shower Compartment Size and Clearance

608.2.2.1 Clearance. A 30 inch (760 mm) wide minimum by 60 inch (1,525 mm) long minimum clearance shall be provided adjacent to the open face of the shower compartment.

> **EXCEPTION:** A lavatory complying with 606 shall be permitted on one 30 inch (760 mm) wide minimum side of the clearance provided that it is not on the side of the clearance adjacent to the controls or, where provided, not on the side of the clearance adjacent to the shower seat.

608.2.3 Alternate Roll-In Type Shower Compartments. Alternate roll-in type shower compartments shall be 36 inches (915 mm) wide and 60 inches (1525 mm) deep minimum clear inside dimensions measured at center points of opposing sides. A 36 inch (915 mm) wide minimum entry shall be provided at one end of the long side of the compartment.

608.3 Grab Bars. Grab bars shall comply with 609 and shall be provided in accordance with 608.3. Where multiple grab bars are used, required horizontal grab bars shall be installed at the same height above the finish floor.

> **EXCEPTIONS:**
> 1. Grab bars shall not be required to be installed in a shower located in a bathing facility for a single occupant accessed only through a private office, and not for common use or public use provided that reinforcement has been installed in walls and located so as to permit the installation of grab bars complying with 608.3.
> 2. In residential dwelling units, grab bars shall not be required to be installed in showers located in bathing facilities provided that reinforcement has been installed in walls and located so as to permit the installation of grab bars complying with 608.3.

608.3.1 Transfer-Type Shower Compartments.

In transfer-type compartments, grab bars shall be provided across the control wall and back wall to a point 18 inches (455 mm) from the control wall.

608.3.2 Standard Roll-In Type Shower Compartments.

Where a seat is provided in standard roll-in type shower compartments, grab bars shall be provided on the back wall and the side wall opposite the seat. Grab bars shall not be provided above the seat. Where a seat is not provided in standard roll-in type shower compartments, grab bars shall be provided on three walls. Grab bars shall be installed 6 inches (150 mm) maximum from adjacent walls.

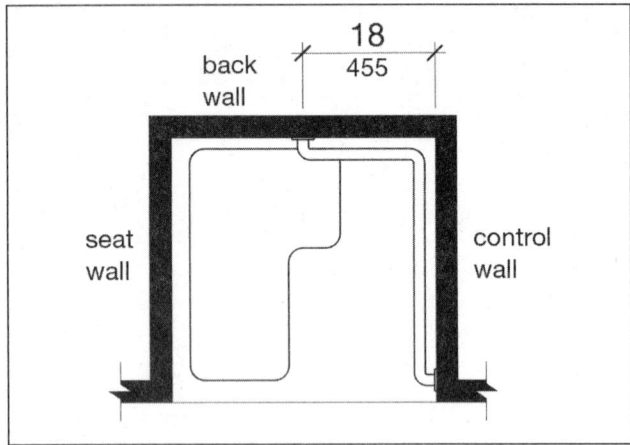

Figure 608.3.1 Grab Bars for Transfer-Type Showers

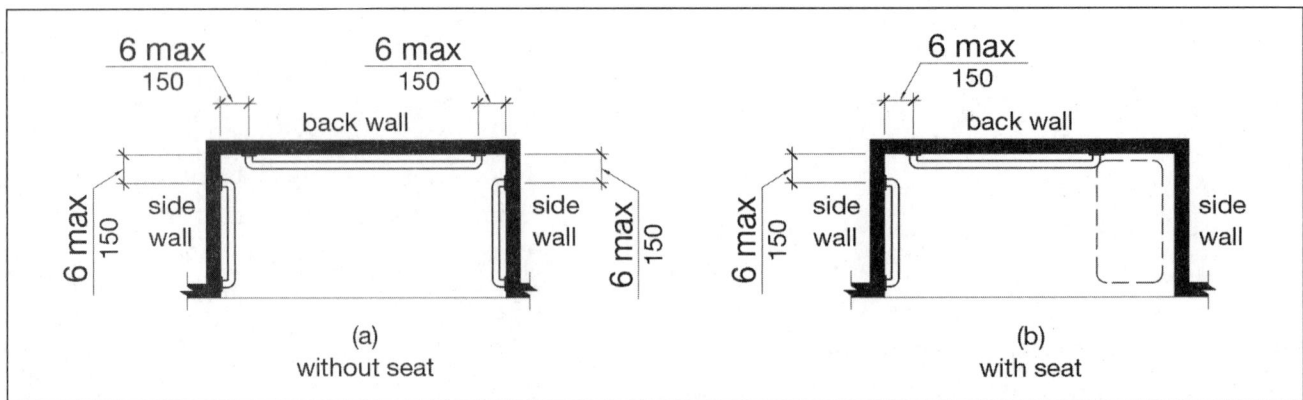

Figure 608.3.2 Grab Bars for Standard Roll-In Type Showers

APPENDIX C

608.3.3 Alternate Roll-In Type Shower Compartments. In alternate roll-in type shower compartments, grab bars shall be provided on the back wall and the side wall farthest from the compartment entry. Grab bars shall not be provided above the seat. Grab bars shall be installed 6 inches (150 mm) maximum from adjacent walls.

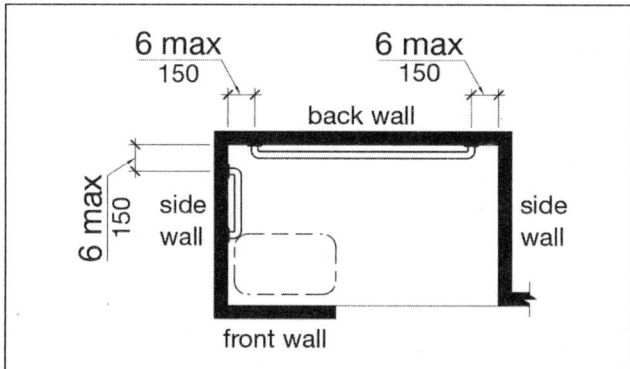

Figure 608.3.3 Grab Bars for Alternate Roll-In Type Showers

608.4 Seats. A folding or nonfolding seat shall be provided in transfer type shower compartments. A folding seat shall be provided in roll-in type showers required in transient lodging guest rooms with mobility features complying with 806.2. Seats shall comply with 610.

> **EXCEPTION:** In residential dwelling units, seats shall not be required in transfer type shower compartments provided that reinforcement has been installed in walls so as to permit the installation of seats complying with 608.4.

608.5 Controls. Controls, faucets, and shower spray units shall comply with 309.4.

608.5.1 Transfer-Type Shower Compartments. In transfer-type shower compartments, the controls, faucets, and shower spray unit shall be installed on the side wall opposite the seat 38 inches (965 mm) minimum and 48 inches (1,220 mm) maximum above the shower floor and shall be located on the control wall 15 inches (380 mm) maximum from the centerline of the seat toward the shower opening.

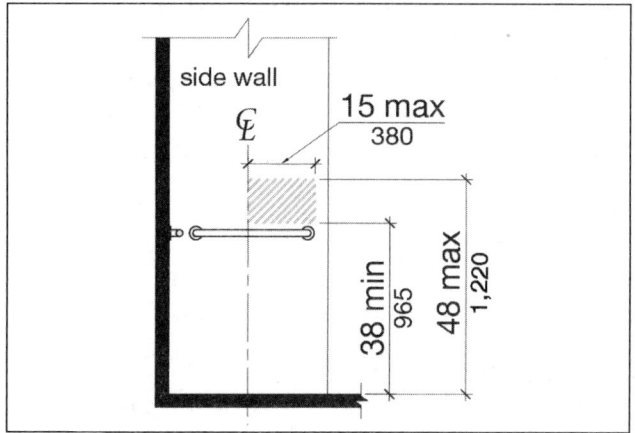

Figure 608.5.1 Transfer-Type Shower Compartment Control Location

608.5.2 Standard Roll-In Type Shower Compartments. In standard roll-in type shower compartments, the controls, faucets, and shower spray unit shall be located above the grab bar, but no higher than 48 inches (1,220 mm) above the shower floor. Where a seat is provided, the controls, faucets, and shower spray unit shall be installed on the back wall adjacent to the seat wall and shall be located 27 inches (685 mm) maximum from the seat wall.

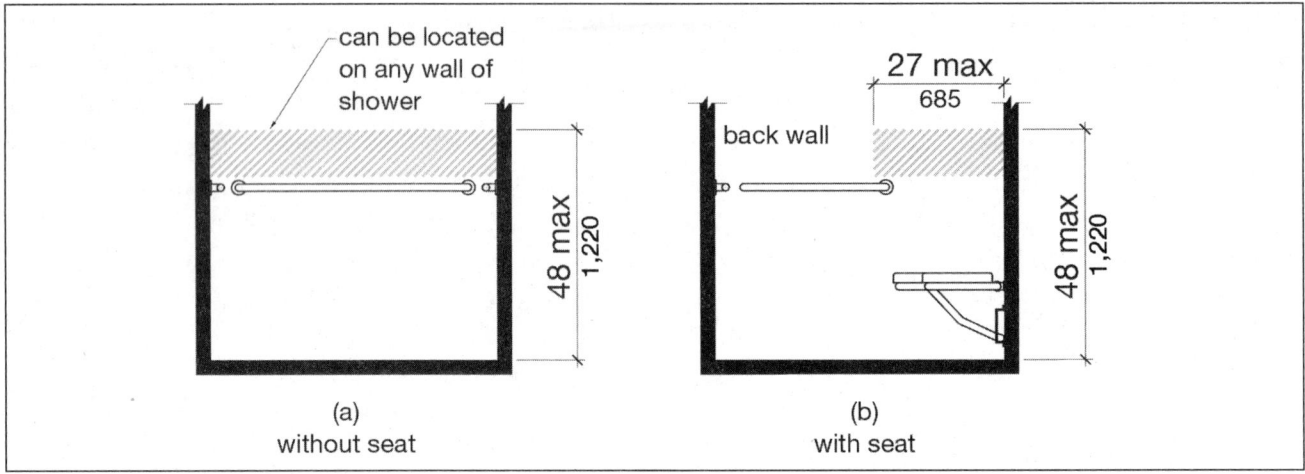

Figure 608.5.2 Standard Roll-In Type Shower Compartment Control Location

> **ADVISORY**
> **608.5.2 Standard Roll-in Type Shower Compartments.** In standard roll-in type showers without seats, the shower head and operable parts can be located on any of the three walls of the shower without adversely affecting accessibility.

608.5.3 Alternate Roll-In Type Shower Compartments. In alternate roll-in type shower compartments, the controls, faucets, and shower spray unit shall be located above the grab bar, but no higher than 48 inches (1,220 mm) above the shower floor. Where a seat is provided, the controls, faucets, and shower spray unit shall be located on the side wall adjacent to the seat 27 inches (685 mm) maximum from the side wall behind the seat or shall be located on the back wall opposite the seat 15 inches (380 mm) maximum, left or right, of the centerline of the seat. Where a seat is not provided, the controls, faucets, and shower spray unit shall be installed on the side wall farthest from the compartment entry.

APPENDIX C

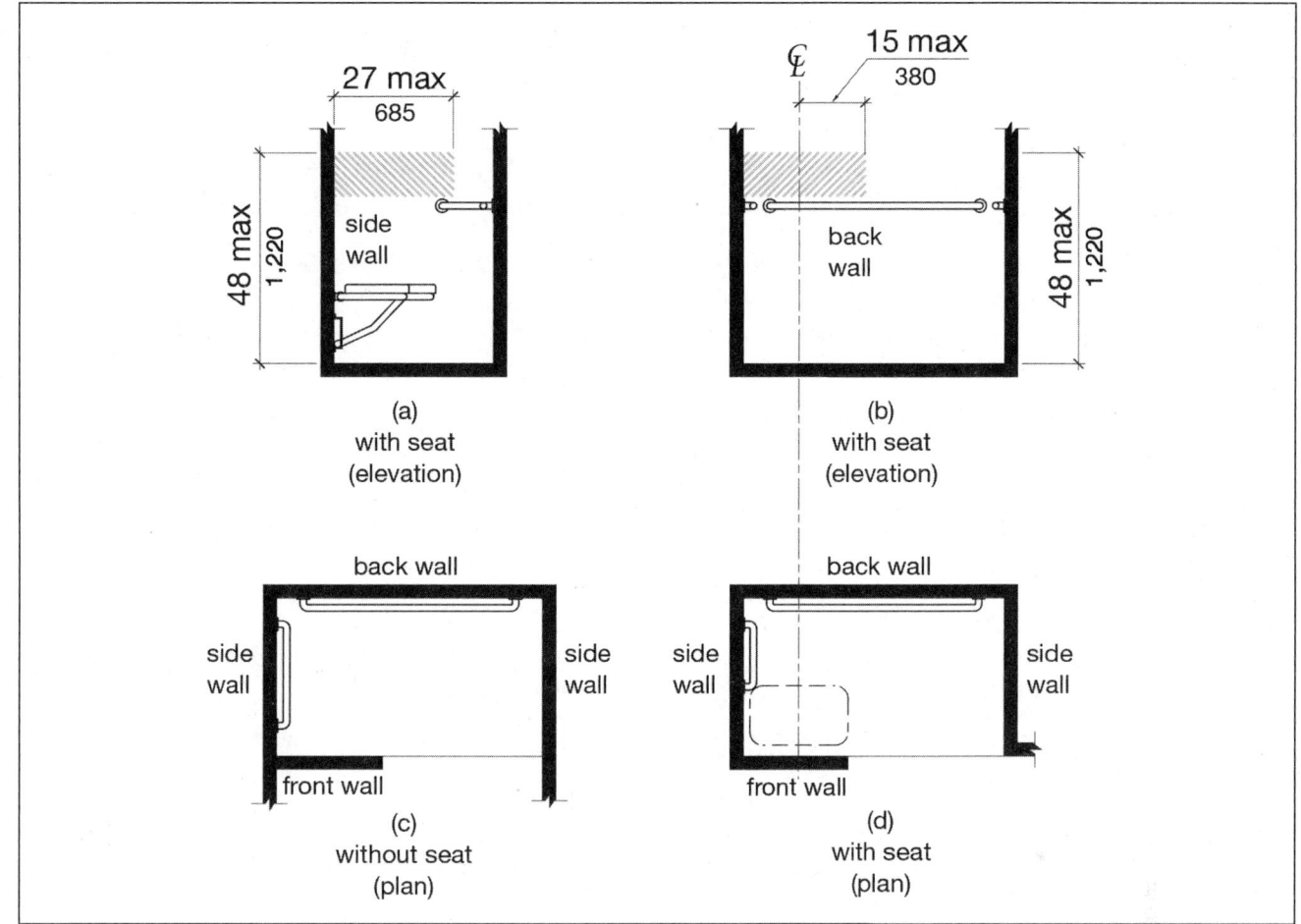

Figure 608.5.3 Alternate Roll-In Type Shower Compartment Control Location

608.6 Shower Spray Unit and Water. A shower spray unit with a hose 59 inches (1,500 mm) long minimum that can be used both as a fixed-position shower head and as a handheld shower shall be provided. The shower spray unit shall have an on/off control with a nonpositive shutoff. If an adjustable-height shower head on a vertical bar is used, the bar shall be installed so as not to obstruct the use of grab bars. Shower spray units shall deliver water that is 120°F (49°C) maximum.

EXCEPTION: A fixed shower head located at 48 inches (1,220 mm) maximum above the shower finish floor shall be permitted instead of a handheld spray unit in facilities that are not medical care facilities, long-term care facilities, transient lodging guest rooms, or residential dwelling units.

> **ADVISORY**
> **608.6 Shower Spray Unit and Water.** Ensure that handheld shower spray units are capable of delivering water pressure substantially equivalent to fixed shower heads.

608.7 Thresholds. Thresholds in roll-in type shower compartments shall be $\frac{1}{2}$ inch (13 mm) high maximum in accordance with 303. In transfer type shower compartments, thresholds $\frac{1}{2}$ inch (13 mm) high maximum shall be beveled, rounded, or vertical.

> **EXCEPTION:** A threshold 2 inches (51 mm) high maximum shall be permitted in transfer type shower compartments in existing facilities where provision of a $\frac{1}{2}$ inch (13 mm) high threshold would disturb the structural reinforcement of the floor slab.

608.8 Shower Enclosures. Enclosures for shower compartments shall not obstruct controls, faucets, and shower spray units or obstruct transfer from wheelchairs onto shower seats.

609 Grab Bars

609.1 General. Grab bars in toilet facilities and bathing facilities shall comply with 609.

609.2 Cross Section. Grab bars shall have a cross section complying with 609.2.1 or 609.2.2.

609.2.1 Circular Cross Section. Grab bars with circular cross sections shall have an outside diameter of $1\frac{1}{4}$ inches (32 mm) minimum and 2 inches (51 mm) maximum.

609.2.2 Noncircular Cross Section. Grab bars with noncircular cross sections shall have a cross-section dimension of 2 inches (51 mm) maximum and a perimeter dimension of 4 inches (100 mm) minimum and 4.8 inches (120 mm) maximum.

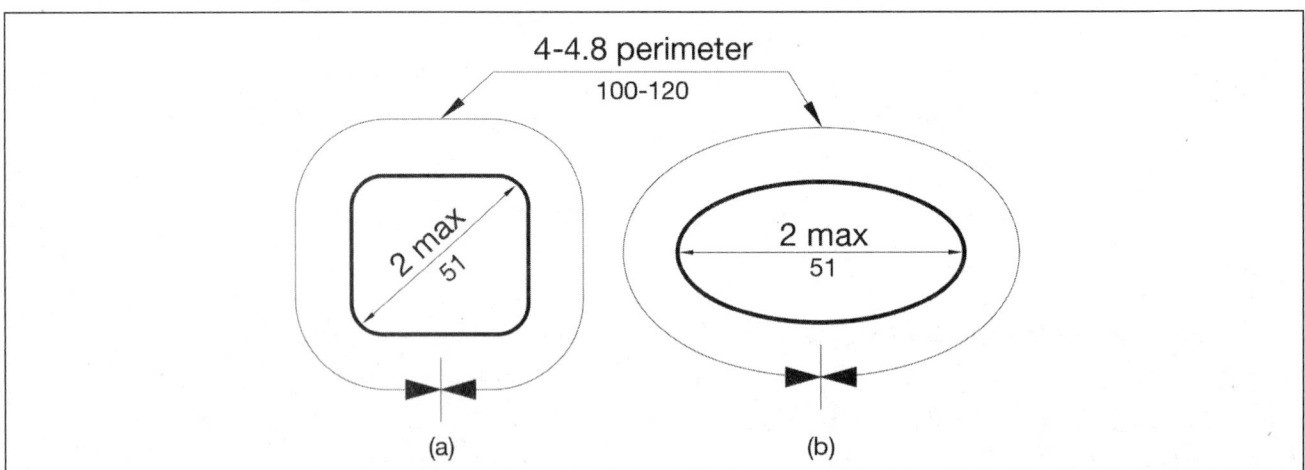

Figure 609.2.2   Grab Bar Noncircular Cross Section

609.3 Spacing. The space between the wall and the grab bar shall be $1\frac{1}{2}$ inches (38 mm). The space between the grab bar and projecting objects below and at the ends shall be $1\frac{1}{2}$ inches (38 mm) minimum. The space between the grab bar and projecting objects above shall be 12 inches (305 mm) minimum.

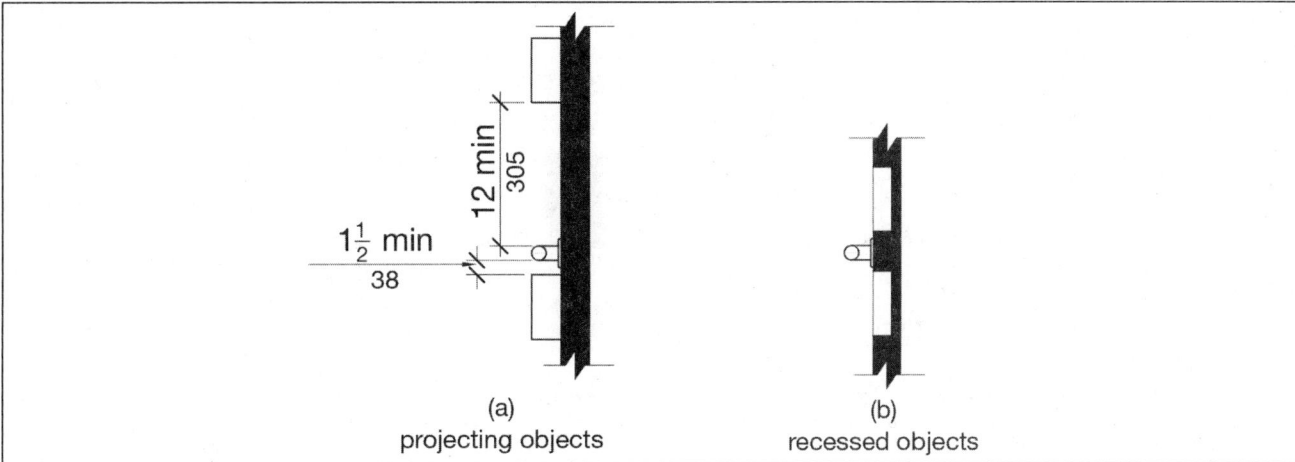

Figure 609.3 Spacing of Grab Bars

EXCEPTION: The space between the grab bars and shower controls, shower fittings, and other grab bars above shall be permitted to be 1½ inches (38 mm) minimum.

609.4 Position of Grab Bars. Grab bars shall be installed in a horizontal position, 33 inches (840 mm) minimum and 36 inches (915 mm) maximum above the finish floor measured to the top of the gripping surface, except that at water closets for children's use complying with 604.9, grab bars shall be installed in a horizontal position 18 inches (455 mm) minimum and 27 inches (685 mm) maximum above the finish floor measured to the top of the gripping surface. The height of the lower grab bar on the back wall of a bathtub shall comply with 607.4.1.1 or 607.4.2.1.

609.5 Surface Hazards. Grab bars and any wall or other surfaces adjacent to grab bars shall be free of sharp or abrasive elements and shall have rounded edges.

609.6 Fittings. Grab bars shall not rotate within their fittings.

609.7 Installation. Grab bars shall be installed in any manner that provides a gripping surface at the specified locations and that does not obstruct the required clear floor space.

609.8 Structural Strength. Allowable stresses shall not be exceeded for materials used when a vertical or horizontal force of 250 pounds (1,112 N) is applied at any point on the grab bar, fastener, mounting device, or supporting structure.

610 Seats

610.1 General. Seats in bathtubs and shower compartments shall comply with 610.

610.2 Bathtub Seats. The top of bathtub seats shall be 17 inches (430 mm) minimum and 19 inches (485 mm) maximum above the bathroom finish floor. The depth of a removable in-tub seat shall be 15 inches (380 mm) minimum and 16 inches (405 mm) maximum. The seat shall be capable of secure placement. Permanent seats at the head end of the bathtub shall be 15 inches (380 mm) deep minimum and shall extend from the back wall to or beyond the outer edge of the bathtub.

APPENDIX C

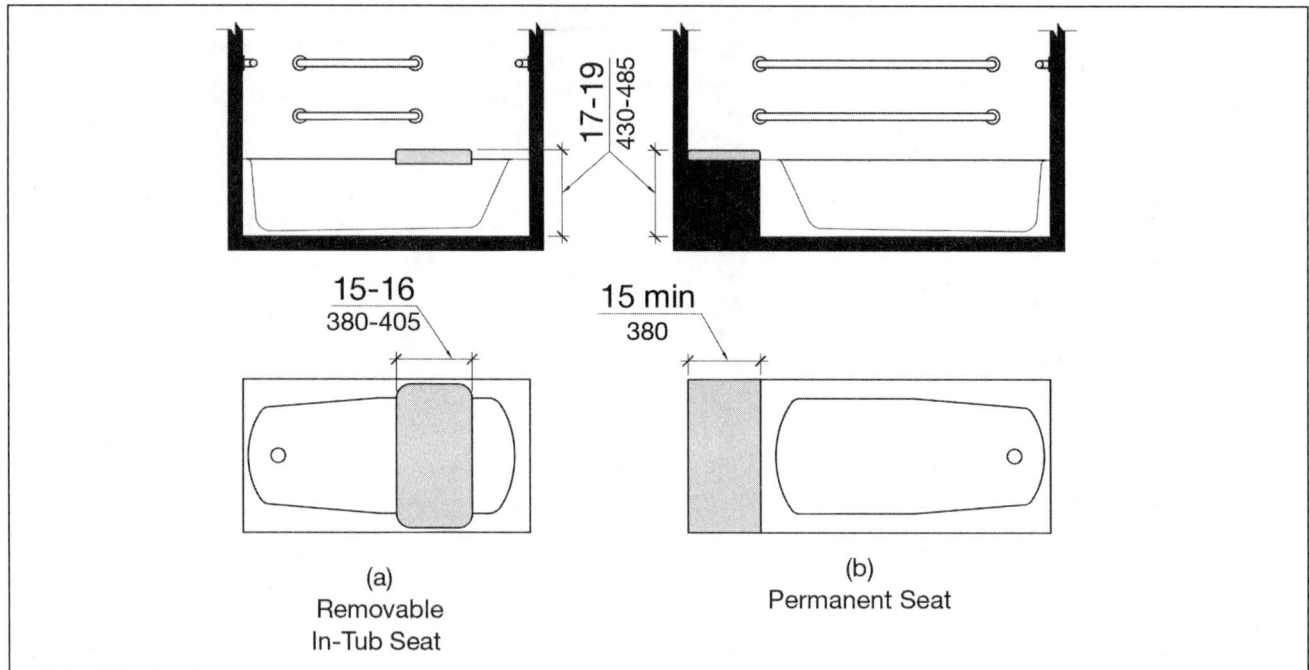

Figure 610.2 Bathtub Seats

610.3 Shower Compartment Seats. Where a seat is provided in a standard roll-in shower compartment, it shall be a folding type, shall be installed on the side wall adjacent to the controls, and shall extend from the back wall to a point within 3 inches (75 mm) of the compartment entry. Where a seat is provided in an alternate roll-in type shower compartment, it shall be a folding type, shall be installed on the front wall opposite the back wall, and shall extend from the adjacent side wall to a point within 3 inches (75 mm) of the compartment entry. In transfer-type showers, the seat shall extend from the back wall to a point within 3 inches (75 mm) of the compartment entry. The top of the seat shall be 17 inches (430 mm) minimum and 19 inches (485 mm) maximum above the bathroom finish floor. Seats shall comply with 610.3.1 or 610.3.2.

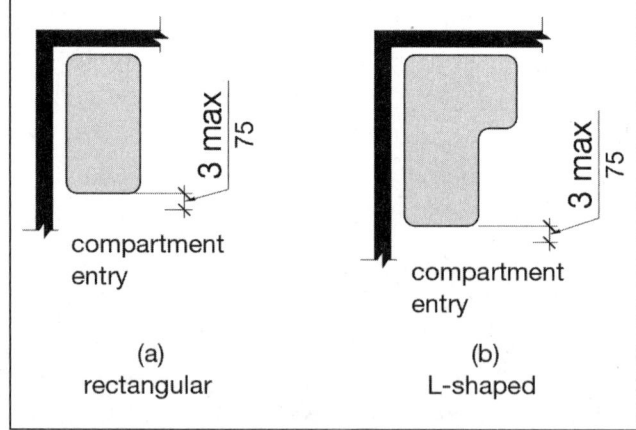

Figure 610.3 Extent of Seat

610.3.1 Rectangular Seats. The rear edge of a rectangular seat shall be $2\frac{1}{2}$ inches (64 mm) maximum and the front edge 15 inches (380 mm) minimum and 16 inches (405 mm) maximum from the seat wall. The side edge of the seat shall be $1\frac{1}{2}$ inches (38 mm) maximum from the adjacent wall.

APPENDIX C

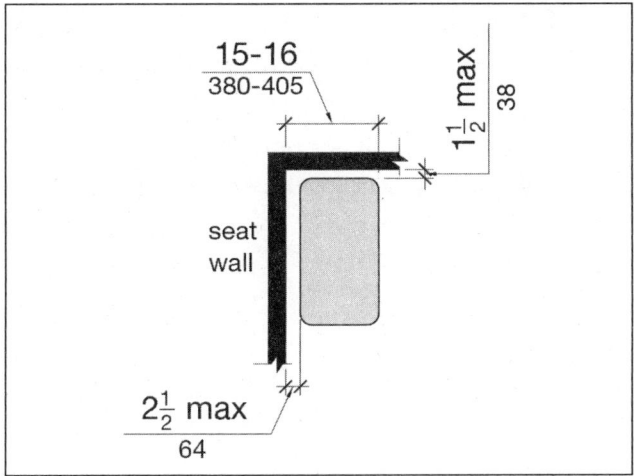

Figure 610.3.1 Rectangular Shower Seat

610.3.2 L-Shaped Seats. The rear edge of an L-shaped seat shall be 2½ inches (64 mm) maximum and the front edge 15 inches (380 mm) minimum and 16 inches (405 mm) maximum from the seat wall. The rear edge of the "L" portion of the seat shall be 1½ inches (38 mm) maximum from the wall and the front edge shall be 14 inches (355 mm) minimum and 15 inches (380 mm) maximum from the wall. The end of the "L" shall be 22 inches (560 mm) minimum and 23 inches maximum (585 mm) from the main seat wall.

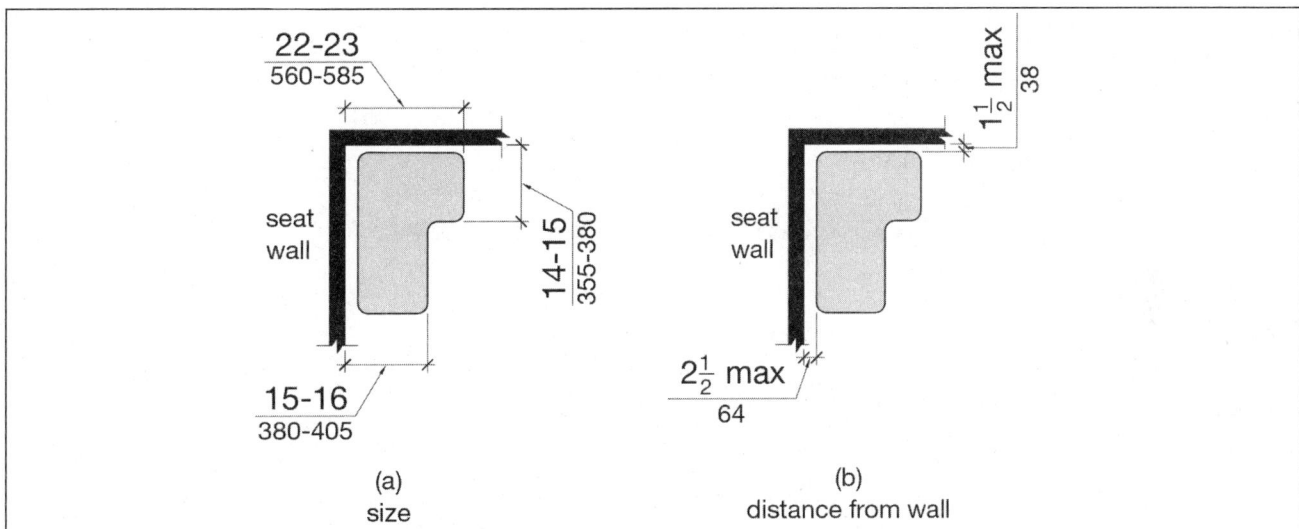

Figure 610.3.2 L-Shaped Shower Seat

610.4 Structural Strength. Allowable stresses shall not be exceeded for materials used when a vertical or horizontal force of 250 pounds (1,112 N) is applied at any point on the seat, fastener, mounting device, or supporting structure.

611 Washing Machines and Clothes Dryers

611.1 General. Washing machines and clothes dryers shall comply with 611.

611.2 Clear Floor Space. A clear floor or ground space complying with 305 positioned for parallel approach shall be provided. The clear floor or ground space shall be centered on the appliance.

611.3 Operable Parts. Operable parts, including doors, lint screens, and detergent and bleach compartments shall comply with 309.

611.4 Height. Top-loading machines shall have the door to the laundry compartment located 36 inches (915 mm) maximum above the finish floor. Front-loading machines shall have the bottom of the opening to the laundry compartment located 15 inches (380 mm) minimum and 36 inches (915 mm) maximum above the finish floor.

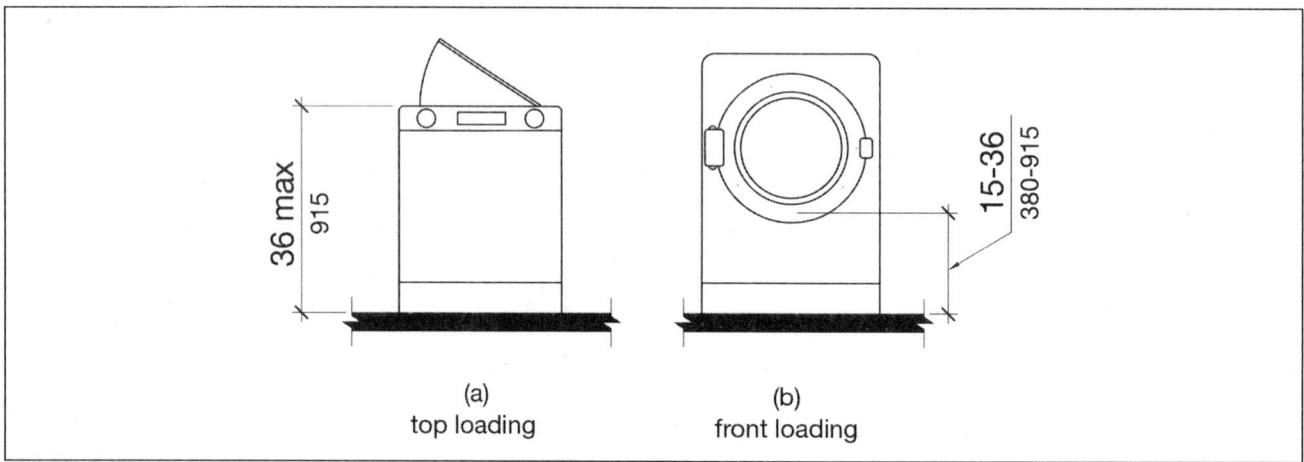

Figure 611.4 Height of Laundry Compartment Opening

612 Saunas and Steam Rooms

612.1 General. Saunas and steam rooms shall comply with 612.

612.2 Bench. Where seating is provided in saunas and steam rooms, at least one bench shall comply with 903. Doors shall not swing into the clear floor space required by 903.2.

EXCEPTION: A readily removable bench shall be permitted to obstruct the turning space required by 612.3 and the clear floor or ground space required by 903.2.

612.3 Turning Space. A turning space complying with 304 shall be provided within saunas and steam rooms.

APPENDIX D
STATE LICENSING RESOURCE LIST

This list of state contractor license boards is provided to help you find more information about licensing in a particular state.

Alabama
General Contractors Board
2525 Fairlane Drive
Montgomery, AL 36116
334-272-5030
http://www.genconbd.state.al.us

Alaska
Division of Occupational Licensing
Box 110806
Juneau, AK 99811-0806
907-465-2546
http://www.dced.state.ak.us/occ

Arizona
Arizona Registrar of Contractors
800 W. Washington, 6th Floor
Phoenix, AZ 85007
602-542-1525
http://www.rc.state.az.us

Arkansas
Contractors License Board
621 E. Capitol Avenue
Little Rock, AR 72202
501-372-4661
http://www.state.ar.us/clb

California
Contractors State License Board
P.O. Box 26000
Sacramento, CA 95826
800-321-CSLB
http://www.cslb.ca.gov

Colorado
Division of Registrations
1560 Broadway, Suite 1300
Denver, CO 80202
303-894-7690
http://www.dora.state.co.us/registrations/index.htm

APPENDIX D

Connecticut
Department of Consumer Protection
165 Capitol Avenue, Room 110
Hartford, CT 06106
860-713-6000
http://www.dcp.state.ct.us/licensing/

Delaware
Division of Revenue
Carvel State Building
820 N. French Street
Wilmington, DE 19801
302-577-8200
http://www.state.de.us/revenue

District of Columbia
No information available.

Florida
Dept. of Business and Professional Regulation
1940 N. Monroe
Tallahassee, FL 3299-0783
850-487-1395
http://www.state.fl.us/dbpr/pro/cilb/cilb_index.shtml

Georgia
Construction Industry Board
237 Coliseum Drive
Macon, GA 31217
478-207-1416
http://www.sos.state.ga.us/plb/construct

Hawaii
P.O. Box 3469
Honolulu, HI 96801
808-586-3000
http://www.state.hi.us/dcca/pvl

Idaho
Division of Building Safety
1090 East Waterfront Street
Meriman, ID 83642
208-334-3951
http://www.2.state.id.us/dbs/dbs_index.html

Illinois
Department of Professional Regulation
320 W. Washington Street
Springfield, IL 62786
217-782-0458
http://www.dpr.state.il.us

Indiana
Indiana Professional Licensing Agency
302 West Washington Street, Room E034
Indianapolis, IN 46204-2700
317-232-2980
http://www.IN.gov/pla/

Iowa
Division of Labor
1000 East Grand Avenue
Des Moines, IA 50319
515-242-5871
http://www.iowaworkforce.org/labor/index.html

Kansas
Dept. of Revenue—Division of Taxation
Robert B. Docking State Building
Topeka, KS 66625
913-296-0222
http://www.ksrevenue.org/

Kentucky
Department of Housing, Buildings,
    and Construction
1047 U.S. Highway 127 South, Suite 1
Frankfort, KY 40601
502-564-3580
http://hbc.ppr.ky.gov/

APPENDIX D

Louisiana
Contractors Board
P.O. Box 14419
Baton Rouge, LA 70898
225-765-2301

Maine
Department of Environmental Protection
17 State House Station
Augusta, ME 04333
207-287-2651
http://www.state.me.us/dep/index.shtml

Maryland
Division of Occupational and Professional Licensing
500 N. Calvert Street
Baltimore, MD 21202
410-230-6270
http://www.dllr.state.md.us

Massachusetts
McCormack State Office Building
One Ashburton Place, Room 1301
Boston, MA 02108
617-727-3200
http://www.state.ma.us/bbrs/hic.htm

Michigan
Bureau of Commercial Services
P.O. Box 30245
Lansing, MI 48909
517-241-9254
http://www.michigan.gov/cis

Minnesota
Department of Commerce
133 East 7th Street
St. Paul, MN 55101
800-657-3978
http://www.state.mn.us/cgi-bin/portal/mn/jsp/
    home.do?agency=Commerce

Mississippi
Mississippi Contractors License Board
215 Woodline Drive, Suite B
Jackson, MS 39232
601-354-6161
http://www.msboc.state.ms.us/

Missouri
Business Services
P.O. Box 778
Jefferson City, MO 65102
573-751-4153
http://www.sos.state.mo.us/

Montana
Department of Labor and Industry
P.O. Box 8011
Helena, MT 59604-8011
406-444-7734
http://erd.dli.state.mt.us/

Nebraska
Nebraska Workforce Development—Department
    of Labor
5404 Cedar Street, 3rd Floor
Omaha, NE 68106
402-595-3183
http://www.dol.state.ne.us/

Nevada
State of Nevada Contractors Board—Reno Office
9670 Gateway Drive, Suite 100
Reno, NV 89511
775-688-1141
http://nscb.state.nv.us/

New Hampshire
Secretary of State
State House, Room 204
107 North Main Street
Concord, NH 03301-4989
603-217-3246
http://www.state.nh.us/sos/

New Jersey
Department of Community Affairs
Bureau of Homeowner Protection
New Home Warranty Program
P.O. Box 805
Trenton, NJ 08625-0805
609-530-8800
http://www.state.nj.us/dca/index.html

New Mexico
Regulation and Licensing Department
725 St. Michael's Drive
P.O. Box 25101
Santa Fe, NM 87504
505-827-7000
http://www.rld.state.nm.us

New York
NYS Department of State Division of Corporations
41 State Street
Albany, NY 12231
518-473-2492
http://www.dos.state.ny.us/

North Carolina
North Carolina Licensing Board for General
    Contractors
P.O. Box 17187
Raleigh, NC 27619
919-571-4183
http://www.nclbgc.org

North Dakota
Secretary of State
600 East Boulevard Avenue, Dept. 108
Bismarck, ND 58505
701-328-3556
http://www.state.nd.us/sec

Ohio
Ohio Construction Industry Examining Board
P.O. Box 4409
Reynoldsburg, OH 43068
614-644-3493
http://www.com.state.oh.us/ODOC/dic/dicocieb.htm

Oklahoma
Oklahoma Tax Commission
2501 Lincoln Blvd.
Oklahoma City, OK 73194
405-521-4437
http://oktax.state.ok.us/oktax/

Oregon
Construction Contractors Board
P.O. Box 14140
Salem, OR 97309-5052
503-378-4621
http://www.ccb.state.or.us

Pennsylvania
Department of General Services
18th and Herr Streets, Room 103
Harrisburg, PA 17125
717-783-7610
http://www.dgs.state.pa.us/

Rhode Island
Department of Administration
Contractor's Registration Board
One Capitol Hill
Providence, RI 02908-5859
401-222-1268
http://www.crb.state.ri.us

APPENDIX D

South Carolina
South Carolina Department of Labor, Licensing,
    and Regulation
Synergy Business Park, Kingtree Building
110 Centerview Dr., Suite 102
Columbia, SC 29210
803-896-4696
http://www.llr.state.sc.us/POL/ResidentialBuilders/

South Dakota
Professional and Occupational Licensing
118 West Capitol
Pierre, SD 57501
605-773-3153
http://www.state.sd.us/sos/sos.htm

Tennessee
Board for Licensing Contractors
500 James Robertson Parkway, Suite 100
Nashville, TN 37243-1150
615-741-8307
http://www.state.tn.us/commerce/boards/
    contractors/index.html

Texas
Corporations Section
Office of the Secretary of State
P.O. Box 13697
Austin, TX 78711-3697
512-463-5555
http://www.license.state.tx.us/

Utah
Division of Occupational and Professional Licensing
160 East 300 South, 4th Floor
Salt Lake City, UT 84145
http://www.commerce.state.ut.us

Vermont
Office of the Vermont Secretary of State
Corporations/UCC Division
81 River Street Drawer 09
Montpelier, VT 05609
802-828-2386
http://www.sec.state.vt.us

Virginia
Department of Professional and Occupational
    Regulation
Board for Contractors
3600 West Broad Street, P.O. Box 11066
Richmond, VA 23230-1066
http://www.state.va.us/dpor/indexie.html

Washington
Department of Labor & Industries, Contractors
Regulation Section
P.O. Box 44450
Olympia, WA 98504-4450
360-902-5226
https://wws2.wa.gov/lni/bbip/contractor.asp

Wisconsin
Department of Financial Institutions
345 West Washington Avenue
P.O. Box 7846
Madison, WI
608-261-7577
http://www.commerce.state.wi.us/SB/SB-Div
ProgramsListed.html

Wyoming
State of Wyoming, Electrical Board
Department of Fire Prevention & Electrical Safety
Cheyenne, WY 82002
307-777-7288
http://wyofire.state.wy.us/

GLOSSARY

The following glossary is meant as a tool to prepare you for the plumber's licensing exam. You will not be asked any vocabulary questions on the exam, so there is no need to memorize any of these terms or definitions. However, reading through this list will familiarize you with general words and concepts, as well as terms you may encounter in the exam questions.

A

ADA Americans with Disabilities Act. Enacted in 1990, this federal act created regulations that guarantee access of public areas to people with various disabilities. Some of the requirements include ramps for wheelchairs, grab bars, and larger toilet areas to accommodate wheelchairs.

adhesion an attraction between different materials on a molecular basis.

air admittance valve a one-way valve that closes by gravity. This allows air to be drawn into a drainage system when negative pressure exists, such as that created when a slug of water travels down it, but does not allow air or sewer gasses into the building through the valve.

air break an indirect waste that discharges into a receptor below the flood-level rim of the receptor.

air gap an indirect waste that discharges into a receptor above the flood-level rim of the receptor.

ambient the area surrounding something; often referred to as room temperature for plumbing purposes.

aquastat a device that controls the water temperature in a vessel or in piping. An aquastat has a sensing probe, either inside the water vessel or mounted on the surface. This actuates the manually set control that turns the heating appliance on or off according to the water temperature desired. Aquastats are always in hydronic (heat distributed by water) systems.

GLOSSARY

ASME American Society of Mechanical Engineers. The ASME has researched, created, and published many engineering standards, including those for piping and storage systems for medical gasses.

B

backflow preventer a system of check valves of various configurations that allow flow in one direction and securely prevent flow in the opposite direction.

black water water that contains human bodily waste.

BTU British Thermal Unit, a measurement of heat. The amount of heat required to raise one pound of liquid water 1°F is 1 BTU.

building drain the lowest (with the exception of subdrains) horizontal part of a drain system within a building that receives all the drainage from the fixtures within the building and extends to 30 inches outside the exterior wall of the building.

building sewer the final part of a building drain system that begins where the building drain ends, 30 inches beyond the exterior wall of the building. It then extends to the sewage disposal system, be it the public sewer main, a septic system, or other sewage disposal system.

C

closet carrier a cast iron and steel structure that is concealed in the wall and is used to support a wall-mounted closet.

concealed fouling surface from the definitions section of the IPC, a concealed fouling surface is "any surface of a plumbing fixture which is not readily visible and is not scoured or cleansed with each fixture operation." For our use in this chapter, the definition also includes the surface of the wall or floor adjacent to and covered by the fixture body.

conductor a pipe inside a building that carries storm water from a roof drain to the storm drain or a combined drain.

constant a number that is always the same in a particular mathematical equation.

D

DFU drainage fixture unit. A number used to measure the load impact of a fixture on the drainage system.

dielectric unions a type of three-part union that is usually half galvanized and half brass. Between the galvanized and brass sections is a plastic insert that prevents galvanic corrosion between the two metals. Any time two pipes of dissimilar metals are joined, it's a good idea to install a dielectric union to reduce corrosion.

domestic hot water potable water supplied by the building water distribution system, which has been heated for residential use at sinks, lavatories, baths, showers, a laundry, or any other residential heated water application.

E

effluent the liquid waste discharged from a drainage pipe.

F

flashing a system of sealing layers around a pipe or device that passes through a surface, which is designed to prevent water from entering a roof or wall through the penetration.

flow pressure the pressure in the water supply pipe near the faucet or water outlet while the faucet or water outlet is wide open and flowing.

flush tank a tank equipped with a valve that automatically refills it. When actuated, water is released from the tank by gravity into the fixture to be flushed.

GLOSSARY

flushometer tank a tank equipped with another pressure tank that fills with water, compressing the air captured inside it until an equal pressure of water and air is achieved. When actuated, the compressed air forces the water out of the tank and into the fixture to be flushed.

flushometer valve a valve that is connected to a water line under pressure and uses that water pressure to flush a fixture. The flushometer automatically closes after the prescribed amount of water needed for the flush has passed through it.

G

GPF gallon(s) per flush.

grade for our purposes, the level of the earth where it meets a building.

gray water water drained from showers, bathtubs, automatic clothes washers, laundry trays, and lavatories. This is water that has no human bodily waste or food waste.

green environmentally friendly.

H

head the pressure at the bottom of a certain height of a column of a liquid.

hoar frost a buildup of frost over a period of time when condensation contacts a cold surface and freezes.

hydraulic jump water rises to a depth greater than a half pipe in the horizontal pipe at a distance within 10 pipe diameters of the vertical stack. This is caused by the sudden slowing of the velocity of the water from the approximately 10 feet per second it attains in the vertical pipe to the pace of 2 feet per second in the horizontal pipe sloped at $\frac{1}{4}$-inch per foot. As the water slows, it "stacks up" like traffic at a tollbooth until it can resume a steady flow at a slower pace.

I

ID inside diameter or nominal diameter.

indirect waste a waste pipe that is not directly connected to the drainage system of the building. It ends above a receptor such as a floor drain or floor sink. The end of the indirect waste pipe is suspended above the receptor a minimum of 1 inch to prevent a backup in the drainage system from entering the indirect waste pipe.

individual water supply a water supply that is not supplied by or approved by a public entity. Individual water supplies can be from wells, springs, streams, or cisterns. If surface water is utilized as the water supply, it must be appropriately treated to ensure its potability.

invert the bottom portion of a drainage pipe. The part of the pipe where the waste flows in a stream.

L

local vent a vent that connects to a fixture on the building side of the fixture trap to vent foul odors from the air in the fixture and extends through the roof. Local vents are commonly found on bedpan washers.

N

natural gas a naturally occurring combination of flammable gasses found underground in pockets of certain rocks. It can be used as is, and it is lighter than air. Natural gas is customarily delivered to the point of use by underground pipelines and is metered at the service entrance to the building. If there is a natural gas leak in a building, it will not settle in the lower portions of the building, but will rise and usually exit the building through the roof venting system.

GLOSSARY

$97\frac{1}{2}$% design temperature the temperature in a particular area will be above this temperature $97\frac{1}{2}$% of the time. If the outdoor winter design temperature is 30°F, this means that the temperature in the winter will be 30°F or higher $97\frac{1}{2}$% of the time. Charts have been created by the National Oceanic and Atmospheric Administration (NOAA) that provide the $97\frac{1}{2}$% temperature for a number of localities in every state of the United States.

O

offset a combination of fittings that change the location of a pipe, but ends with it continuing in the same original direction. This is commonly done to move around an obstruction such as a beam or post.

oxidation the chemical process that results when an element combines with oxygen, such as rusting or burning.

P

plenum a space or box, up to room size, that is used for distribution and pressure balance of air throughout a building that has been conditioned by heating, cooling, humidification, dehumidification, filtering, or other treatment. A plenum is not intended to be for habitation. A distribution system of ducts is generally the distribution means attached to a plenum.

potable water that is free from impurities that would cause harm if ingested and is therefore suitable for drinking.

primary roof drain a grate in a roof at a low point that is connected to a conductor with the purpose of collecting rainwater and transporting it to the storm drainage system.

propane gas a by-product of the refining process of oil. It is burned as a gas, but it is usually compressed into a liquid (liquefied petroleum gas, LPG) for easy transport and delivery. Propane is most often stored in tanks at the site where it is consumed. Leaking propane can collect in explosive pools in crawl spaces, basements, and other low portions of a building since it is heavier than air.

psi pounds per square inch, or, the force exerted on 1 square inch of a surface, such as that created on the inside of a water tank by the pressure from the water supply pipe or the added pressure exerted by the expansion of water being heated in the water heater.

psig pounds per square inch, gauge. This is pressure expressed in pounds per square inch above normal atmospheric pressure. Normal atmospheric pressure is 14.7 psi. Psig is a measurement of pressure greater than 14.7 psi, using 14.7 psi as zero.

S

sanitary drainage system a drainage system that carries sewage. It does not carry any surface water, storm runoff, or groundwater. This system receives its supply from the plumbing fixtures in a building and nowhere else.

scupper rectangular holes in the wall at a height demanded by the load-bearing ability of the roof for the purpose of removing rainwater that has collected on the roof when the primary roof drain cannot carry all the water away.

secondary roof drain a construction method, usually accomplished with scuppers, that conducts rainwater off a roof to a location that is obvious to an observer, who can recognize that the primary roof drain is not removing all the water.

stack vent the vertical extension of a soil stack above the point of any fixture branch. The stack vent conducts no waste.

GLOSSARY

station outlet the point, usually mounted on a wall, where the user makes connections to use the medical gas.

subsoil drain a buried drainage system, usually around the footing of a building, designed to carry water in the soil near the footing away from the building.

sump a pit designed for the collection of water.

sump pump a pump placed inside a sump for the removal of the water collected in the sump.

T

thermostat a device that regulates the temperature in a building or a section of a building. A thermostat does this by turning a heating or cooling system on or off according to a manually determined setting. Temperature in the room is sensed by various means, including thermistors, bimetallic strips, bimetallic discs, gas expansion, thermo couples, and thermopiles.

Transite pipe a brand name of asbestos cement pipe manufactured by the Johns-Manville Company. The name "Transite" has become synonymous with asbestos cement pipe, since Johns-Manville was the largest manufacturer of the product.

trap seal the distance vertically from the trap weir, or point of overflow into the drain, to the top of the dip of the trap. As long as there is water in this area, no sewer gas can enter the building through the trap.

V

vacuum drainage system a vacuum pump connected to a receiving tank creates a vacuum condition in the drainage system. Fixtures are emptied by opening a valve in the drain, which allows atmospheric pressure to push the waste and water down the drain to the receiving tank.

vacuum relief valve also known as a vacuum breaker, this device consists of a one-way check valve along with a means for allowing air into the outlet side of the valve when the water supply is off. This construction prevents backflow into the supply piping and prevents a siphon from occurring in the outlet pipe.

vent stack a vertical pipe connecting at or near the base of the stack it serves and extending through the roof of the building, used only for conducting air and sewer gas. No waste flows in a vent stack.

W

WC water column. This is the pressure required to create a differential of a certain number of inches in a U-shaped tube. It is used primarily for measuring small pressures similar to those found in end-user gas supplies.

weir the point in a trap where the water overflows the trap and continues down the drainage pipe.

ADDITIONAL ONLINE PRACTICE

Whether you need help building basic skills or preparing for an exam, visit the **LearningExpress Practice Center**! On this site, you can access FREE additional practice materials. This online practice will also provide you with:

- **Immediate Scoring**
- **Detailed answer explanations**
- **Personalized recommendations for further practice and study**

Follow the directions below to access your free additional practice:

- E-mail **LXHub@learningexpresshub.com** for your access code. Please include your name, contact information, and complete book title in your e-mail.
- Write "Free online practice code" in your e-mail subject line.
- You will receive your access code by e-mail.
- Go to **www.learningexpresshub.com/affiliate** and follow the easy registration steps. Be sure to have your access code handy!
    ○ If this is your first time registering, be sure to register as a **new user**.
    ○ If you've registered before with another product, be sure to follow the steps as a **returning user**.

The e-mail address you register with will become your username. You will also be prompted to create a password. For easy reference, record them here:

Username: _____

Password: _____

With your username and password, you can log in and access your additional practice. If you have any questions or problems, please contact LearningExpress customer service at 1-800-295-9556 ext. 2, or e-mail us at **customerservice@learningexpressllc.com**.

NOTES

NOTES

NOTES

NOTES

Plumbing Code Resources from the ICC PMG Resource Center

The International Code Council® (ICC®), a membership association dedicated to building safety and fire prevention, develops the codes used to construct residential and commercial buildings, including homes and schools. Most U.S. cities, counties, and states that adopt codes choose the International Codes®, developed by the International Code Council. The ICC PMG (Plumbing, Mechanical and Fuel Gas) Resource Center is the complete source for all your plumbing, mechanical, and fuel gas code needs.

Products—ICC codebooks and electronic media, references, training tools, and other industry publications provide everything you need to know about Plumbing, Mechanical and Fuel Gas Codes.

Training—ICC offers industry-leading, customized training programs that can help keep you current on the latest codes and code changes. These programs are offered where and when you need them—on-site, online, or on the phone.

Certification—ICC certification services are nationally recognized, with easy, web-based registration and results reporting. Keep certifications up-to-date and your career on track with ICC Certifications.

Membership—Join now and save on all ICC products and services. Benefits include free code opinions and discounts.

Plumbing product listings—Whether you are installing, purchasing, inspecting or specifying, plumbing products, you can be confident that they meet requirements in codes and standards by ensuring they are listed with the ICC Evaluation Service® (ICC-ES®) PMG Listing Service. An ICC-ES PMG listing is a document issued on a product that has demonstrated compliance with the specified standard as well as compliance with one or more of the following codes: *International Plumbing Code®*, *International Mechanical Code®*, *International Fuel Gas Code®*, *Uniform Plumbing Code®*, *Uniform Mechanical Code®*, or *International Residential Code®*. You can verify that products have an ICC-ES PMG listing by looking for the product at www.icc-es.org/pmg.

Find out more about ICC Plumbing, Mechanical and Fuel Gas (PMG) products and services at the PMG Resource Center.
www.iccsafe.org/PMG 1-888-ICC-SAFE (422-7233), x4PMG Email: PMGResourcecenter@iccsafe.org

31901060565514